'Nothing matches the reach of this volume! Learn what politicizes a census; produces questions on racial identities; expands the use of third party data. These insights instruct us in whether globalization of census-taking is in our reach, with far-reaching consequences'.

Kenneth Prewitt, *Columbia University, New York; Director of the 2000 Census, USA*

'This truly international edited volume offers highly competent, nuanced, and empirically well-supported hypotheses to show how census making represents and enacts the classification of citizens; how it strives for autonomy while being part of politics and international standardization; and how the digitization of population registers might eventually make it superfluous'.

Richard Rottenburg, *Wits University, Johannesburg*

'In an increasingly globalized and standardized production of numbers, this book offers an outstanding contribution to both a political epistemology as well as an institutional and methodological framing of census taking, making sense of what the state sees or avoids to see when counting its population'.

Patrick Simon, *National Institute for Demographic Studies, Paris*

I0031547

The Global Politics of Census Taking

This book examines in detail the state of the art on census taking to spark a more vivid debate on what some may see as a rather technical – and hence uncontroversial – field of inquiry. Against the backdrop of controversy between instrumental and performative theoretical stances towards census taking, it analyses the historical trajectories and political implications of seemingly technical decisions made during the quantification process by focusing on the 2020 round of censuses, which have been particularly revealing as activities have been affected by the ongoing COVID-19 pandemic and the ensuing containment policies. Through case studies of countries from the Global North and the Global South, the book highlights the consequences of, and innovations and challenges in census taking focusing on three particular areas of concern – the politics of the census in terms of identity politics; the institutional autonomy of the census; and significant and transformative methodological innovations.

This book will be of key interest to scholars, students and practitioners of quantification studies, and social demography and more broadly to public policy, governance, comparative politics and the broader social sciences.

Walter Bartl is Senior Lecturer of Sociology (Privatdozent) at Martin Luther University Halle-Wittenberg, Germany, and President of the International Sociological Association's Research Committee 41, Sociology of Population.

Christian Suter is Professor Emeritus of Sociology at the University of Neuchâtel, Switzerland, and President of the International Sociological Association's Research Committee 55, Social Indicators.

Alberto Veira-Ramos is Professor of Demography and Population Theory at the Carlos III University of Madrid, Spain, Treasurer of International Sociological Association's Research Committee 41, Sociology of Population, and Vice President of the Research Network on Economic Sociology of the European Sociological Association.

Routledge Studies in Governance and Public Policy

The Politics of Local Innovation
Conditions for the Development of Innovations
Edited by Hubert Heinelt, Björn Egner and Nikolaos-Komninos Hlepas

Municipal Territorial Reforms of the 21st Century in Europe
Paweł Swianiewicz, Adam Gendźwiłł, Kurt Houlberg and Jan Erling Klausen

Coping with Migrants and Refugees
Multilevel Governance across the EU
Edited by Tiziana Caponio and Irene Ponzo

Policy Styles and Trust in the Age of Pandemics
Global Threat, National Responses
Edited by Nikolaos Zahariadis, Evangelia Petridou,
Theofanis Exadaktylos and Jörgen Sparf

The Culture of Accountability
A Democratic Virtue
Gianfranco Pasquino and Riccardo Pelizzo

Expertise, Policy-making and Democracy
Johan Christensen, Cathrine Holst and Anders Molander

The Structure of Policy Evolution
Painting an Integrated Picture of Change in Policy and Institutional Systems
Oldrich Bubak

The Global Politics of Census Taking
Quantifying Populations, Institutional Autonomy, Innovation
Edited by Walter Bartl, Christian Suter and Alberto Veira-Ramos

For more information about this series, please visit: https://www.routledge.com/Routledge-Studies-in-Governance-and-Public-Policy/book-series/GPP

The Global Politics of Census Taking

Quantifying Populations, Institutional Autonomy, Innovation

Edited by
Walter Bartl, Christian Suter
and Alberto Veira-Ramos

Routledge
Taylor & Francis Group
LONDON AND NEW YORK

First published 2024
by Routledge
4 Park Square, Milton Park, Abingdon, Oxon OX14 4RN

and by Routledge
605 Third Avenue, New York, NY 10158

Routledge is an imprint of the Taylor & Francis Group, an informa business

British Library Cataloguing-in-Publication Data
A catalogue record for this book is available from the British Library

ISBN: 978-1-032-19546-9 (hbk)
ISBN: 978-1-032-19551-3 (pbk)
ISBN: 978-1-003-25974-9 (ebk)

DOI: 10.4324/9781003259749

Typeset in Times New Roman
by codeMantra

Open Access version funded by World Society Foundation (www.worldsociety.ch), Martin Luther University Halle-Wittenberg, and the Berkeley Research Impact Initiative (BRII) sponsored by the University of California at Berkeley Library.

Contents

Contributors

Walter Bartl is at Martin Luther University Halle-Wittenberg, Germany.

Ram B. Bhagat is at International Institute for Population Sciences, Mumbai, India.

Daniel Capistrano is at University College Dublin, Ireland.

Eva Grimm is at Federal Statistical Office, Germany.

Thomas Körner is at Federal Statistical Office, Germany.

Mara Loveman is at the University of California at Berkeley, USA.

Gabriel Mendes Borges is at the Brazilian Institute of Geography and Statistics, Brazil.

Temitope J. Owolabi is at the University of Lagos, Nigeria.

Rachel Pereira Rabelo is at the National Institute for Educational Research and Studies, Brazil.

Walter J. Radermacher is at Ludwig Maximilian University of Munich, Germany.

Nicolás Sacco is at Pennsylvania State University, USA.

Christyne C. Silva is at the National Institute for Educational Research and Studies, Brazil.

Ilona Sologoub is at VoxUkraine, Ukraine.

Christian Suter is at the University of Neuchâtel, Switzerland.

Teke Johnson Takwa is at the Central Bureau for Censuses and Population Studies, Cameroon.

Alena Thiel is at Martin Luther University Halle-Wittenberg, Germany.

Tetyana Tyshchuk is at VoxUkraine, Ukraine.

Alberto Veira-Ramos is at Universidad Carlos III de Madrid, Spain.

Byron Villacís is at the Bowdoin College, USA.

Preface

The idea of this book dates back to 2019, when the International Sociological Association (ISA) called for session proposals for the Fourth ISA Forum of Sociology, which was originally to be held in Porto Alegre, Brazil, in mid-2020. Soon afterwards, things developed differently than planned, the COVID-19 pandemic forced the organizing committee at first to postpone the conference and then to host it completely online. An initial version of most of the chapters of this book was presented in one of the overall three sessions on census taking organized by the editors at the Fourth ISA Forum of Sociology in February 2021. Two sessions were hosted by Research Committee (RC) 41, Sociology of Population, and RC55, Social Indicators. In addition to these two research committees, the third session was co-hosted by RC02, Economy and Society, RC09 Social Transformations and Sociology of Development, and RC11, Sociology of Aging. The presentations in these sessions sparked vivid debate, which may have been even more fruitful in a conventional conference with the possibility of continuing conversations in an informal setting afterwards. As organizational sociologist Stefan Kühl once observed: what people really missed about not being able to attend work-related meetings in person during the pandemic were informal talks, which are so important for creative exchange. Looking at the bright side of things, the fact that the Fourth ISA Forum of Sociology took place in a virtual setting has the benefit that recordings of the sessions are available online.[1] The authors of Chapters 4 and 10 presented their thoughts originally at a session organized by Walter Bartl and Rainer Diaz-Bone at the annual conference of the German Sociological Association in September 2020, which was also held online.

The book project benefited from very constructive comments and suggestions by three anonymous reviewers, based on our proposal to Routledge. Furthermore, Reinhold Sackmann from Martin Luther University Halle-Wittenberg, and the members of his working group, especially, Sten Becker, Jakob Hartl, Katja Klebig, Melanie Olczyk, Fabian Schmid, Anne Vatter, Oliver Winkler, as well as Katarzyna Kopycka from the University of Warsaw, were so kind to comment on early and more mature drafts of the book proposal, the Introduction, Chapter 10 and the concluding chapter. We truly appreciate their unconditional openness to sometimes quite specialized ramifications of the topic and the insightful and measured feedback they provided. We also thank Jon Anson from Ben Gurion University of the Negev and Andrzej Kulczycki from the University of Alabama at Birmingham who generously provided editorial advice during the revision process of one of the chapters in the book.

There are two types of books: perfect ones and published ones. While the contributors to this volume and we as editors did our best to produce an innovative and interesting book, it is certainly not perfect. We strived to bring together scholars from the Global North as well as from the Global South, and we aimed at including both academics and practitioners. Against this backdrop, we would be happy if it could be seen as a valuable contribution towards globally more inclusive research on census taking as well as towards framing this research in a slightly more transdisciplinary way.

Furthermore, the idea of inclusive research led us to seek possibilities for making the book available in Open Access. This endeavor was finally supported by funds from the World Society Foundation in Zurich, the publication fund of Martin Luther University Halle-Wittenberg and the Berkeley Research Impact Initiative (BRII) sponsored by the University of California at Berkeley Library. We are indeed thankful for this encouragement.

Halle, Neuchâtel, and Madrid, January 2023

Note

1 https://isaconf.confex.com/isaconf/forum2020/meetingapp.cgi/Session/14810, https://isaconf.confex.com/isaconf/forum2020/meetingapp.cgi/Session/13695, https://isaconf.confex.com/isaconf/forum2020/meetingapp.cgi/Session/15368.

The global politics of census taking in the 2020 census round

An introduction

Walter Bartl, Alberto Veira-Ramos and Christian Suter

0.1 Introduction

The current slogan of the United Nations Statistical Commission "Better Data, Better Lives" captures the longstanding and widespread belief that better information will lead not only to better collective decision-making but also to better individual results. In line with this instrumental view, the 2020 World Population and Housing Census Program of the United Nations (UN) regards nationally implemented population censuses as a primary source of data needed for policy formulation, implementation, monitoring, and evaluation. This view has become even more salient since the 2030 Agenda for Sustainable Development relies on population and housing censuses in order to develop many of the indicators operationalizing the Sustainable Development Goals (SDGs). While today, more territories than ever participate in the Census Program, censuses as a political instrument of information gathering have come under various forms of pressure, which increases the challenge of orchestrating such a massive global endeavor extending over a period of ten years. Among the current pressures on the census are:

1 New (and old) political controversies about who should be counted and which identity categories to apply (if any);
2 The increasing relevance of numbers as part of the indispensable epistemic infrastructure of global governance, modern statehood, and public critique, which has also contributed to making the production of official statistics a target of political interference, a problem that had never vanished from the UN agenda on statistical capacity building in the Global South, but has reemerged as an issue of concern in countries of the Global North as well;
3 Critique directed at the tangible and intangible costs associated with the traditional census together with new developments in information technology, which has recently increased the pressure for methodological innovations in census taking, creating a very open situation, in which the dominant global model is increasingly facing competition from new forms, which are put to the test in different parts of the world;
4 The COVID-19 pandemic, which resulted in logistical problems and often the postponement of national census implementation. The pandemic might prove

DOI: 10.4324/9781003259749-1

to be a critical juncture for national statistical systems, one that could catalyze methodological innovations, but also one that could hamper census taking in significant ways.

While there is little doubt that the census is a principal source of information for global governance and modern statehood, the instrumental view exposed by official UN documents comes up for discussion when we look at the performative effects of counting populations. By producing key indicators on the populations to be governed, censuses shape a state's knowledge in important ways. In fact, without the cognitive infrastructure of the census and related instruments, the sociodemographic reality of a state would not be known, and it would not be "legible" for policymakers (Scott, 1998). The process of objectifying demographic identities does not begin when data are published, but already when they are constructed (Kreager, 2004). The systematic construction and description of standardized features of the population transforms the object itself by defining those that are to be included and those that are to be left out, as well as by classifying subgroups that might not have previously existed. The numerically generated visibility of otherwise latent subgroups very often is the starting point for politically contentious identity politics or affirmative action programs. On the other hand, ignoring existing social classifications, such as ethnicity, race, or religion, when producing population indicators, could contribute not only to a blurring of social boundaries but also to a neglect of social problems related to particular subgroups. Similarly, while methodological innovations of census taking are often discussed in purely technical terms, there is a political dimension to them, since they typically produce new forms of inclusion and exclusion at the same time. Making these hidden selectivities of methodological innovations visible is an important aspect of performativity research. In contrast to this reflexive stance toward census taking, policy formulation and implementation usually treat the results of censuses as black boxes representing reality. This paradoxical nature of public statistics led Alain Desrosières (1993/1998) to conceptualize them in a dual way, as constructed and real at the same time. While situations of reflection and critique emphasize the contingency of methodological decisions, situations of intervention presuppose their objective character.

Against this backdrop of controversy between instrumental and performative theoretical stances toward census taking, the proposed book aims to expose the current state-of-the-art on the topic and spark a globally more integrative and more transdisciplinary debate on what some may see as a rather technical – and hence uncontroversial – field of inquiry. We aim to analyze the political implications of seemingly technical decisions made during the quantification process by focusing on analyses of the 2020 global census round from different world regions. Even though international organizations have promoted population censuses worldwide since the 19th century and standardization efforts of the UN have increased since the 1950s (Ventresca, 1995), the census is by no means a global model that travels without translations. Varieties of census taking can hardly be explained without careful attention to local practices of categorization and quantification as well as social histories of state-society interactions with respect to official statistics (Emigh, Riley, & Ahmed, 2016a, 2016b). Our book highlights the tensions between international coordination efforts on the one hand and (path-dependent) trajectories of statehood on the other hand by analyzing controversies and translation processes mediating these tensions. Therefore, a general research question, addressed throughout the book, is: *How can we describe and explain the trajectory of national censuses?* We follow this general research interest focusing on three more specific issues that have become particularly salient in the 2020 census round, and, which, according to our view, make global census taking a more reflective endeavor than it was some decades ago.

The first issue refers to representing collective identities. While quantifying populations is known to have important implications for the formation and representation of collective identities, the categories of quantification have become increasingly reflexive and less self-evident. Ethnic, racial, and religious census categories have become more politicized, and their use or non-use increasingly requires explicit justifications (Simon, 2015, 2017). A crucial research question addressed in this respect is: *How can we explain the (non-)use of ethnoracial or religious questions in the census?* We analyze the politics of cognition and recognition as they manifest themselves during the 2020 census round through comparative and historical case studies on Latin America and India in particular (Chapters 1–3), but the cases of Nigeria and Ukraine offer some insight in that respect as well (Chapters 7 and 8).

The second issue attracting attention during the 2020 census round is the institutional autonomy of the census in the national statistical system of a country and the related political culture. Census taking has been considered the backbone of national statistical systems and as pivotal for democratic politics (Egeler, Dinsenbacher, & Kleber, 2013; Prewitt, 1987). At the same time, there is considerable variety in the institutional design of national statistical systems (Prévost, 2019). A crucial research question in this regard is: *How can the relative institutional autonomy of census taking be adequately described and explained?* The institutional design of national statistical systems is actually not coincidental with the politicization or depoliticization of census taking. In many countries – not only in the Global South but also in the Global North – attempts at political interference have been made in recent years. The cases covered in our book range from micro politics of data collection (USA, Brazil, Ecuador, Chapter 5) and outright attempts at manipulating census data (Nigeria, Chapter 7) to even deciding not to implement the census at all (Ukraine, Chapter 8). Chapters 4 and 6, in contrast, focus more on possible ways to deal with a lack of trust in public statistics.

The third salient issue in the current global census round consists of significant methodological innovations, which rely substantially on new forms of digital information gathering and processing and are about to transform the classic model of census taking in forms that have barely been explored systematically. Our guiding question in this respect is: *How can socio-material and methodological innovations of census taking be explained?* While scholars in the tradition of world polity theory emphasize the importance of global scripts and international organizations for explaining census taking as a globally coordinated endeavor, this approach remains fairly silent about how methodological and organizational innovations of census taking can be explained when they do not conform to global standards. In this respect, more conflict-sensitive and practice-oriented approaches might prove to be fruitful (Emigh, Riley, & Ahmed, 2020; Diaz-Bone & Horvath, 2021; Didier, 2021; Ruppert & Scheel, 2021a). Our introduction gives an overview of the current state of methodological innovations in census taking, as well as problems due to the COVID-19 pandemic. Individual chapters in the third book section analyze particular challenges, as well as intended and non-intended consequences of the innovative forms of producing key population indicators through case studies of countries from the Global North and the Global South. Innovations highlighted by these case studies range from the transition to a register-based census (Spain, Germany, Chapters 9 and 10), to the utilization of biometric data and mobile devices (Ghana, Cameroon, Chapters 11 and 12).

With regard to possible consequences of census taking, a relevant question related to the use or non-use of ethnoracial categories is the way in which it affects categorical inequality (Piketty, 2022, pp. 175–202; Tilly, 1999). While research on categorical inequality is crucial for creating an empirical base for debates about social justice, using data based on census categories might be ambivalent in the sense that it runs the risk of reifying state categories instead of creating epistemologically more sophisticated research designs (Monk, 2022). The issues of institutional autonomy and methodological innovation have in common that they both are likely to affect trust in public statistics, a resource not to be underestimated for democratic governance (Radermacher, 2020).

0.2   Theoretical approaches and empirical variety of census taking

Social studies of quantification are a growing field (Bartl, Papilloud, & Terracher-Lipinski, 2019; de Paiva Rio Camargo & Daniel, 2021; Di Fiore, Kuc-Czarnecka, Lo Piano, Puy, & Saltelli, 2022; Diaz-Bone & Didier, 2016; Espeland & Stevens, 2008; Henneguelle, 2023; Ringel, Espeland, Sauder, & Werron, 2021; Rottenburg, Merry, Park, & Mugler, 2015; Sætnan, Lomell, & Hammer, 2011; Vollmer, 2007). In a recent review article on sociological approaches to the study of quantification, Andrea Mennicken and Wendy Espeland observed a parallel growth in and mistrust of the production of numbers during the last decades (Mennicken & Espeland, 2019). As regards explanations for these phenomena, they cautioned against overgeneralization, but nevertheless came to six conclusions about the nature of quantification processes, which they deduced from different research traditions on the subject: 1. Quantification is itself an intervention in the world because it constitutes a new reality that did not exist before. 2. Counting on a large scale, such as in the census, requires cognitive, monetary, and material investments. 3. Quantification typically addresses problems of uncertainty, control, distrust, and coordination. 4. Once created, numbers are often hard to abandon again because they become built into routine practices and institutions. 5. Quantification is selective and things that are hard to measure are often left out, which gives the impression that they might be less valuable. 6. The observed coincidence of increasing quantification and mistrust of it seems to be due to our limited capacity to check the accuracy of numbers, due to a lack of time, skill or access (Mennicken & Espeland, 2019, pp. 238–239). While it is certainly true that no unifying theory of quantification processes has emerged, other authors have called for more rigorous research in order to push theory development further by conducting research more systematically (Emigh et al., 2020). An important reservation against attempts at purification comes from the fact that quantification studies address very different substantial fields. Since we focus on the narrower field of census taking, in the following, we will highlight the first steps toward theory building, which often takes the form of developing typologies for comparative purposes. Against such a background, we will also discuss the variety of empirical evidence, which has often been gathered without direct reference to these typologies. Therefore, only a limited degree of consistency will be possible.

0.2.1   *Modern census taking*

Since we focus on current census taking as a globally coordinated endeavor, it seems worthwhile to point out how modern censuses differ from earlier approaches to counting populations. A starting point for doing so is the typology constructed by Marc Ventresca (1995, p. 32) (Table I.1). Examples of historical counts date back to ancient times. Hence, taking together cases from a very long time period obviously is a simplification, but it highlights the particular features of current census activities in an analytical way. Historical counts usually had a practical purpose, were often extractive, and could be conducted by diverse authorities. When, during the 19th-century, the census became basically an exclusive activity of the state, it formalized older social practices of counting, which had provided the numerical competencies and cultural techniques that predated official state practices (Emigh, 2002). In 19th century Europe, population censuses were increasingly conducted to gain knowledge about the population itself; they fed into the governmentality of the state, which began to see the population as the ultimate end of government (Foucault, 1991). Colonial censuses in contrast imagined the populations to be enumerated as essentially different from the metropolis, corresponding to exploitative interests in often complicated ways (Appadurai, 1993; Renard, 2021). When censuses became a separate state activity, it was organized according to formal rules, and it was conducted by specialized staff of then newly founded statistical offices (Ho, 2019). These stylized facts should not let us forget that official census taking required considerable cognitive, material, and organizational investments, which is, for example, why the first attempts at census taking after the French revolution failed and could only be realized after the "Adunation" of the country had been completed, e.g. the metric system had been introduced (Desrosières, 1993/1998, p. 31). During state-building efforts in other world regions, such as Latin America, census taking similarly took several attempts before it succeeded (Loveman, 2014). Furthermore, even if we feel inclined to view census taking as an evolutionary achievement, the success of census projects is not a given. On the

Table 1.1 Features of historical counts and modern census activity

Dimension	Count	Census
Purpose		
Purpose	Extractive; Monitoring	Scientific
Authority	Religious; Commercial; State	State
Organization		
Organization	Informal; undifferentiated	Formal; specialized
Personnel	Non-specialists	Professional staff
Content and conduct		
Scope	Narrow and idiosyncratic	Broad and consistent
Population	Selective	Inclusive, citizen-based
Technique	Estimates, samples	Individual enumeration
Method	Convenience	Rational, scientific
Publication		
Status of data	Secret, private	Public
Dissemination	Local	Global

Source: Ventresca (1995, p. 32).

contrary, examples of contemporary failures (e.g. Pavez, 2020) emphasize the importance of continued investments in form (Thévenot, 1984).

With the official organization of census taking, content and methods changed as a consequence of the increasing professional and scientific "theorization" of census taking (Strang & Meyer, 1993). The scope of the census was broadened to include the entire population of a given territory with experts deliberating about the consequences of employing the concept of *de jure* or *de facto* population (e.g. Fabricius, 1868). The dispute of how to deal with temporarily absent persons has not been resolved yet in a globally standardized manner (Thorvaldsen, 2018, p. 159). An inclusive perspective on all residents of a delimited territory implied a cognitive equalization of the population. While individuals had entered population counts selectively based on certain characteristics before, they were now seen as comparable at least in terms of their personhood: "It makes no sense to count people if their common personhood is not seen as somehow more significant than their differences" (Porter, 1986, p. 25). From the 1850s onwards, methods of census taking were increasingly standardized by efforts of internationally active statisticians who began to meet at International Statistical Congresses in 1853 in Brussels (Ventresca, 1995, p. 61). This professional standardization led to formal definitions of criteria for inclusion as well as clearly demarcated census districts, and as a consequence also to more precise census results. During the early 19th century, the use of census results began to change dramatically as well. Ian Hacking (1990, p. 2) felicitously captured this aspect of the then new counting practices of state administrations in the phrase "an avalanche of printed numbers". While most official counting had been kept secret to administrators before Napoleonic times, a vast amount of it was printed and published afterward. The publicness of statistics can be seen as a crucial ingredient of trust in a liberal democracy because it allows for public control and critique of policymakers and administrators (Porter,

1995; Prewitt, 1987; Rose, 1991). The American constitution was the first to make a census the base of legitimate political representation. Today, the results of census taking are disseminated at a global scale, implying not only a domestic but also an international public as a source of state legitimacy (Ventresca, 1995).

This characterization of the modern census can be refined by further distinguishing between descriptive and interventionist purposes of census taking (Emigh et al., 2016a, 2016b). While older forms of counting populations were typically extractive in purpose, early modern censuses strived to describe the population in a demographic way, because "populous" territories were seen as inherently advantageous (Curtis, 2002). Descriptive censuses still bore the intellectual traces of sovereignty and mercantilism, the ideologies that attributed absolute political power to the state and aimed to promote the state's economic power vis-à-vis other states based on its population number. However, these early censuses bore little sense that the governed themselves were changeable through social intervention. Later, in contrast, state action became linked to the concept of "population" as an aggregate unit of analyses for which it was possible to identify statistical regularities and design public policies defining the needs and the welfare of their object (Emigh et al., 2016a, pp. 49–50; cf. Foucault, 1991). Censuses with interventionist intentions became effective only after the Second World War, when they were linked to policies that had broad popular support (Emigh et al., 2016b, p. 52). Comparative case studies show that censuses vary in their linkage to interventionist public policy according to their politicization in public debates, with the cases of the United States of America (USA), the United Kingdom (UK), and Italy representing a decreasing order in this sense (Emigh et al., 2016a, 2016b). Broader accounts of the historical co-evolution of statistical devices and public policy point out that, in the long run, the relevance of census taking diminished vis-à-vis other statistical techniques (Desrosières, 2011; Radermacher, 2020), an observation that increases even the need to adequately describe and explain the empirical variety of census taking.

Simplifying an increasingly unsurmountable body of research to more digestible dimensions, theoretical accounts of census taking can be classified as falling into three main categories, depending on which actors are attributed with crucial agency[1]: The first are accounts that embed census taking in national projects of state building or imperial projects of colonization (Anderson, 1983/2006). In these accounts, census taking is mainly instrumental to the power of the state. Bourdieu's notion of the "information capital" of the state is an example of such a perspective (Bourdieu, 1994) and the Foucauldian framing of census taking has been interpreted similarly, with limits of bureaucratic knowledge having remained rather unexplored (Szreter & Breckenridge, 2012, p. 7). A second account portrays census taking as part of a cultural script of World Society, which developed gradually since the mid-19th century and became a global standard after the Second World War (Ventresca, 1995). In this account, international organizations – first and foremost the UN and their methodological manuals, but later also global financial institutions such as the World Bank (Loveman, 2014) – have been decisive agents of global-local translation of World Society scripts and practices (Michalopoulou, 2016; Thompson, 2016). The third account has been articulated most recently and was directed against the first two theoretical approaches, which from that perspective both appear to be excessively state-centered. Instead, so the argument goes, it is not possible to explain the historical contingencies of individual census trajectories without paying due attention to the social history of societies and to the interaction of asymmetrically endowed societal actors, among which the state is merely one and not necessarily the most influential one (Emigh et al., 2020, p. 301). Building upon Boltanski and Thévenot (1991/2006), Emigh and colleagues argue that actors can construct categories for themselves and can criticize imposed categories. Furthermore, instead of taking the symbolic power of the state as a given, it is necessary to also investigate the – sometimes contentious – processes involved in building up symbolic state power (Loveman, 2005). The differences between the first two accounts have been leveled maybe too quickly when they have both been subsumed under the label of a state-centered approach (even if the didactic intention is understandable). Especially, the manifold translations of data practices between the global and the national level get lost (Desrosières, 2000), when the commonalities between the two perspectives are overly emphasized. Instead, it seems to be worthwhile to study the "debates, struggles, tensions, discourses, techniques, material devices, logics, rationalities, values, assumptions" involved in the international data practices of census taking (cf. Diaz-Bone & Horvath, 2021; Ruppert & Scheel, 2021b, p. 15).

We will bear these three types of theoretical accounts in mind when we discuss coordination efforts and controversies around census taking in the early 21st century. The following subsections will address more specifically the issues of ethnoracial categories, institutional autonomy, and political interference in census taking, as well as socio-technical innovations.

0.2.2   *The use and non-use of ethnoracial categories*

Official statistics, and especially censuses, reflect legitimate representations of the population of the societies in which they are conceived and produced. There are at least three elements of categorization through which the census constructs group identities. In the first instance, the enumeration of people requires a (seemingly) clear setting of social boundaries for inclusion as well as for further distinction of subgroups. A second aspect is that census enumeration reveals a community's size, growth, and often also socio-economic characteristics. Knowledge of these attributes enters public discourse and shapes relations between social groups (Alba, 2020; Bhagat, 2022; Morning & Rodríguez-Muñiz, 2016). Thirdly, this aspect can be further stepped up in a democratic political system, in which group size indicates power and access to social and economic resources. Then, the knowledge generated by the census allows not only for quantitative comparisons, but might also fuel formal political competition between communities (cf. Maktabi, 1999; Weiss, 1999).

Like other passive technologies for governing populations, such as registers or identity cards, the census draws boundaries of belonging. The baseline of census taking is the definition of the national population as a community of equals as distinct from other state populations (Anderson, 1983/2006, pp. 164–170). While social boundaries of subgroups might not be clearly defined or might overlap in mundane interactions (Brubaker, 2006/2008), the census enables neat social and geographical boundaries to be drawn, especially if it uses mutually exclusive categories that are claimed to be an inherited feature of individuals. In these cases, the census not only sets quasi-natural social boundaries over time and space but also contributes to the creation of communities. By creating particular sets of categories for classifying the population, the census shapes our shared experience (Watson, 2005, p. 249, 329). In order to highlight the constructedness of the categorization process of populations that census taking performs, we follow Mara Loveman (2014, p. 37) in using the umbrella term *ethnoracial categories* for "any categorical distinction that names or delimits sets of human beings who are construed to belong together naturally, as a collectivity or community, due to some source of heritable similarity".

Despite repeated efforts to standardize the content of censuses in guidelines promoted by the UN, some topics remain country-specific. Of this data, ethnicity, and race stand with religion as the most controversial and least standardized categories in population censuses (Bennani & Müller, 2021). Countries can be divided into those that collect colorblind statistics and those that produce ethnoracial statistics. The strategies behind the choices of whether or not to collect ethnoracial statistics can be linked to ideologies of equality and diversity, which are grounded in national histories and political mobilization. One of the first attempts to capture the use and non-use of ethnoracial categories in census taking from a comparative perspective observed quite a variety of categories across nation-states. In order to systematize this variety, Rallu, Piché, and Simon (2006, p. 534; Simon, 2015) distinguished six ideological frameworks legitimizing the form and content of ethnoracial categories in censuses.

- First, "not counting in the name of national integration" emerged in the context of West European nation-states that imagined their populations as monoethnic and assumed that existing ethnic identifications would inevitably be assimilated. Furthermore, many postcolonial states, such as in western Africa, tried to monopolize the modernization of their societies by negating ethnic awareness.
- Second, "not counting in the name of multiculturalism" counterintuitively describes reasons for not gathering ethnoracial statistics that are grounded in a positive valuation of cultural mixing and multiculturalism.
- Third, "counting to dominate" mainly originated in colonial settings in Asia, Africa, and the Soviet Union, where the imperial power strived to dominate subjugated groups, but also in large immigration countries, such as the USA and Canada, where Blacks and Asians were to be excluded from full citizenship rights until the 1960s.
- Fourth, "counting in the name of multiculturalism" is associated with the increased appreciation of cultural diversity and ethnoracial mixing. This ideology is particularly present in Latin America, where positive race relations are emphasized.
- Fifth, "counting for survival" refers to relatively small national minorities using ethnic statistics to demand more power and support for maintaining their

cultural specificity. This type was particularly strengthened through the global institutionalization of the category "indigenous people" (Bennani, 2017; Walter, Kukutai, Carroll, & Rodriguez-Lonebear, 2020).

• Sixth, "counting to justify positive action" refers to the political aim of compensating socio-economic disadvantages of ethnoracial groups; it can therefore be seen as an outright reversal of the discriminatory perspective of the third case. Social demand for ethnoracial statistics for the purpose of positive action has been greatest in the USA, Canada, and the UK.

Despite noticing various contradictions in state practices, the authors were able to assign particular countries and historical periods to each type (Rallu et al., 2006, pp. 534–536). However, this should not lead us to expect overly homogeneous national cultures using quantitative data in the public sphere. For example, although the USA uses ethnoracial statistics in the census, these coexist with color-blind explanations of racial inequality (Beaman & Petts, 2020). As a consequence, others, who build on the typology of Rallu et al. (2006) as well as on Desrosières' (2000) notion of a "statistical production chain", suggested paying closer attention to policy fields in which ethnoracial data are to be used rather than to national patterns of quantification (Supik & Spielhaus, 2019). In anti-discrimination policy, ethnoracial statistics are essential (cf. Potvin, 2005).

A different analytical contribution was made by Kertzer and Arel (2006). First, they proposed an interactionist view on the link between identity categories in the census and respondents. Second, they outlined three different models of linking the ethnoracial categorization of the population to political power.

• In the Affirmative Action Model, the state creates an identity category whose members benefit from special programs, justified as a way of making up for past discrimination against this group. In this model, census politics operates at two levels. First, it creates an incentive for political entrepreneurs to get their favored group category on the list of beneficiaries. Second, individuals tend to see the census as a site of identity recognition – in the event that their identity category is offered in the census questionnaire.
• In the Territorial Threshold Model, political power is assigned to members of a particular identity category based on the group surpassing a given statistical threshold on a particular territory. Being the majority, for example, might give rise to claims for devolved political power, or being a significant minority might legitimate claims for minority-language schooling.
• In the Power-sharing Model, political or administrative offices are allocated according to a formula among members of two or more identity categories. Such a division is almost always justified according to a principal of equality or proportionality. Political offices are likely to equate with actual numerical divisions within a population. This model is inherently unstable because it leads to power relations being questioned in terms of demographic change (Kertzer & Arel, 2006, pp. 670–673).

Kertzer and Arel emphasize not only the political choices involved in creating the repertoire offered in the census questionnaire but also that the affirmation of a categorical identity by an individual is a matter of choice. In the Soviet Union, the margin of slack of respondents awakened the suspicion of party officials and attempts to control (Mespoulet, 2022). More recently, a similar idea was framed in terms of practice theory as the "subjectivation" of census taking (Cakici & Ruppert, 2021). The construction of national identity in post-Yugoslav censuses, for example, met the challenge that many individuals refused to ethnically identify themselves (Bieber, 2015). Despite incentives for standardization from the EU, the politicization of the census debate led to a delay in Bosnia and to its abortion in Macedonia (Hoh, 2018). Such an interactive approach seems to be particularly well-suited as a lens for exploring cases of "national indifference" (Labbé, 2019) that limit the power of the census to shape national identities.

Since the 1990s an increasing global pressure to recognize minorities and the demand for ethnoracial statistics in order to document discriminatory practices has been observed (Lieberman & Singh, 2017, pp. 15–16; Simon, 2005, 2021). Based on a global dataset of 141 countries in the 2000 census round compiled by the United Nations Statistical Division (UNSD), Morning (2008) finds that 63% of the national censuses studied incorporate some form of ethnoracial enumeration. However, there is substantial variation regarding the content of ethnoracial categorization (race, phenotype, ethnicity, religion, language, or combinations of these categories) following world-regional patterns. Nonetheless, Morning concludes that "the variety of approaches can be grouped into a basic taxonomy of ethnic classification approaches, suggesting greater commonality in worldwide manifestations of the ethnicity concept than some have recognized" (Morning, 2008, p. 239). In a historical analysis of census questionnaires between 1800 and 2005, Lieberman and Singh (2017) found that ethnoracial enumeration has been the modal approach across time, used in 74.9% of their cases.

> Throughout the nineteenth century, we find a steady climb toward increased ethnic enumeration, such that in the first four decades of the twentieth century, more than 85% of censuses enumerated ethnic cleavages. Following the Second World War, however, with a proliferation of new states and perhaps amidst some international resistance to ethnic categorization, we find a reverse trend, reaching the nadir in the 1990s, when 74 of 113 censuses (65.5%) asked questions about at least one set of ethnic categories. This trend was reversed again in the first decade of the new millennium.
>
> (Lieberman & Singh, 2017, p. 15)

Despite this rather broad recent development, monographic studies on ethnoracial census categories exist only for a minority of countries or world regions, e.g. for Nazi Germany (Aly & Roth, 1984/2004), the USA (Anderson, 1988/2015; Schor, 2009/2017), Brazil and the USA (Nobles, 2000), the UK (Supik, 2014), Latin America (Loveman, 2014; Telles, 2014), Rwanda (Tesfaye, 2014), Brazil and Colombia (Paschel, 2016), USA, Canada, and UK (Thompson, 2016), USA, UK, and Italy (Emigh et al., 2016b), Eastern Europe (Surdu, 2016; critical: Mirga, 2018), and India (Bhagat, 2022).

Based on data from the 2000 census round, Kukutai and Thompson (2015) attribute the recently increased use of ethnoracial categorization to the integration of countries into global civil society measured, first, as the expression of state commitment to human rights instruments on the world stage, and second, as state participation in international non-governmental organizations. The Organization for Economic Cooperation and Development (OECD), for example, has also started to pay attention to practices of ethnoracial categorization (Balestra & Fleischer, 2018). In terms of national pressures, they identified rights claims by established ethnic minorities challenging the state as well as structural pressures resulting from the presence of an ethnic contender group and higher levels of net immigration. From a world polity perspective, older states were less likely to enumerate by ethnoracial categories because of historical path dependencies of colorblind census policies. Also, states with a high Gross Domestic Product may "be less likely to receive opposition from ethnic groups within their states, given their economic position and relative lack of deprivation" (Kukutai & Thompson, 2015, p. 59). Yet historical trajectories are much more complex than the theorem of path dependency suggests. Loveman (2014, p. 4) points out that categorization practices in one historical period might indeed give rise to political mobilization in order to achieve a different symbolic order in the census in the next period. The German "colorblind" policy of census taking, for example, seems hard to explain without accounting for the "dark side of numbers" during the National Socialist regime in the Third Reich (Seltzer & Anderson, 2008). In contrast to that policy-turnaround after democratization, ethnoracial categorizations in many postcolonial states date back to colonial times (Appadurai, 1993). Rather than representing a simple continuity of categorization practices, they are sometimes still in use, because they have acquired new social functions (Balaton-Chrimes & Cooley, 2022). Therefore, while statistical analyses of ethnoracial enumeration at a global level are valuable and could still be pushed further, comparative case studies of national trajectories will remain indispensable if more systematic accounts of historical processes are to be developed (Emigh et al., 2016b).

Yet, the political and ethical question remains, whether ethnoracial classification should be used in the census or not. When deliberating about that question, we cannot assume that ethnoracial categories are time-constant variables. This becomes especially evident in the changeability of racial self-identification, e.g. compared to perceived skin color (Dixon & Telles, 2017, p. 409). In Ukraine, for example, many respondents changed their declared nationality between the 1989 and 2001 censuses from Russian to Ukrainian (Stebelsky, 2009), a trend that has clearly continued afterward (Veira-Ramos & Liubyva, 2020, p. 211). This change was driven to a considerable extent by descendants from mixed-nationality couples. International migration and generational change have increased the interest in the question of how individuals born within mixed marriages should be asked to report themselves (Alba, 2020; DaCosta, 2020; Rocha & Aspinall, 2020; Song, 2021). How many categorical distinctions should be enumerated for the sake of fitting all nuances? The historical study of Lieberman and Singh (2017) has shown that representing ethnoracial cleavages in the census increases the likelihood of group competition and also violent conflict. However, these findings do not lend themselves to straightforward policy recommendations for the abolition of ethnoracial categories in the census. Proponents argue that, without ethnoracial statistics, it would not be possible to assess the amount of ethnoracial inequality in society and, hence, colorblind policies are implicitly contributing to the reproduction of these inequalities and to discrimination (Beaman & Petts, 2020). Opponents argue that data gathering of ethnoracial categories is itself discriminating and should be reduced or abolished altogether (cf. Blum, 2002; Surdu, 2019). "Migration background" in Germany, for example, a category that according to the official colorblind data policy of the state is not openly ethnic, has been criticized for perpetuating the othering of immigrants across generations rather than being a useful analytical instrument for fighting discrimination (Elrick & Schwartzman, 2015; Will, 2019). In response to that situation, a government expert commission on integration proposed a more narrow definition of "migration background" that comes closer to international standards but also displayed significant disagreement among commission members (Fachkommission Integrationsfähigkeit, 2021, pp. 222–227). Since "migration background" left experiences of discrimination by Black Germans under the radar of official statistics, activists initiated an "Afrocensus" to become visible as political subjects (Aikins, Bremberger, Aikins, Gyamerah, & Yıldırım-Caliman, 2021). This exemplary initiative to introduce ethnoracial categorization is merely to demonstrate that the debate on this issue is still ongoing and controversial (Des Neiges Léonard, 2015; cf. Simon, 2017).

Chapters 1 and 3 analyze the use of ethnoracial categories in Latin America and Chapter 2 presents a case study on their use in India.

0.2.3 Politics, the institutional autonomy of census taking and trust

From the viewpoint of world polity theory, transnational scientific and professional theorizing produces global scripts on how to institutionalize census taking (Strang & Meyer, 1993; Ventresca, 1995). Diffusion of practices among states is not only facilitated by abstraction but is also likely to invoke merely ritual adoption. What is at stake in the relationship between census taking and politics is trust in numbers and the legitimacy of the state. In the following, we discuss three concepts that theorize the relevance of census taking for (democratic) politics and facilitate the comparison of national variation: political interference, statistical capacity, and institutional autonomy. Furthermore, we briefly characterize the professionalization efforts of international organizations to ensure a high quality of census taking.

The census is often considered a crucial instrument and information infrastructure of democratic politics (Egeler et al., 2013; Prewitt, 1987; Rose, 1991; Sullivan, 2020b). In this view, statistical and political representation are akin (Didier, 2002); statistical representation of individuals is a precondition for their adequate political representation and for their acquisition of social rights – as is aptly captured in the notion of "statistical citizenship" (Hannah, 2001). Politicians, in turn, fund, regulate, and supervise national statistical offices that gather information on the population, their activities and the territory. This information is required for policy formulation, implementation, and evaluation. At the same time, public statistics in a democracy should not be perceived as prioritizing questions of political power over scientific methods (Barlösius, 2005, pp. 119–134). In the Soviet Union, official statistics often came under political pressure to prioritize party ideology of the future society to be achieved over a more accurate representation of social reality (Mespoulet, 2022). Since information gathering is based to a considerable extent on the cooperation of individuals when filling in questionnaires, trust in public statistics and the quality of the products of national statistical offices is crucial for a major statistical endeavor such as census taking. Walter Radermacher (2020, p. 140), the former director of the European Statistical Office declares: "Trust is the main and overarching goal of statistical governance." While census taking of colonial powers created suspicion from the colonized populations, for example, in East Africa (Goldthorpe, 1952), mistrust toward census taking has also been documented for postcolonial environments (Gil & Omaboe, 2005) and established democracies (Hauser, 1973; Seltzer, 1994; Sullivan, 2020b). In a nutshell, the problem of trust boils down to the credibility of the institution in charge of census taking. Taking a discussion about the differentiation between politics and statistics in Germany as an indication, the incipient consciousness of this problem dates back to the 1860s and 1870s, when statistical offices began to professionalize (Maier, 1959, p. 15). That debate took place two to three decades before the first broader discourse on the value-neutrality of social sciences unfolded (Beck, 1974). Today, national statistical offices find themselves in the paradoxical situation of acknowledging the constructedness of data while, at the same time, claiming the scientific objectivity of their results.

National statistical offices tend to resolve this contradiction by emphasizing the objectivity of their methods, beneficial to maintaining public trust and dissipating possible suspicion about the malleability of statistical results (Desrosières, 2009). In this view, any doubts about the political neutrality of the statistical agency in charge are to be avoided. According to the former director of the U.S. Census Bureau, Kenneth Prewitt, *political interference* can be defined as:

> The attempt to gain partisan or regional advantage by shaping the production of a statistical product against the judgment of a nonpartisan and apolitical statistical agency. [...] This definition focuses on the production of statistical products, not their use or application.
>
> (Prewitt, 2010, p. 228)

More recently, definitions of quality in public statistics have aimed to focus on the adherence of production processes to professional standards and the adequacy of products for serving legitimate – not particularistic – political purposes (Radermacher, 2020). Risks of political interference are particularly evident where data from authoritarian regimes are concerned (Carlitz & McLellan, 2021; von der Lippe, 1999; Mespoulet, 2022). However, several cases of political interference and data manipulation have also been documented in democratic polities (Agrawal & Kumar, 2020; Aragão & Linsi, 2022; Foremny, Jofre-Monseny, & Solé-Ollé, 2017).

While there are conceptual doubts that a technocratic census, isolated from political influence and public debate, really is more instrumental for governing societies (Emigh et al., 2016b), the exact line of such a distinction between legitimate political demands and illegitimate political interferences is often difficult to draw. The possible benefit of a certain politicization of state information gathering seems to lie in its creation of a demand for accountability and a healthy sense of mistrust toward statistical activities (Lehtonen, 2019). In the 2020 census of the USA, for example, the inclusion of a citizenship question as proposed by the Trump administration fueled public debate, because it was argued that it could deter undocumented immigrants from participation and, hence, weaken the political power of the respective states in the apportioning of seats in the House of Representatives (Sullivan, 2020a). In the end, the question was withdrawn. Had the question been included, ironically, heavily Republican states would have lost most from its assumed deterrence effect (Poston, 2020).

In order to ensure the quality of official statistics, the International Statistical Institute (ISI) formulated a Declaration on Professional Ethics in 1985 (Jowell, 1986) and the UN adopted Fundamental Principles of Official Statistics in 1994 (Ljones, 2015; Seltzer, 1994). The latter are based on voluntary self-assessments and peer review. Although operational indicators for the UN Fundamental Principles were suggested early on (Vries, 1999), a satisfactory global governance structure has not been implemented yet (Georgiou, 2017). Parallel to these initiatives, the concept of *statistical capacity* developed by the World Bank has aimed to capture the quality of national statistical systems on a structural level, which has become even more important as a consequence of the 2030 Agenda for Sustainable Development and the transformative capacity ascribed to the related SDG indicators (Biermann et al., 2022; Merry, 2019). Statistical capacity deteriorates when censuses or the publication of their results are delayed or are even entirely suppressed. Both have significant consequences for the efficiency of public policy. Budget allocation formulas in intergovernmental transfers, for example, often refer to population numbers when they are to serve general purposes. Hence, erroneous figures lead to democratically unintended misallocations. A study by the Inter-American Development Bank estimated that, in El Salvador, due to inaccurate municipal population numbers, approximately 92 million U.S. dollars were sent to municipalities mistakenly between 2000 and 2007, equaling roughly 700% of the cost of the previous census and around 27 times the annual budget of the statistical office (Roseth, Reyes, & Amézaga, 2019).

Recently, Jean-Guy Prévost suggested integrating the action-oriented notion of political interference and the more structural notion of statistical capacity into a two-dimensional concept of *institutional autonomy* (Prévost, 2019, p. 158). Applied to census taking, the negative dimension of institutional autonomy would then refer to the institutional protection of census authorities from political interference. The positive dimension would refer to the professional, organizational, and infrastructural capacity to conduct a census. The positive dimension of the concept of institutional autonomy is a tricky one: Even if the national organization in charge is independent in general, they usually depend on the government or the parliament for the periodical activity of census taking. Parliament, government, and administration provide the legal framework to conduct the census as well as earmarked funding and logistical support. Local authorities provide not only logistical support but sometimes also quality control of census taking. When the census is perceived by the population as an administrative operation rather than a statistical one, trust from the population in the use of the data and the confidentiality of the information provided may be jeopardized (Durr, 2020). Social scientists working in the tradition of Michael Mann (1984) and James Scott (1998) have further elaborated the statistical capacity concept and related indicators, in which census taking is an important dimension (Brambor, Goenaga, Lindvall, & Teorell, 2020; Cameron, Dang, Dinc, Foster, & Lokshin, 2021; Lee & Zhang, 2016). While this scholarship can be interpreted as theorizing that is likely to facilitate the standardized diffusion of census taking (Strang & Meyer, 1993), how exactly does that work?

Coordinating the 2020 census round, several (parts of) UN handbooks have been devoted to the governance and management of census taking (UN, 2016, 2017, pp. 44–78, 2021, pp. 11–104). The United Nations Fund for Population Activities (UNFPA), the leading institution in support of census taking at the global level, is present in 155 countries and provided specific support to 135 countries during the 2010 census round (IDWG-Census, 2017/2019, p. 6). The role of situated interaction and materiality has to remain open at this point but will be explored by some of the case studies in this book. They show that the national capacity for census taking is much more wide-ranging than is evident when looking at the almost universal participation of countries. Furthermore, they elucidate the conflicts involved in balancing the relationship between census information and politics. Some states have never managed to implement a consistent system for population statistics (e.g. Ghana, Chapter 11), while others have faced more or less obvious political interference (e.g. Brazil, Ecuador, and USA, Chapter 5; Nigeria, Chapter 7), and yet others have halted census activities for a range of political reasons (e.g. Ukraine, Chapter 8).

0.2.4   *Socio-technical innovation, privacy and trust*

Regarding methodological and socio-technical innovations, world polity theory again urges us to look for abstract deliberations on this topic (Strang & Meyer, 1993), while more interactive approaches direct our attention to the role of socio-technical devices employed in concrete practices of census taking as well as related problematizations and conflicts among participants. A good part of the innovations we observe seem to have been developed in response to rather longstanding problems of census taking. Yet, current socio-technical innovations of information gathering in the census also create new problems, because they are at tension with the data practices of census stakeholders (cf. Diaz-Bone & Horvath, 2021). Methodological innovations, while aiming to restore or stabilize trust in public statistics, can also put this trust at risk, because they create new or previously ignored uncertainties. In this section, we cannot cover the entire range of methodological innovations (more comprehensive: Baffour, King, & Valente, 2013; Skinner, 2018), but focus on the problem of undercoverage and strategies to overcome it, register-based censuses and differential privacy a new method for protecting individual privacy that was first applied to census taking in the 2020 US census.[2]

Although censuses are but one of several instruments for gathering demographic information (Tabutin, 1997), they are typically considered the backbone of national statistical systems. Up to the 1970s, conventional census taking was compulsory, comprehensive, total, periodical, instantaneous, and household-based (Coleman, 2013). Currently, in contrast, there is a key concern of statisticians: "that official population statistics such as those generated by traditional methods like questionnaire-based censuses … are going through a transition and are at a crossroads as methods and data sources are being innovated and diversified" (Ruppert & Scheel, 2021b, p. 1). Radermacher regards the management of Big Data as a critical current stage in the historical development of official statistics – on equal footing with former key methodological developments, such as the growth of statistics as a science in the 19th century and the introduction of major scientific innovations such as inferential statistics in the early 20th century and the computerization of official statistics since the 1970s (Radermacher, 2020, p. 3; Thorvaldsen, 2018, pp. 250–286). New sociodemographic data sources and technological possibilities for their analysis are an issue for countries in the Global North as well as in the Global South (Gbadebo, 2021; Kashyap, 2021; Loveman, 2016; Moultrie, 2016; Prewitt, 2022; van der Brakel, 2022).

A longstanding problem of census taking has been achieving universal coverage (Skinner, 2018). Despite the current claim of the UNFPA that "everyone counts", it happens that some (vulnerable) groups of a *de facto* population of a state are not adequately represented, sometimes even due to outright statistical exclusion (Cobham, 2020; Davis, 2020). A few states have never conducted a full census of their population. In Afghanistan for instance, the first census of the country was started in 1979, but was not completed, because of security issues (Askar, 2019; Simonsen, 2004). Censuses in a number of other countries have left some (especially vulnerable) subpopulations uncounted or underrepresented. In Colombia, close to 3 million internally displaced people have remained officially invisible for a long time (Villaveces-Izquierdo, 2004). Not uncommonly, (ethnic) power relations are behind the will not to count. While it is obvious that improvements in census coverage will require a commitment to further investments in information infrastructure, we also have to bear in mind that being counted also means becoming visible, which is not in the interest of every vulnerable group (Herzog, 2021; Moultrie, 2016). Hence, who defines the categories of those who are not to be left behind?

Differential undercount is not only an issue in developing but also in highly developed countries. In the run-up to the 2000 census, the USA for instance, had faced declining participation rates and differential undercount of urban minorities

(Hillygus, Nie, Prewitt, & Pals, 2006). The Census Bureau strived to tackle these issues with an unprecedented civic mobilization campaign and a dual system estimation that was meant to correct census results using sample surveys of hard-to-count populations, such as undocumented immigrants (Hannah, 2001; cf. Herzog, 2021). While the methodological correction of census results was not applied in the end, the mobilization campaign was evaluated as successful in reducing the differential undercount (Hillygus et al., 2006). Contributing to a society-centered approach to census taking, Michael Rodríguez-Muñiz (2017) analyzed how local leaders have campaigned among unregistered Hispanic immigrants for their participation in the 2010 US census. For that census, it was shown that net undercount affected young children, while young adults and those above 60 are overrepresented (O'Hare, 2019). Undercounted are also males, while females display a net overcount. Whites are counted more accurately than those of other ethnoracial groups, and homeowners more accurately than renters. Ethnographic research revealed that the undercount of children was to a large degree attributable to them living in complex households with nonrelatives, distant relatives, and multigenerational households, a situation that had not been taken into account in the questionnaire so far (Schwede, 2022). In order to address potential undercoverage of vulnerable subgroups in the 2020 census round, the UNFPA committed to advocating for the inclusion of new questions on migration, disability, and marriage (IDWG-Census, 2017/2019, p. 25). The UN published, for example, a handbook covering the issue of how to measure international migration in censuses (UNSD, 2022).

While several of the sociodemographic variables affecting differential undercount are intersectional, the persons represented by them are often concentrated in certain areas. Yet, space is relevant on a much more fundamental level to census taking. Space represents an irreducible dimension of differential undercount because the modern concept of population is inherently tied to a particular territory. It implies a sedentary population with identifiable individuals and an address system identifying housing units (cf. Manderscheid, 2021). It is no coincidence that the population census is typically combined with a housing census. A consequence of the sedentary population concept is that the homeless require concerted counting strategies focusing on shelters and particular urban places in order to be represented (Chamberlain & MacKenzie, 2003; Ruppert & Ustek-Spilda, 2021; Wright & Devine, 1992). Another consequence is that unregistered housing constitutes a problem to legibility: One reason why attempts at conducting a census in Afghanistan have proved futile is that the country lacks a consistent address system (Askar, 2019).

When unregistered housing is limited to certain areas of a country, such as slums in urban areas, geographic information systems (GIS) tools and digital satellite imagery combined with more traditional fieldwork methodologies can be used to gather information and put these areas on the census map (Angeles et al., 2009). Aerial photography allows for the estimation of the population living in these areas (Tabutin, 1997, p. 383). Some of the most innovative approaches to mapping informal settlements use satellite images and learning algorithms (Neal, Seth, Watmough, & Diallo, 2022). In contrast, grassroots citizen science approaches rely on building trust among local actors and the state in order to map areas formerly uncovered by the census (Carranza-Torres, 2021). Recently, the potential of citizen science projects to contribute to SDG indicators has been especially highlighted (Fraisl et al., 2020; MacFeely & Nastav, 2019). The UNFPA promoted hybrid census approaches for also counting populations living in hard-to-reach areas (IDWG-Census, 2017/2019). Recommendations in the "Handbook on Geospatial Infrastructure in Support of Census Activities" are especially crucial for census-related mapping activities in hard-to-reach areas (UN, 2009).

Once slums are officially recognized by the state as a problem, that very category creates an incentive to optimize quantified indicators while questions of quality of access to public services in these areas as well as problems of urban poverty beyond these areas may remain neglected (Bhan & Jana, 2013). In other cases, it is the inhabitants of informal settlements themselves, who take bottom-up civic action to make their area legible to the state in order to get access to public services (Motta, 2019). Nevertheless, the process of quantifying these areas is not linear, and the resulting numbers often remain controversial. In Romania, for example, the enumeration of Roma, who often live concentrated in particular settlements, has been a longstanding problem. Some authors claim that the state has rather successfully counted this group, which indicates also the growing trust of Roma toward the state (Csata, Hlatky, & Liu, 2021). This assessment is based on an external validation of census results by an expert survey in 2,800 municipalities. At the same time, others maintain that census figures are fictitious because they hide several methodological decisions that are questionable and respond to interests favoring overreporting (Surdu, 2019). In 2011, the category Roma was made up of 19 subcategories, most of which referred to occupations traditionally considered to be specific to Roma. Hence, the numbers did not result from individual self-identification with the label Roma, but from post-census coding work by administrators.

The high costs of traditional censuses are of concern to most governments. In a survey on the 2010 census round, census cost was the most commonly reported challenge for governments (67%) (UNSC, 2012, p. 6). Innovative technologies can play a role in reducing the cost of census taking as well as the time burden they put on citizens. Handheld devices and the internet are the main technological innovations relevant to traditional census taking (UNSD, 2019). The 2000 and 2010 rounds saw a decline in the number of traditional censuses from 30 to 24 in the Europe and Central Asia region, and an increase from 9 to 19 in the number of censuses using either registry data alone or in combined methodologies (IDWG-Census, 2017/2019, p. 8). Ten years ago, 23 of 30 European countries disposed of local population registers, and 20 (except Germany, Italy, and Switzerland) were developing central population registers – mostly in order to use them in the census (Poulain & Herm, 2013). In Europe, this shift has been encouraged by the European Union. However, the diversification of census taking has also been observed beyond Europe:

In the 1990 round, only eight countries undertook an alternative census and all were located within Europe. By the 2010 round, this had increased nearly fivefold to 39 countries and included 14 countries outside of Europe, of which half were French Overseas Departments and Territories (DOMTOMs).

(Kukutai, Thompson, & McMillan, 2015, p. 13)

Reliance on population registers for census taking outside of Europe comprised just five countries in the 2010 round: Israel, Bahrain, Turkey, Greenland, and Singapore (Kukutai et al., 2015, p. 16), underlining the infrastructural preconditions required for such a transition. Case studies on register-based censuses in our book cover the cases of Spain and Germany (see Chapters 9 and 10).

The United Nations Economic Commission for Europe has promoted register-based censuses by studying best practices in Scandinavian countries (UN-ECE, 2007) and by developing guidelines for standardizing international practices (UNECE, 2018). Register-based censuses do not rely on questionnaires, but link information from a variety of administrative registers. Once these registers have been put into place, a census can be conducted at only 2% of the cost of a traditional census (UNECE, 2018, p. 2). Furthermore, data gathering would not be bound to decennial intervals, but become available at relatively short notice. Population registers can be considered as a part of state-craft that emerged in the north of Europe and has traveled from there (Szreter & Breckenridge, 2012; Watson, 2018). Today, the right to identity registration at birth is considered a global human right, which nevertheless requires individual states to be realized. The civil registration and vital statistics systems (CRVS) in Europe, North America and Australasia perform better than those in Asia, Latin America, and Africa (Mikkelsen et al., 2015). The coverage of birth registration is generally higher than the level of death registration. Between 2000 and 2012, in 148 countries and territories assessed, the percentage of children aged less than five years whose birth has been registered increased from 58% to 65%, and the percentage of deaths registered increased from 36% to 38%. Various international organizations such as the UN, the World Bank, and the Inter-American Development Bank are advocating for investments in CRVS, which would empower especially women and children as a key to sustainable development (Atick, Dahan, Gelb, & Harbitz, 2016; Harbitz & Kentala, 2015; Silva, Snow, Andreev, Mitra, & Abo-Omar, 2019; UNSD, 2014). At the same time, academic attention to CRVS has increased significantly during the last 30 years (Silva, 2022, p. 3). Research into how CRVS can be integrated with other data sources, such as the census, has also brought to light inconsistencies between them, which is reflected not only in technical terms but also regarding the social processes producing these inconsistencies and their consequences for access to civil rights. Civil registers are increasingly complemented by biometric technologies. Biometric technologies of identification and registration can be understood as "the automated recognition of individuals based on precisely measured features of the body" (Breckenridge, 2014, p. 12). They are today under the most rapid, and systematic, development in countries in Africa and Asia formerly colonized by European empires (cf. Chapter 11; Eyenga, Omgba Mimboe, & Bindzi, 2022). A prominent example of such a biometric database used for the administration of social policy in the first place is the Aadhaar system in India (Rao, 2019). Although the proposal was there already, Aadhaar has not been used to assist census taking so far (Bhagat, 2022, p. 32; see also Chapter 2).

New technologies and methodological innovations in census taking also pose particular challenges to individual privacy. Three risks to privacy result especially from data mining practices that have become available during the last 20 years: re-identification, reconstruction, and tracing (Dwork, Smith, Steinke, & Ullman, 2017). While the notion of anonymization rests on the assumption that an explicit identifier that is contained in a single database is removed, the first type of attack results from the combination of several seemingly unrelated databases that allow for the matching of cases based on the combination of information in each row. Re-construction refers to an attack on released statistical data that contains nonprivate identifying information. The goal of a reconstruction attack is to determine secret bits of information for nearly all individuals in the dataset by using a large amount of nonprivate information. The goal of a tracing attack is to determine the membership of particular individuals in a given dataset. The risk of re-identification sometimes emerges because formal regulation lags behind technological and administrative innovation. Linking administrative data in register-based censuses, for example, requires particular precaution against the risk of re-identifying individuals after data have technically been anonymized (Christen, Ranbaduge, & Schnell, 2020). While in India, for example, the data collected from the census is protected by law and is to be kept confidential, hardly any provision of confidentiality exists regarding the personal information in the National Population Register created as a household list in preparation of the 2011 census (Bhagat, 2022, p. 29).

In the 2020 census round, however, one of the most controversial debates on privacy protection did not focus on a lack of institutional provisions, but on a new technical solution to the risk of privacy breaches adopted by the U.S. Census Bureau: differential privacy (Sullivan, 2020a). Differential privacy means basically adding noise, and random data, to census products such as tables or (public) use files (Hotz & Salvo, 2022). Innovative methods of census technology have historically contributed to broader transformations of societal practices (Ruggles & Magnuson, 2020). In this case, however, the U.S. Census Bureau is an innovation taker from Big Tech companies. The U.S. Census Bureau has applied different methods of privacy protection since the 1980s. They became aware that these were insufficient when the first successful attacks on seemingly anonymous databases were carried out. The only mechanism that seemed to be safe was the differential privacy mechanism hitherto mainly applied in Big Tech companies, but on a much lower scale (Sarathy, 2022). When the U.S. Census Bureau announced the introduction of this new method (Abowd, 2018) and asked its stakeholders about their data needs in order to design the system in the most responsible way, "all hell broke loose", involving lawsuits being filed against the U.S. Census Bureau (Boyd & Sarathy, 2022, p. 8). Differential privacy is highly controversial because it raises doubts about the accuracy and usability of data for science and policymaking (Hauer & Santos-Lozada, 2021; Mervis, 2019). It seems to be clear that many stakeholders held a statistical image of the accuracy of census data that was basically unrealistic. While former privacy protection mechanisms had also deliberately introduced uncertainty, these procedures had never been made explicit to the extent of the differential privacy mechanism (Abowd et al., 2022). The strategic use of uncertainty within the crucial information infrastructure of the state poses a serious challenge to its legitimacy. A satisfactory closure of this debate in the USA has yet to develop (Boyd & Sarathy, 2022).

Comparable levels of conflict, involving constitutional court decisions, were witnessed in Germany, revolving around questions of privacy and the methodological accuracy of the census (Hannah, 2009; Scholz & Kreyenfeld, 2016). While the political conflicts of the 1980s related to questions of privacy, the latest political conflicts related to the methodology of the census in 2011, combining administrative registers and a sample survey of the population (Chapters 4 and 10). Although the court decision confirmed the legitimacy of a combined census in Germany, the theoretical foundations of register-based censuses are still in question (MacDonald, 2020; Wallgren & Wallgren, 2014/2022). In contrast to traditional censuses grappling with undercoverage, register-based censuses face the methodological challenge of overcoverage (Monti, Drefahl, Mussino, & Härkönen, 2020). In Spain, the transition toward a register-based census has been completed already in the 2020 census round. Critique has remained limited in scope and restricted to a quite narrow circle of experts (Treviño Maruri & Domingo, 2020; Chapter 9; Vega Valle, Argüeso Jiménez, & Pérez Julián, 2020). It might be that stakeholders still don't fully grasp the practical implications of this new census type.

At a global level, diversification of census methods creates conditions for the development of multi-partner cooperation between National Statistical Offices (NSOs), UN institutions, and private practitioners such as providers of satellite imagery or GIS specialists. Collaboration between countries is also considered critical. Exchange of information on experiences when testing different methodologies can help countries to choose the most appropriate method given the context and nature of national-specific challenges. This is particularly true for South-South collaboration cases between countries (Chapter 12), but also for cooperation among public, private, and civil society actors (see Chapters 6 and 13).

It is not easy to summarize our discussion of theoretical approaches to census taking and the empirical variety described. For the moment, it might suffice to say that during our discussion we found the analytical tensions we were able to create by referring to world polity theory on the one hand and to typologies aiming for international comparisons or micro-sociological approaches aiming for more processual accounts on the other hand rather productive. This productive tension will be reiterated in the concluding chapter of the book. While we have striven to strike a balance between these analytical perspectives during our hitherto discussion, we will now turn toward giving a brief overview of the 2020 census round.

0.3   The 2020 census round: UN-strategy and the impact of the COVID-19 pandemic

The purpose of the present section is to shed light on UN strategies for coordinating 2020 national censuses, to give an overview of the current state of the census round and the impact of the COVID-19 pandemic as well as describe the applied methods of census taking. While some countries of the Global North also engage in bilateral international cooperation and capacity building (Fischer, 2020; see also Chapter 12), we will focus on the activities of international organizations.

As a consequence of recommendations from an evaluation report of the UNF-PA's support to NSOs and national governments during the 2010 census round (UNFPA Evaluation Office, 2016), the UNFPA for the first time developed an explicit and formal "Strategy for the 2020 Round of Population & Housing Census" (Jhamba, Juran, Jones, & Snow, 2020). Supporting NSOs in capacity building for census taking is the overarching goal of this strategy, which is broken down into several strategic methodological goals that are pursued through established forms of assistance tools and cooperative actions (Table I.2). The strategy is embedded in the 2030 Agenda for Sustainable Development, and the corresponding 17 SDGs. SDGs indicator 17.19.2 underscores the specific importance of census implementation because population data are needed for 98 of the 232 SDG indicators (IDWG-Census, 2017/2019, p. 26).

In following the statistical production chain all the way through, one strategic goal that has been highlighted by the UNFPA strategy for the 2020 census round (Table I.2) stresses the importance of strategic dissemination of data produced by censuses. Data should be freely available not only to policymakers, but also to research and a wider public to favor good governance and innovation. In particular, census data should be used to elaborate indicators such as those designed to measure the level of progress made on SDGs and provide Small Area Estimates. However, such an instrumental use of census data is not always the case, as several chapters of this book show. Chapter 6 stands out regarding the issue of data

Table I.2 UNFPA strategic goals and means for their achievement for the 2020 census round.

Strategic goals	Tools and actions
• Advocacy for use of Handheld devices with Global Navigation Satellite System (GNSS) • Integrating GIS in census and digital mapping • Advocating for greater/wider use of georeferenced census data • Advocating for questions to leave no one behind • Advocating for questions on civil registration • Promoting wider dissemination of census data, while protecting privacy • Generating modeled population estimates, where needed • Supporting the transition to register-based censuses, where relevant • Strengthening governance and quality assurance	• Providing tools and guidelines • Regional workshops • Ongoing technical advice • Chief technical advisors on census • Population data fellows • Operational support • Leveraging institutional partnerships for more effective and coordinated census support • Brokering south-to-south support

Source: Jhamba et al. (2020).

dissemination because it describes a project at the interface of science, policy, and civic engagement that aims at building a network of statistically literate actors for using census data.

In the 2010 census round, 214 of 235 territories conducted census operations covering 93% of the world population (IDWG-Census, 2017/2019, p. 5). The fact that the number of territories participating in the census is higher than the number of sovereign nation-states in the world is due to the separate enumeration of non- or semi-sovereign territories, a heritage of empires that still creates some ambiguity in international statistics (Heintz, 2012). Compared to the 2000 census round, the number of territories not participating was reduced from 26 to 21. Social and political instability was the main factor behind the lack of coverage of the missing 7% of the world population. According to the UNFPA (2022), the following eleven countries held their last census before the 2010 round: Lebanon (1932), Afghanistan (1979), Democratic Republic of Congo (1984), Somalia (1987), Uzbekistan (1989), Iraq (1997), Ukraine (2001), Central African Republic (2003), Haiti (2003), Syrian Arab Republic (2004), Yemen (2004). While some of these countries announced their participation in the 2020 census round (Democratic Republic of Congo, Iraq, Somalia, Ukraine, Uzbekistan), this is not the case for others (Afghanistan, Lebanon, Syrian Arab Republic, Yemen); the announcement by Ukraine that it will conduct a census in 2023 does not seem to be realistic (Appendix, Table I.4). Often, political or violent conflicts lead to humanitarian emergencies that make planning and fieldwork unviable.

The 2020 census round is scheduled to extend over the period from 2015 to 2024 (Appendix, Table I.4)[3], therefore, a final assessment of census activities will be possible only in a few years. Nevertheless, some patterns can be described already. Four types of censuses and four modes of data collection were planned to be employed in 218 territories listed in the Global Census Tracker of the UNFPA (2022). Traditional censuses still comprised the standard type of census taking (72.5%) and around 22% of the countries have relied either completely on it or in combination with a survey on administrative registers (Figure I.1; Appendix, Table I.3). The rolling census is a type employed only in France and in Martinique. In terms of data collection, Computer-Assisted Personal Interviews (CAPI) has been the dominant mode (45.4%), followed by mixed mode (15.1%), Pen-and-Paper Personal Interview (PAPI) (8.7%) and Computer-Assisted Web-Interviewing (CAWI) (3.2%).

Following the UN recommendations for increasing the international comparability of census data in the 2020 census round, the peak years of census taking were planned to be 2020 and 2021 (Mrkić, 2020, p. 36). However, the World Health Organization declared the COVID-19 infection caused by the SARS-CoV-2 virus first a Public Health Emergency of International Concern (30 January 2020) and later a pandemic (11 March 2020). As a consequence, the disease itself and the related containment measures affected census taking in unforeseen ways. Only 16.5% of the countries had completed their census before 2020. Due to delays in the implementation of the census, 2022 became the modal year (31%). Considering that the WHO announced in September 2022 that the end of the pandemic was in sight (Mishra, 2022), one can estimate that the effects of the pandemic will

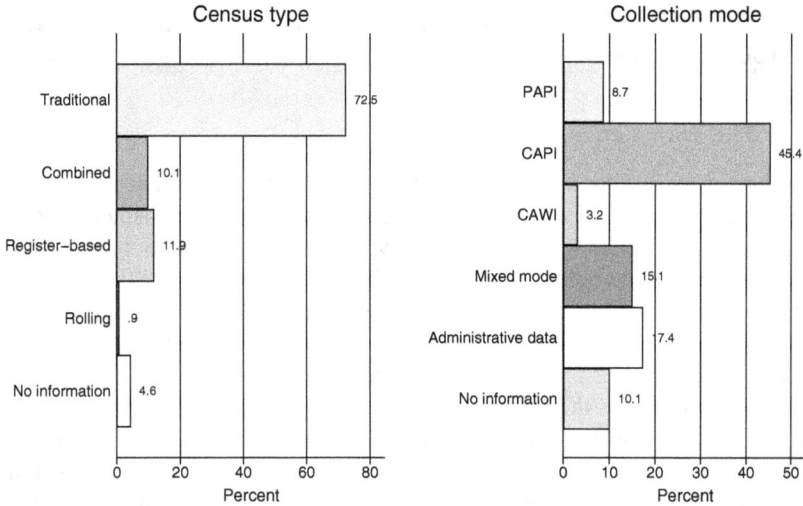

Census type

Traditional	72.5
Combined	10.1
Register–based	11.9
Rolling	.9
No information	4.6

Collection mode

PAPI	8.7
CAPI	45.4
CAWI	3.2
Mixed mode	15.1
Administrative data	17.4
No information	10.1

N=218

Figure I.1 The 2020 global census round by census type and data collection mode.
Source: UNFPA (2022), author's calculations.

be concentrated in the years 2020–2022. In conclusion, 64% of the census round will have been conducted during the pandemic (UNFPA, 2022, own calculations), which has implications for the quality of census taking, comparability with previous censuses, interpretation of results, and maybe also the future of census taking (Mrkic, 2021, p. 483). According to a UN survey conducted among those 121 countries that had scheduled their censuses originally for the years 2020 and 2021, 75% of the 111 countries that responded to the questionnaire answered that the pandemic affected census taking (UNSD, 2021, own calculations). Today it is clear that in 32.6% of the cases registered in the Global Census Tracker, COVID-19 has implied a postponement of implementation (UNFPA, 2022). Four countries have not yet announced new data for census taking (Bosnia and Herzegovina, India, Namibia, and Thailand). According to the UN survey, 41% changed the mode of data collection: 24% resorted to Computer-Assisted Telephone Interviews (CATI), 20% to Computer-Assisted Web-Interviewing (CAWI), 13% used administrative data and 12% changed to self-enumeration with paper questionnaire (UNSD, 2021). The three main difficulties encountered in conducting the census during the COVID-19 pandemic were the need to reduce face-to-face interaction, funding limitations, and mobility restrictions. Among the modes of data collection, CAWI proved to be fairly robust: the census was postponed in only one of seven cases applying that mode. It is noteworthy that the register-based censuses were the least affected by the pandemic. Except for South Korea and some Gulf States in the Middle East, all countries conducting register-based census are European. As a consequence of that and also due to the low cost of this census type, the UN announced that it would increase its methodological support for register-based censuses on a global scale (Mrkic, 2021). Hence, a further increase in this census type is likely to be seen in the 2030 global census round.

0.4   The contributions to this volume

The chapters of this book have been organized into three sections depending on the main topic addressed. The first book section includes the three chapters that focus on issues related to ethnoracial categorization. The second section includes five chapters, which revolve around the tension between professionally rigorous census implementation and the ambivalent relationship to politics. The third section includes another five chapters where methodological and technological innovations are described and their implications discussed.

In order to address the research questions of our book, analyses of changes in census taking are of strategic importance. Chapter 1 by Mara Loveman analyzes the dramatic transformation of the information infrastructure for the production of ethnoracial statistics in Latin America. This transformation is most strikingly evident in national censuses in the region. A few decades ago, almost no Latin American country included questions about race or ethnicity in their national census. By the 2010 census round, almost every country in the region collected information about ethnic or racial identification. Her chapter argues that ethnoracial statistics contribute to shaping the terrain they seem to merely describe, but not always as predicted or intended. Placing the Brazilian experience in a regional comparative perspective illuminates how official ethnoracial statistics can produce outcomes that are simultaneously productive and counterproductive to the aims of those who struggle for their production in the first place. Loveman takes her empirical analysis as an occasion to reflect more broadly on the rationale of using ethnoracial categories in the census.

Another important example of ethnoracial categorization, if we take that notion to include religion, is that of India. The case of India is particularly interesting because the country has realized a highly ambitious, but nevertheless contradictory approach to positive discrimination (Piketty, 2022). In Chapter 2, Ram B. Bhagat stresses the idea that the way in which a state views its people and their characteristics is very much a political phenomenon, which changes according to the nature of the state and its strategy to maintain power. The census makes people legible to the state by converting them into a population divided into mutually exclusive ethnic, religious, racial, and caste categories. In contrast to this practice, historical records show the interwoven and inclusive nature of social identities in many spheres of life in India. Furthermore, there are people identifying with multiple categories, but more often than not, there are no multiple-choice options in census questionnaires, including the census of India. Bhagat argues that this creates a contrived social reality and identity politics in the shadow of the state. While the colonial regime petrified originally rather fluid caste categories through census taking and administrative measures, this process did not change much after independence; instead, it was reinforced and has reemerged in various new forms.

Chapter 3 brings the reader back to the Latin American reality, this time to focus on the categorization of indigenous populations in education censuses. Daniel Capistrano, Christyne Carvalho da Silva, and Rachel Pereira Rabelo describe national regulations as well as data collection practices from ten Latin American countries. Using the past three decades as a reference, they observe that the political context of increasing tensions between neoliberal education reforms and recognition movements has been decisive for the design of statistical systems for education in Latin America. The results of their work indicate that regular data collection on indigenous peoples via education censuses is related to a rather global political commitment by national governments to recognize these populations and their cultural needs (cf. Bennani, 2017). However, further analysis of national education plans suggests that, despite this rather generalized recognition, the political use of these data differs significantly among countries. The authors identify one group that proposes the use of ethnicity data to reinforce traditional practices in educational planning, and another group that proposes a postcolonial revision of these practices.

The five chapters in the second book section revolve around the question of how the relationship between official statistics – as exemplified by the census and policymakers – affects trust in statistics, but also the legitimacy of the state more in general. Census data often conditions governmental policies and can serve as an evaluation of the state of the country. Hence, it is important to have an independent, respected, and trusted institution in charge of its implementation and a vigilant civil society aware of the potential negative implications of particularistic government intervention. Against the backdrop of the crucial role of statistics in the COVID-19 pandemic, in Chapter 4, Walter Radermacher reminds us that official statistics are the backbone of policy in democratic societies. But how can statistical facts be simultaneously constructed and yet objective? Official statistics are more than an assembly of specific surveys; they are a coherent system with an architecture that also corresponds to the respective conditions of a time and a country. Three areas have a particular role to play in several respects (not least because of their expense and budget); they are something like the cornerstones of the business architecture that are important for the statics of the entire system: national accounts and external trade for economic statistics and population censuses for demographic and social statistics. In this sense, the population census and its perspectives are interrelated with the system of official statistics, which in turn is closely linked with social developments and social issues. Patterns and questions of development of population censuses, whether methodological, technical, or political, are therefore usually of importance to the whole system and vice versa. Radermacher points out cases where trust in statistics has been at stake, escalating up to the level of constitutional lawsuits being filed against NSOs. Furthermore, he sets out the reasons why two professional standards could act as potential safeguards against the loss of trust in statistics: the UN Fundamental Statistical Principles and the EU Statistics Code of Practice.

Complementary to that argument, Byron Villacís basically shows in Chapter 5, that professional standards alone are not sufficient for assuring trust in statistics. Villacís presents the results of a comparative analysis of the relationship between census taking and politics in the USA, Brazil, and Ecuador. His research is based on a content analysis of documents from archives, media reports, and in-depth interviews. According to the author, censuses are far from being objective or unbiased; in fact, they are sensitive devices, providing ample opportunities for political intervention that hardly become visible to a larger lay public. Results from his research show that a functional and effective legal system is essential for defending a census that is capable of standing above particularistic interests. Furthermore, an organized society aware of the scope and implications of government interventions is crucial for activating a response, when governments intend to manipulate the process of census data collection for their own convenience.

Adding a third layer to the arguments of the previous two chapters on how to assure the institutional autonomy of census taking, Chapter 6 describes the aims and experiences of the "Latin American Observatory of Population Censuses" (OLAC), a self-financed, autonomous, and collaborative scientific-civic network created by researchers from different countries in this world region. While the informational practices of lay users find increasing interest in research (Agrawal & Kumar, 2020, p. 13), the group of experts explores census processes, socializes information and news that focus on methodological resources, promotes dialogue on demographic and public policy issues, seeks to monitor the processes around national census taking and explores ways to improve statistical literacy both among census takers and (potential) data users. Five years after its foundation, Nicolás Sacco-Zeballos, Gabriel Mendes-Borges, and Byron Villacís review the group's aims, achievements as well as future challenges and opportunities.

Temitope Owolabi maintains in Chapter 7 that Nigeria's developmental problems are inextricably intertwined with the type of census politics that have been repeatedly documented for this country (Obono & Omoluabi, 2014; Olorunfemi & Fashagba, 2021; Serra & Jerven, 2021; Udo, 1998; Yesufu, 1968). His analysis is mainly based on the Nigerian Federation's 2006 population census because the census in the 2020 round is still to be conducted. According to Owolabi, data generated from the population census is used in determining who gets what, when, and why in the country's federal polity. Consequently, there has been an unending drive among Nigerian states to inflate census figures, with the aim of obtaining the advantages arising from having higher population figures in the country. This scenario has created a situation of distributive imbalance and, subsequently, territorial injustice in the allocation of funds and other resources in the federal polity.

The Ukrainian case, discussed in Chapter 8 by Tetyana Tyshchuk and Ilona Sologub, constitutes a peculiar example of politically motivated postponement of census implementation – long before the first Russian invasion of the country and the beginning of the full-scale war on February 24, 2022. After its independence from the Soviet Union, Ukraine had only one census in 2001. Since then, the State Statistics Service of Ukraine (Ukrstat) has been updating the population data using administrative records and living conditions or labor force data using sample surveys. To find out why the government did not have a sense of urgency for the census, the authors performed 20 in-depth interviews with central and local officials. From their research, they concluded that officials do not perceive a necessity for census data since they do not develop policy solutions based on the data and they do not believe in the ability of the national statistical office to collect high-quality data. At the same time, many of them have developed their own sector-specific infrastructure for information gathering.

The third book section focuses on how methodological and technical innovations are shaping census taking around the world. Chapters 9 and 10, by Alberto Veira-Ramos and Walter Bartl, respectively Thomas Körner and Eva Grimm, describe the transitions from a traditional census into a register-based census in Spain and Germany. While this transition has been completed in Spain for the 2020 round, in Germany it is planned to be completed for the 2030 round. In the case of Germany, the transition takes place in two steps: In the 2010 round from a traditional to a combined census, and in the 2030 round from a combined to a register-based model. Such a transition seems to be gaining momentum in European countries, where digitalization of administrative records has advanced remarkably since the beginning of the century. Due to positive experiences with this type of data gathering during the COVID-19 pandemic, it is also actively advocated for by the UN in other world regions (Mrkic, 2021). The EU aims to obtain data more frequently (the ultimate objective being an annual census), more timely, and in more detailed regional breakdowns, while insisting on harmonization. The harmonization of results is achieved mainly via output harmonization, which allows the member states to combine a range of different data sources and to correct for conceptual differences in administrative data. At the same time, the EU and national statistical offices are concerned with privacy issues resulting from record linkage from different parts of the administration. While censuses in the past required the collaboration of the population, the process now becomes an administrative routine depending to a greater extent on the technological performance of administrative registers involved in the provision of various data and their interoperability. Although this shift lowers the financial burden on public budgets and the time burden on the population, it also implies stronger closure of the process of census taking from lay inquiries and a potential gain in power for the state and its experts. The challenges and opportunities of register-based censuses are discussed in greater depth in each of these chapters.

The last two chapters of the third section discuss the complexities and nuances when introducing new technologies in the process of census implementation in two developing countries: Ghana, and Cameroon. Chapter 11 presents results from a combined archival and ethnographic research conducted by Alena Thiel. The author traces the Ghanaian census history from its colonial past to the current ambitions of developing new measurement practices based on digital population data infrastructures. Approaching Ghana's census history from the perspective of data infrastructures and their global circulations, the paper explores the political effects of the continuous translations and adaptations that have shaped the Ghanaian census model since the middle of the 19th century.

Teke Johnson Takwa describes in Chapter 12 the unexpected challenges faced by census enumerators in Cameroon when it was decided that handheld digital devices (CAPI) were to be used instead of the traditional pen-and-paper based (PAPI) questionnaire during the pilot survey for the census in 2023. Problems such as low internet and electricity coverage and respondents' fear of providing information that would be registered in an unfamiliar format added to frequent field mistakes committed by enumerators when using handheld devices for data collection. However, the author claims that despite these challenges, the use of digital devices for data collection in Cameroon can be regarded as a real improvement over the traditional PAPI.

Notes

1 The three categories used here do not match the more elaborate typology of Riley, Ahmed and Emigh (2021). In the epistemological dimension of their typology they distinguish between empiricism, transcendental idealism and praxis. We set aside empiricist accounts of census taking because they usually do not focus on explicating their heuristic assumptions. Hence, our first two categories would count as being rooted epistemologically in transcendental idealism and our third category in an epistemology of praxis.
2 We are grateful to an anonymous reviewer for pointing out this public issue in the USA.
3 We are truly grateful to Tetiana Liubyva for her kind cooperation in producing Table I.4 from UNFPA (2022).

References

Abowd, J. M. (2018). The U.S. Census Bureau adopts differential privacy. In Y. Guo & F. Farooq (Eds.), *Proceedings of the 24th ACM SIGKDD International Conference on Knowledge Discovery & Data Mining* (p. 2867). New York, NY: ACM. doi:10.1145/3219819.3226070

Abowd, J. M., Ashmead, R., Cumings-Menon, R., Garfinkel, S., Heineck, M., Heiss, C., … Zhuravlev, P. (2022). The 2020 Census disclosure avoidance system TopDown algorithm [Special issue]. *Harvard Data Science Review, (2)*. doi:10.1162/99608f92.529e3cb9

Agrawal, A., & Kumar, V. (2020). *Numbers in India's periphery. The political economy of government statistics*. Cambridge: Cambridge University Press. doi:10.1017/9781108762229

Aikins, M. A., Bremberger, T., Aikins, J. K., Gyamerah, D., & Yıldırım-Caliman, D. (2021). *Afrozensus 2020: Perspektiven, Anti-Schwarze Rassismuserfahrungen und Engagement Schwarzer, afrikanischer und afrodiasporischer Menschen in Deutschland.* Berlin, Germany. www.afrozensus.de

Alba, R. (2020). *The great demographic illusion: Majority, minority, and the expanding American mainstream*. Princeton, NJ: Princeton University Press. doi:10.1515/9780691202112

Aly, G., & Roth, K. H. (2004). *The Nazi census: Identification and control in the Third Reich*. Philadelphia, PA: Temple University Press (Original work published 1984).

Anderson, B. (2006). *Imagined communities: Reflections on the origin and spread of nationalism* (2nd ed.). London, England: Verso (Original work published 1983).

Anderson, M. J. (2015). *The American census: A social history* (2nd ed.). New Haven, CT: Yale University Press (Original work published 1988).

Angeles, G., Lance, P., Barden-O'Fallon, J., Islam, N., Mahbub, A. Q., & Nazem, N. I. (2009). The 2005 census and mapping of slums in Bangladesh: Design, select results and application. *International Journal of Health Geographics, 8*(1), 32. doi:10.1186/1476-072X-8-32

Appadurai, A. (1993). Number in the colonial imagination. In C. A. Breckenridge & P. van der Veer (Eds.), *Orientalism and the postcolonial predicament* (pp. 314–339). Philadelphia, PA: University of Pennsylvania Press.

Aragão, R., & Linsi, L. (2022). Many shades of wrong: What governments do when they manipulate statistics. *Review of International Political Economy, 29*(1), 88–113. doi:10.1 080/09692290.2020.1769704

Askar, M. (2019). *The will not to count: Technologies of calculation and the quest to govern Afghanistan* (PhD thesis). McGill University, Montreal, Quebec, Canada. Retrieved from https://www.proquest.com/docview/2605170010

Atick, J., Dahan, M., Gelb, A., & Harbitz, M. (2016). Enabling digital development. Digital identity. In World Bank (Ed.), *World development report 2016: Digital dividends* (pp. 194–197). Washington, DC: The World Bank.

Baffour, B., King, T., & Valente, P. (2013). The modern census: Evolution, examples and evaluation. *International Statistical Review, 81*(3), 407–425. doi:10.1111/insr.12036

Balaton-Chrimes, S., & Cooley, L. (2022). To count or not to count? Insights from Kenya for global debates about enumerating ethnicity in national censuses. *Ethnicities, 22*(3), 404–424. doi:10.1177/14687968211056379

Balestra, C., & Fleischer, L. (2018). *Diversity statistics in the OECD: How do OECD countries collect data on ethnic, racial and indigenous identity?* (OECD Statistics Working Papers No. 2018/09). Paris. doi:10.1787/89bae654-en

Barlösius, E. (2005). *Die Macht der Repräsentation: Common Sense über soziale Ungleichheiten.* Wiesbaden, Germany: VS Verlag für Sozialwissenschaften.

Bartl, W., Papilloud, C., & Terracher-Lipinski, A. (2019). Governing by numbers: Key indicators and the politics of expectations. An introduction [Special issue]. *Historical Social Research, 44*(2), 7–43. doi:10.12759/hsr.44.2019.2.7-43

Beaman, J., & Petts, A. (2020). Towards a global theory of colorblindness: Comparing colorblind racial ideology in France and the United States. *Sociology Compass, 14*(4), e12774. doi:10.1111/soc4.12774

Beck, U. (1974). *Objektivität und Normativität: die Theorie-Praxis-Debatte in der modernen deutschen und amerikanischen Soziologie.* Reinbek bei Hamburg, Germany: Rowohlt.

Bennani, H. (2017). *Die Einheit der Vielfalt: zur Institutionalisierung der globalen Kategorie "indigene Völker".* Frankfurt (Main), Germany: Campus.

Bennani, H., & Müller, M. (2021). "Who are we and how many?" – Zur statistischen Konstruktion globaler Personenkategorien. *KZfSS Kölner Zeitschrift für Soziologie und Sozialpsychologie, 73*(S1), 223–252. doi:10.1007/s11577-021-00747-x

Bhagat, R. B. (2022). *Population and the political imagination: Census, register and citizenship in India.* London, England: Routledge.

Bhan, G., & Jana, A. (2013). Of slums or poverty: Notes of caution from census 2011. *Economic and Political Weekly, 48*(18), 13–16.

Bieber, F. (2015). The construction of national identity and its challenges in post-Yugoslav censuses. *Social Science Quarterly, 96*(3), 873–903. doi:10.1111/ssqu.12195

Biermann, F., Hickmann, T., Sénit, C.-A., Beisheim, M., Bernstein, S., Chasek, P., . . . Nilsson, M. (2022). Scientific evidence on the political impact of the sustainable development goals. *Nature Sustainability, 5*(9), 795–800. doi:10.1038/s41893-022-00909-5

Blum, A. (2002). Resistance to identity categorization in France. In D. I. Kertzer & D. Arel (Eds.), *Census and identity: The politics of race, ethnicity, and language in national censuses* (pp. 121–147). Cambridge, England: Cambridge University Press.

Boltanski, L., & Thévenot, L. (2006). *On justification: Economies of worth.* Princeton, NJ: Princeton University Press (Original work published 1991).

Bourdieu, P. (1994). Rethinking the state: Genesis and structure of the bureaucratic field. *Sociological Theory, 12*(1), 1–18.

Boyd, D., & Sarathy, J. (2022). Differential perspectives: Epistemic disconnects surrounding the U.S. Census Bureau's use of differential privacy [Special issue]. *Harvard Data Science Review, 4*(Issue 2). doi:10.1162/99608f92.66882f0e

Brambor, T., Goenaga, A., Lindvall, J., & Teorell, J. (2020). The lay of the land: Information capacity and the modern state. *Comparative Political Studies, 53*(2), 175–213. doi:10.1177/0010414019843432

Breckenridge, K. (2014). *Biometric state: The global politics of identification and surveillance in South Africa, 1850 to the present.* Cambridge, England: Cambridge University Press. doi:10.1017/CBO9781139939546

Brubaker, R. (2008). *Nationalist politics and everyday ethnicity in a Transylvanian town* (2nd ed.). Princeton, NJ: Princeton University Press (Original work published 2006).

Cakici, B., & Ruppert, E. (2021). Data subjects: Calibrating and sieving. In E. Ruppert & S. Scheel (Eds.), *Data practices: Making up a European people* (pp. 205–235). London, England: Goldsmith Press.

Cameron, G. J., Dang, H.-A. H., Dinc, M., Foster, J., & Lokshin, M. M. (2021). Measuring the statistical capacity of nations. *Oxford Bulletin of Economics and Statistics, 83*(4), 870–896. doi:10.1111/obes.12421

Carlitz, R. D., & McLellan, R. (2021). Open data from authoritarian regimes: New opportunities, new challenges. *Perspectives on Politics, 19*(1), 160–170. doi:10.1017/S1537592720001346

Carranza-Torres, J. A. (2021). How can traditional statistical relationships be redefined through citizen to government partnerships? *Statistical Journal of the IAOS, 37*, 229–243. doi:10.3233/SJI-190578

Chamberlain, C., & MacKenzie, D. (2003). *Australian census analytic program: Counting the homeless, 2001.* Canberra, Australia: Australian Bureau of Statistics.

Christen, P., Ranbaduge, T., & Schnell, R. (2020). *Linking sensitive data. Methods and techniques for practical privacy-preserving information sharing.* Cham, Switzerland: Springer. doi:10.1007/978-3-030-59706-1_14

Cobham, A. (2020). *The uncounted.* Cambridge, MA: Polity Press.

Coleman, D. (2013). The twilight of the census. *Population and Development Review, 38*, 334–351. doi:10.1111/j.1728-4457.2013.00568.x

Csata, Z., Hlatky, R., & Liu, A. H. (2021). How to head count ethnic minorities: Validity of census surveys versus other identification strategies. *East European Politics, 37*(3), 572–92. doi:10.1080/21599165.2020.1843439

Curtis, B. (2002). Foucault on governmentality and population: The impossible discovery. *The Canadian Journal of Sociology / Cahiers canadiens de sociologie, 27*(4), 505–533. doi:10.2307/3341588

DaCosta, K. A. (2020). Multiracial categorization, identity, and policy in (mixed) racial formations. *Annual Review of Sociology, 46*(1), 335–353. doi:10.1146/annurev-soc-121919-054649

Davis, S. L. M. (2020). *The uncounted: Politics of data in global health.* Cambridge, England: Cambridge University Press. doi:10.1017/9781108649544

de Paiva Rio Camargo, A., & Daniel, C. J. (2021). Social studies of quantification and its implications in sociology. *Sociologias, 23*(56), 42–81. doi:10.1590/15174522-109768

de Vries, W. (1999). Are we measuring up...? Questions on the performance of national statistical systems. *International Statistical Review/Revue Internationale de Statistique, 67*(1), 63–77.

Des Neiges Léonard, M. (2015). Who counts in the census? Racial and ethnic categories in France. In R. Sáenz, N. P. Rodríguez, & D. G. Embrick (Eds.), *International handbooks* on population: Vol. 4. The international handbook of the demography of race and ethnicity* (pp. 537–552). Dordrecht, The Netherlands: Springer.

Desrosières, A. (1998). *The politics of large numbers: A history of statistical reasoning.* Cambridge, MA: Harvard University Press (Original work published 1993).

Desrosières, A. (2000). Measurement and its uses: Harmonization and quality in social statistics. *International Statistical Review = Revue internationale de statistique, 68*(2), 173–187. doi:10.1111/j.1751-5823.2000.tb00320.x

Desrosières, A. (2009). How to be real and conventional: A discussion of the quality criteria of official statistics. *Minerva, 47*(3), 307–322.

Desrosières, A. (2011). Words and numbers. For a sociology of the statistical argument. In A. R. Sætnan, H. M. Lomell, & S. Hammer (Eds.), *The mutual construction of statistics and society* (pp. 41–63). New York, NY: Routledge.

Diaz-Bone, R., & Didier, E. (2016). The sociology of quantification - Perspectives on an emerging field in the social sciences. *Historical Social Research, 41*(2), 7–26. doi:10.12759/hsr.41.2016.2.7-26

Diaz-Bone, R., & Horvath, K. (2021). Official statistics, big data and civil society. Introducing the approach of "economics of convention" for understanding the rise of new data worlds and their implications. *Statistical Journal of the IAOS, 37*, 219–228. doi:10.3233/SJI-200733

Didier, E. (2002). Sampling and democracy. Representativeness in the first United States surveys. *Science in Context, 15*(3), 427–445. doi:10.1017/S0269889702000558

Didier, E. (2021). *Quantitative marbling: New conceptual tools for the socio-history of quantification* (Anton Wilhelm Amo Lectures No. 7). Halle (Saale). Retrieved from Martin-Luther-Universität Halle-Wittenberg website: https://www.scm.uni-halle.de/amo_lecture/

Di Fiore, M., Kuc-Czarnecka, M., Lo Piano, S., Puy, A., & Saltelli, A. (2022). The challenge of quantification: An interdisciplinary reading. *Minerva, 61*(1), 53–70. doi:10.1007/s11024-022-09481-w

Dixon, A. R., & Telles, E. E. (2017). Skin color and colorism: Global research, concepts, and measurement. *Annual Review of Sociology, 43*(1), 405–424. doi:10.1146/annurev-soc-060116-053315

Durr, J.-M. (2020). The balance between governance support needed and influence avoided: The case of population censuses. *Statistical Journal of the IAOS, 36*(1), 211–215. doi:10.3233/SJI-190513

Dwork, C., Smith, A., Steinke, T., & Ullman, J. (2017). Exposed! A survey of attacks on private data. *Annual Review of Statistics and Its Application, 4*(1), 61–84. doi:10.1146/annurev-statistics-060116-054123

Egeler, R., Dinsenbacher, N., & Kleber, B. (2013). The relevance of census results for a modern society. In W. Franz & P. Winker (Eds.), *Jahrbücher für Nationalökonomie und Statistik: 233/3. 150 Years Journal of Economics and Statistics* (pp. 389–405). Stuttgart, Germany: Lucius & Lucius. doi:10.1515/jbnst-2013-0307

Elrick, J., & Schwartzman, L. F. (2015). From statistical category to social category: Organized politics and official categorizations of 'persons with a migration background' in Germany. *Ethnic and Racial Studies, 38*(9), 1539–1556. doi:10.1080/01419870.2014.996240

Emigh, R. J. (2002). Numeracy or enumeration? The uses of numbers by states and societies. *Social Science History, 26*(4), 653–698.

Emigh, R. J., Riley, D. J., & Ahmed, P. (2016a). *Antecedents of censuses from medieval to nation states: How societies and states count.* New York, NY: Palgrave Macmillan.

Emigh, R. J., Riley, D. J., & Ahmed, P. (2016b). *Changes in censuses from imperialist to welfare states: How societies and states count.* New York, NY: Palgrave Macmillan. doi:10.1057/9781137485069

Emigh, R. J., Riley, D. J., & Ahmed, P. (2020). The sociology of official information gathering: Enumeration, influence, reactivity, and power of states and societies. In T. Janoski, C. de Leon, J. Misra, & I. W. Martin (Eds.), *The new handbook of political sociology* (pp. 290–320). Cambridge, England: Cambridge University Press.

Espeland, W., & Stevens, M. L. (2008). A sociology of quantification. *European Journal of Sociology, 49*(3), 401–436. doi:10.1017/S0003975609000150

Eyenga, G. M., Omgba Mimboe, G., & Bindzi, J. F. (2022). Être sans-papiers chez soi? Les mésaventures de l'encartement biométrique au Cameroun. *Critique internationale, 97*(4), 113–134. doi:10.3917/crii.097.0116

Fabricius, A. (1868). Ueber factische und rechtliche Bevölkerung. *Jahrbücher für Nationalökonomie und Statistik, 10*, 1–19.

Fachkommission Integrationsfähigkeit (Ed.). (2021). *Gemeinsam die Einwanderungsgesellschaft gestalten. Bericht der Fachkommission der Bundesregierung zu den Rahmenbedingungen der Integrationsfähigkeit.* Berlin, Germany: Bundeskanzleramt.

Fischer, O. P. (2020). Innovative approaches to support the global 2020 round of population and housing censuses. *Statistical Journal of the IAOS, 36*(1), 51–54. doi:10.3233/SJI-190559

Foremny, D., Jofre-Monseny, J., & Solé-Ollé, A. (2017). 'Ghost citizens': Using notches to identify manipulation of population-based grants. *Journal of Public Economics, 154*, 49–66. doi:10.1016/j.jpubeco.2017.08.011

Foucault, M. (1991). Governmentality. In G. Burchell, C. Gordon, & P. Miller (Eds.), *The Foucault effect: Studies in governmentality with two lectures by and an interview with Michel Foucault* (pp. 87–104). Chicago, IL: University of Chicago Press.

Fraisl, D., Campbell, J., See, L., Wehn, U., Wardlaw, J., Gold, M., … Fritz, S. (2020). Mapping citizen science contributions to the UN sustainable development goals. *Sustainability Science, 15*(6), 1735–1751. doi:10.1007/s11625-020-00833-7

Gbadebo, B. M. (2021). Old, contemporary and emerging sources of demographic data in Africa. In C. O. Odimegwu & Y. Adewoyin (Eds.), *The Routledge handbook of African demography* (pp. 111–128). London, England: Routledge.

Georgiou, A. V. (2017). Towards a global system of monitoring the implementation of UN fundamental principles in national official statistics. *Statistical Journal of the IAOS, 33*(2), 387–397. doi:10.3233/SJI-160335

Gil, B., & Omaboe, E. N. (2005). Population censuses and national sample surveys in developing countries. In M. Bulmer & D. P. Warwick (Eds.), *Social research in developing countries: Surveys and censuses in the Third World* (pp. 43–54). London, England: UCL Press.

Goldthorpe, J. E. (1952). Attitudes to the census and vital registration in East Africa. *Population Studies, 6*(2), 163–171.

Hacking, I. (1990). *The taming of chance.* Cambridge, England: Cambridge University Press.

Hannah, M. G. (2001). Sampling and the politics of representation in US census 2000. *Environment and Planning D: Society and Space, 19*(5), 515–534. doi:10.1068/d289

Hannah, M. G. (2009). Calculable territory and the West German census boycott movements of the 1980s. *Political Geography, 28*(1), 66–75. doi:10.1016/j.polgeo.2008.12.001

Harbitz, M., & Kentala, K. (2015). *Dictionary for civil registration and identification.* Washington, DC: Inter-American Development Bank.

Hauer, M. E., & Santos-Lozada, A. R. (2021). Differential privacy in the 2020 census will distort COVID-19 rates. *Socius, 7*, 2378023121994014.

Hauser, P. M. (1973). Statistics and politics. *The American Statistician, 27*(2), 68–71.

Heintz, B. (2012). Welterzeugung durch Zahlen. Modelle politischer Differenzierung in internationalen Statistiken, 1948–2010. *Soziale Systeme, 18*(1+2), 7–39.

Henneguelle, A. (2023). Socio-economics of quantification and value: The perspective of convention theory. In R. Diaz-Bone & G. d. Larquier (Eds.), *Handbook of economics and sociology of conventions* (pp. 1–20). Cham, Switzerland: Springer. doi:10.1007/978-3-0 30-52130-1_55-1

Herzog, B. (2021). Managing invisibility: Theoretical and practical contestations to disrespect. In G. Schweiger (Ed.), *Migration, recognition and critical theory* (pp. 211–227). Cham, Switzerland: Springer. doi:10.1007/978-3-030-72732-1_10

Hillygus, D. S., Nie, N. H., Prewitt, K., & Pals, H. (2006). *The hard count: The political and social challenges of census mobilization.* New York, NY: Russell Sage Foundation.

Ho, J.-M. (2019). *Social statisticalization: Number, state, science* (Dissertation). Cornell University, New York, NY.

Hoh, A.-L. (2018). *Counting for EU enlargement? Census-taking in Croatia, Bosnia and Macedonia* (PhD thesis). Maastricht University, Maastricht, The Netherlands. doi:10.26481/dis.20180926ah

Hotz, V. J., & Salvo, J. (2022). A chronicle of the application of differential privacy to the 2020 census [Special issue]. *Harvard Data Science Review, 4*(2). doi:10.1162/99608f92. ff891fe5

IDWG-Census. (2019). *UNFPA strategy for the 2020 round of population & housing censuses (2015–2024).* New York, NY. Retrieved from UNFPA Inter-Divisional Working Group on Census website: https://www.unfpa.org/sites/default/files/pub-pdf/Census_ Strategy_Final_July.pdf

Jhamba, T., Juran, S., Jones, M., & Snow, R. (2020). UNFPA strategy for the 2020 round of population and housing censuses (2015–2024). *Statistical Journal of the IAOS, 36*(1), 43–50. doi:10.3233/SJI-190600

Jowell, R. (1986). The codification of statistical ethics. *Journal of Official Statistics, 2*(3), 217–253.

Kashyap, R. (2021). Has demography witnessed a data revolution? Promises and pitfalls of a changing data ecosystem. *Population Studies, 75*(sup1), 47–75. doi:10.1080/0032472 8.2021.1969031

Kertzer, D. I., & Arel, D. (2006). Population composition as an object of political struggle. In R. E. Goodin & C. Tilly (Eds.), *The Oxford handbook of contextual political analysis* (pp. 664–677). Oxford, England: Oxford University Press. doi:10.1093/oxfordhb/ 9780199270439.003.0036

Kreager, P. (2004). Objectifying demographic identities. In S. Szreter, H. Sholkamy, & A. Dharmalingam (Eds.), *Categories and contexts: Anthropological and historical studies in critical demography* (pp. 33–56). Oxford, England: Oxford University Press.

Kukutai, T., & Thompson, V. (2015). 'Inside out': The politics of enumerating the nation by ethnicity. In P. Simon, V. Piché, & A. Gagnon (Eds.), *Social statistics and ethnic diversity: Cross-national perspectives in classifications and identity politics* (pp. 39–61). Cham, Switzerland: Springer. doi:10.1007/978-3-319-20095-8_3

Kukutai, T., Thompson, V., & McMillan, R. (2015). Whither the census? Continuity and change in census methodologies worldwide, 1985–2014. *Journal of Population Research, 32*(1), 3–22. doi:10.1007/s12546-014-9139-z

Labbé, M. (2019). National indifference, statistics and the constructivist paradigm. The case of the Tutejsi ('the people from here') in interwar polish censuses. In M. van Ginderachter & J. Fox (Eds.), *National indifference and the history of nationalism in modern Europe* (pp. 161–179). London, England: Routledge.

Lee, M. M., & Zhang, N. (2016). Legibility and the informational foundations of state capacity. *The Journal of Politics, 79*(1), 118–132. doi:10.1086/688053

Lehtonen, M. (2019). The multiple faces of trust in statistics and indicators: A case for healthy mistrust and distrust. *Statistical Journal of the IAOS, 35*(4), 539–548. doi:10.3233/SJI-190579

Lieberman, E. S., & Singh, P. (2017). Census enumeration and group conflict: A global analysis of the consequences of counting. *World Politics, 69*(1), 1–53. doi:10.1017/S0043887116000198

von der Lippe, P. (1999). Die politische Rolle der amtlichen Statistik ein der ehemaligen DDR. In Statistisches Bundesamt (Ed.), *Amtliche Statistik - ein konstitutives Element des demokratischen Staates: 50 Jahre Bundesrepublik Deutschland* (pp. 25–46). Wiesbaden, Germany: Statistisches Bundesamt.

Ljones, O. (2015). Twenty years of the United Nations fundamental principles of official statistics: The present and an outlook into the future. *Statistical Journal of the IAOS, 31*(3), 375–379. doi:10.3233/SJI-150910

Loveman, M. (2005). The modern state and the primitive accumulation of symbolic power. *American Journal of Sociology, 110*(6), 1651–1683. doi:10.1086/428688

Loveman, M. (2014). *National colors. Racial classification and the state in Latin America.* Oxford, England: Oxford University Press.

Loveman, M. (2016). New data, new knowledge, new politics: Race, color, and class inequality in Latin America. In J. Hooker & A. B. Tillery (Eds.), *The double bind: The politics of racial & class inequalities in the Americas* (pp. 47–56). Washington, DC: American Political Science Association.

MacDonald, A. L. (2020). Of science and statistics: The scientific basis of the census. *Statistical Journal of the IAOS, 36*(3), 17–34. doi:10.3233/SJI-190596

MacFeely, S., & Nastav, B. (2019). "You say you want a [data] revolution": A proposal to use unofficial statistics for the SDG global indicator framework. *Statistical Journal of the IAOS, 35*(3), 309–327. doi:10.3233/SJI-180486

Maier, W. (1959). Gedanken über den Brauch und Mißbrauch der Statistik in der Politik. *Allgemeines statistisches Archiv, 43*(1), 12–16.

Maktabi, R. (1999). The Lebanese census of 1932 revisited. Who are the Lebanese? *British Journal of Middle Eastern Studies, 26*(2), 219–241.

Manderscheid, K. (2021). Concepts of society in official statistics. Perspectives from mobilities research and migration studies on the re-figuration of space and cross-cultural comparison. *Forum: Qualitative Sozialforschung/Forum: Qualitative Social Research, 22*(2), Art. 15. doi:10.17169/fqs-22.2.3719

Mann, M. (1984). The autonomous power of the state: Its origins, mechanisms and results. *European Journal of Sociology, 25*(2), 185–213. doi:10.1017/S0003975600004239

Mennicken, A., & Espeland, W. (2019). What's new with numbers? Sociological approaches to the study of quantification. *Annual Review of Sociology, 45*(1), 223–245. doi:10.1146/annurev-soc-073117-041343

Merry, S. E. (2019). The sustainable development goals confront the infrastructure of measurement. *Global Policy, 10*(S1), 146–148. doi:10.1111/1758-5899.12606

Mervis, J. (2019). Researchers object to census privacy measure. *Science, 363*(6423), 114. doi:10.1126/science.363.6423.114

Mespoulet, M. (2022). Creating a socialist society and quantification in the USSR. In A. Mennicken & R. Salais (Eds.), *The new politics of numbers* (pp. 45–70). Cham, Switzerland: Springer.

Michalopoulou, C. (2016). Statistical internationalism: From Quetelet's census uniformity to Kish's cross-national sample survey comparability. *Statistical Journal of the IAOS, 32*(4), 545–554. doi:10.3233/SJI-160960

Mikkelsen, L., Phillips, D. E., AbouZahr, C., Setel, P. W., Savigny, D. de, Lozano, R., & Lopez, A. D. (2015). A global assessment of civil registration and vital statistics systems: Monitoring data quality and progress. *Lancet (London, England), 386*(10001), 1395–1406. doi:10.1016/S0140-6736(15)60171-4

Mirga, A. (2018). Mihai Surdu, those who count: Expert practices of Roma classification, Central European University Press, 2016. *Critical Romani Studies, 1*(1), 114–126. doi:10.29098/crs.v1i1.14

Mishra, M. (2022, September 14). End of COVID pandemic is 'in sight' - WHO chief. *Reuters.* Retrieved from https://www.reuters.com/business/healthcare-pharmaceuticals/who-chief-says-end-sight-covid-19-pandemic-2022-09-14/

Monk, E. P. (2022). Inequality without groups: Contemporary theories of categories, intersectional typicality, and the disaggregation of difference. *Sociological Theory, 40*(1), 3–27. doi:10.1177/07352751221076863

Monti, A., Drefahl, S., Mussino, E., & Härkönen, J. (2020). Over-coverage in population registers leads to bias in demographic estimates. *Population Studies, 74*(3), 451–469. doi:10.1080/00324728.2019.1683219

Morning, A. (2008). Ethnic classification in global perspective: A cross-national survey of the 2000 census round. *Population Research and Policy Review, 27*(2), 239–272.

Morning, A., & Rodríguez-Muñiz, M. (2016). *Race in the demographic imaginary: Population projections and their conceptual foundations* (NYU Population Center Working Paper). New York, NY: New York University.

Motta, E. (2019). Resisting numbers: The favela as an (un)quantifiable reality. In F. Neiburg, J. I. Guyer, G. Onto, M. Bolt, D. James, B. Maurer, … M. Luzzi (Eds.), *The real economy: Essays in ethnographic theory* (pp. 77–102). Chicago, IL: HAU Books.

Moultrie, T. A. (2016). Demography, demographers and the 'data revolution' in Africa. *Afrique Contemporaine, 258*(2), 25–39.

Mrkic, S. (2021). Conducting population and housing censuses during the pandemic: An overview. *Statistical Journal of the IAOS, 37*, 483–493. doi:10.3233/SJI-210820

Mrkić, S. (2020). The 2020 round of population and housing censuses: An overview. *Statistical Journal of the IAOS, 36*(1), 35–42. doi:10.3233/SJI-190574

Neal, I., Seth, S., Watmough, G., & Diallo, M. S. (2022). Census-independent population estimation using representation learning. *Scientific Reports, 12*(1), 5185. doi:10.1038/s41598-022-08935-1

Nobles, M. (2000). *Shades of citizenship: Race and the census in modern politics.* Stanford, CA: Stanford University Press.

Obono, O., & Omoluabi, E. (2014). Technical and political aspects of the 2006 Nigerian population and housing census. *African Population Studies, 27*(2), 249–262. doi:10.11564/27-2-472

O'Hare, W. P. (2019). *Differential undercounts in the U.S. census: Who is missed?* Cham, Switzerland: Springer. doi:10.1007/978-3-030-10973-8

Olorunfemi, J. F., & Fashagba, I. (2021). Population census administration in Nigeria. In R. Ajayi & J. Y. Fashagba (Eds.), *Nigerian politics* (pp. 353–368). Cham, Switzerland: Springer.

Paschel, T. S. (2016). *Becoming black political subjects: Movements and ethno-racial rights in Colombia and Brazil.* Princeton, NJ: Princeton University Press.

Pavez, M. A. (2020). *Using lenses to understand policy failures: The case of the 2012 census in Chile* (PhD dissertation). University of Massachusetts, Boston, MA.

Piketty, T. (2022). *A brief history of equality*. Cambridge, MA: Harvard University Press.

Porter, T. M. (1986). *The rise of statistical thinking: 1820–1900*. Princeton, NJ: Princeton University Press.

Porter, T. M. (1995). *Trust in numbers: The pursuit of objectivity in science and public life*. Princeton, NJ: Princeton University Press.

Poston, J. D. L. (2020). The decennial census and congressional apportionment. *Harvard Data Science Review, 2*(1). doi:10.1162/99608f92.2b99e39a

Potvin, M. (2005). The role of statistics on ethnic origin and 'race' in Canadian anti-discrimination policy. *International Social Science Journal, 57*(183), 27–42. doi:10.1111/j.0020-8701.2005.00529.x

Poulain, M., & Herm, A. (2013). Central population registers as a source of demographic statistics in Europe. *Population, 68*(2), 183–212. doi:10.3917/popu.1302.0215

Prévost, J.-G. (2019). Politics and policies of statistical independence. In M. J. Prutsch (Ed.), *Science, numbers, and politics* (pp. 153–180). Cham, Switzerland: Springer.

Prewitt, K. (1987). Public statistics and democratic politics. In W. Alonso & P. Starr (Eds.), *The politics of numbers* (pp. 261–274). New York, NY: Russell Sage Foundation.

Prewitt, K. (2010). What is political interference in federal statistics? *The AN-NALS of the American Academy of Political and Social Science, 631*(1), 225–238. doi:10.1177/0002716210373737

Prewitt, K. (2022). 2030: A sensible census, in reach. In A. L. Carriquiry, J. M. Tanur, & W. F. Eddy (Eds.), *Statistics in the public interest* (pp. 321–336). Cham, Switzerland: Springer.

Radermacher, W. J. (2020). *Official statistics 4.0: Verified facts for people in the 21st century*. Cham, Switzerland: Springer. doi:10.1007/978-3-030-31492-7

Rallu, J.-L., Piché, V., & Simon, P. (2006). Demography and ethnicity: An ambiguous relationship. In G. Caselli, J. Vallin, & G. J. Wunsch (Eds.), *Demography: Analysis and synthesis. A treatise in population studies* (Vol. 3, pp. 531–549). Boston, MA: Elsevier.

Rao, U. (2019). Population meets database: Aligning personal, documentary and digital identity in aadhaar-enabled India. *South Asia, 42*(3), 537–553. doi:10.1080/00856401.2019.1594065

Renard, L. (2021). Vergleichsverbot? Bevölkerungsstatistiken und die Frage der Vergleichbarkeit in den deutschen Kolonien (1885–1914). *Kölner medizinhistorische Beitrage, 73*, 169–194. doi:10.1007/s11577-021-00745-z

Riley, D., Ahmed, P., & Emigh, R. J. (2021). Getting real: Heuristics in sociological knowledge. *Theory and Society, 50*(2), 315–356. doi:10.1007/s11186-020-09418-w

Ringel, L., Espeland, W., Sauder, M., & Werron, T. (Eds.). (2021). *Research in the sociology of organizations: Vol. 74. Worlds of rankings*. Bingley, England: Emerald.

Rocha, Z. L., & Aspinall, P. J. (2020). Introduction: Measuring mixedness around the world. In Z. L. Rocha & P. J. Aspinall (Eds.), *The Palgrave international handbook of mixed racial and ethnic classification* (pp. 1–25). Cham, Switzerland: Palgrave. doi:10.1007/978-3-030-22874-3_1

Rodríguez-Muñiz, M. (2017). Cultivating consent: Nonstate leaders and the orchestration of state legibility. *American Journal of Sociology, 123*(2), 385–425. doi:10.1086/693045

Rose, N. (1991). Governing by numbers: Figuring out democracy. *Accounting, Organizations and Society, 16*(7), 673–692.

Roseth, B., Reyes, A., & Amézaga, K. Y. (2019). *The value of official statistics* (IDB Technical Note No. 1682). Inter-American Development Bank. doi:10.18235/0001883

Rottenburg, R., Merry, S. E., Park, S.-J., & Mugler, J. (Eds.). (2015). *The world of indicators. The making of governmental knowledge through quantification.* Cambridge, England: Cambridge University Press.

Ruggles, S., & Magnuson, D. L. (2020). Census technology, politics, and institutional change, 1790–2020. *The Journal of American History, 107*(1), 19–51. doi: 10.1093/jahist/jaaa007

Ruppert, E., & Scheel, S. (2021a). Data practices. In E. Ruppert & S. Scheel (Eds.), *Data practices: Making up a European people* (pp. 29–48). London, England: Goldsmith Press.

Ruppert, E., & Scheel, S. (2021b). Introduction: The politics of making up a European people. In E. Ruppert & S. Scheel (Eds.), *Data practices: Making up a European people* (pp. 1–28). London, England: Goldsmith Press.

Ruppert, E., & Ustek-Spilda, F. (2021). Refugees and homeless people: Coordinating and narrating. In E. Ruppert & S. Scheel (Eds.), *Data practices: Making up a European people* (pp. 89–124). London, England: Goldsmith Press.

Sætnan, A. R., Lomell, H. M., & Hammer, S. (Eds.). (2011). *The mutual construction of statistics and society.* New York, NY: Routledge.

Sarathy, J. (2022). From algorithmic to institutional logics: The politics of differential privacy. *SSRN Electronic Journal.* Retrieved from https://privacytools.seas.harvard.edu/files/privacytools/files/ssrn-id4079222.pdf

Scholz, R. D., & Kreyenfeld, M. (2016). The register-based census in Germany: Historical context and relevance for population research. *Comparative Population Studies, 41*(2), 175–204. doi:10.12765/CPoS-2016-08en

Schor, P. (2017). *Counting Americans: How the US census classified the nation.* New York, NY: Oxford University Press (Original work published 2009).

Schwede, L. (2022). Persistent undercounts of race and hispanic minorities and young children in US censuses. In T. M. Redding & C. C. Cheney (Eds.), *Profiles of anthropological praxis: An international casebook* (pp. 260–282). New York, NY: Berghahn.

Scott, J. C. (1998). *Seeing like a state: How certain schemes to improve the human condition have failed.* New Haven, CT: Yale University Press.

Seltzer, W. (1994). *Politics and statistics: Independence, dependence or interaction?* New York, NY. Retrieved from United Nations website: https://unstats.un.org/unsd/statcom/FP-Seltzer.pdf

Seltzer, W., & Anderson, M. (2008). Using population data systems to target vulnerable population subgroups and individuals: Issues and incidents. In J. Asher, D. Banks, & F. J. Scheuren (Eds.), *Statistical methods for human rights* (pp. 273–328). New York, NY: Springer. doi:10.1007/978-0-387-72837-7_13

Serra, G., & Jerven, M. (2021). Contested numbers: Census controversies and the press in 1960s Nigeria. *Journal of African History, 62*(2), 235–253. doi:10.1017/S0021853721000438

Silva, R. (2022). Population perspectives and demographic methods to strengthen CRVS systems: Introduction. *Genus, 78*(1), 1–15. doi:10.1186/s41118-022-00156-8

Silva, R., Snow, R., Andreev, D., Mitra, R., & Abo-Omar, K. (2019). *Strengthening CRVS systems, overcoming barriers and empowering women and children.* Ottawa, Ontario, Canada. Retrieved from International Development Research Centre website: http://hdl.handle.net/10625/60225

Simon, P. (2005). The measurement of racial discrimination: The policy use of statistics. *International Social Science Journal, 57*(183), 9–25. doi:10.1111/j.0020-8701.2005.00528.x

Simon, P. (2015). The making of racial and ethnic categories: Official statistics reconsidered. In P. Simon, V. Piché, & A. Gagnon (Eds.), *Social statistics and ethnic diversity:*

Cross-national perspectives in classifications and identity politics (pp. 1–16). Cham, Switzerland: Springer.

Simon, P. (2017). The failure of the importation of ethno-racial statistics in Europe: Debates and controversies. *Ethnic and Racial Studies, 40*(13), 2326–2332. doi:10.1080/0141987 0.2017.1344278

Simon, P. (2021). Discrimination: Studying the racialized structure of disadvantage. In R. Zapata-Barrero, D. Jacobs, & R. Kastoryano (Eds.), *Contested concepts in migration studies* (pp. 78–94). London, England: Routledge. doi:10.4324/9781003119333-6

Simonsen, S. G. (2004). Ethnicising Afghanistan? Inclusion and exclusion in post-Bonn institution building. *Third World Quarterly, 25*(4), 707–729. doi:10.1080/01436590410 001678942

Skinner, C. (2018). Issues and challenges in census taking. *Annual Review of Statistics and Its Application, 5*(1), 49–63. doi:10.1146/annurev-statistics-041715-033713

Song, M. (2021). Who counts as multiracial? *Ethnic and Racial Studies, 44*(8), 1296–1323. doi:10.1080/01419870.2020.1856905

Stebelsky, I. (2009). Ethnic self-identification in Ukraine, 1989–2001: Why more Ukrainians and fewer Russians? *Canadian Slavonic Papers, 51*(1), 77–100.

Strang, D., & Meyer, J. W. (1993). Institutional conditions for diffusion. *Theory and Society, 22*(4), 487–512.

Sullivan, T. A. (2020a). *Census 2020: Understanding the issues.* Cham, Switzerland: Springer. doi:10.1007/978-3-030-40578-6

Sullivan, T. A. (2020b). Coming to our Census: How social statistics underpin our democracy (and republic). *Harvard Data Science Review, 2*(1), 1–22. doi:10.1162/99608f92.c871f9e0

Supik, L. (2014). *Statistik und Rassismus: das Dilemma der Erfassung von Ethnizität.* Frankfurt (Main), Germany: Campus.

Supik, L., & Spielhaus, R. (2019). Introduction to special issue: Matters of classification and representation: Quantifying ethnicity, religion and migration. *Ethnicities, 19*(3), 455–468. doi:10.1177/1468796819833434

Surdu, M. (2016). *Those who count. Expert practices of Roma classification.* Budapest, Hungary: Central European University Press.

Surdu, M. (2019). Why the "real" numbers on Roma are fictitious: Revisiting practices of ethnic quantification. *Ethnicities, 19*(3), 486–502.

Szreter, S., & Breckenridge, K. (2012). Registration and recognition: The infrastructure of personhood in world history. In K. Breckenridge & S. Szreter (Eds.), *Registration and recognition: Documenting the person in world history* (pp. 1–36). Oxford, England: Oxford University Press.

Tabutin, D. (1997). Sistemas de información en demografía. *Estudios Demográficos y Urbanos, 12*(3), 377–426.

Telles, E. E. (2014). *Pigmentocracies: Ethnicity, race, and color in Latin America.* Chapel Hill, NC: University of North Carolina Press.

Tesfaye, F. (2014). *Statistique(s) et génocide au Rwanda: la genèse d'un système de catégorisation génocidaire.* Paris, France: L'Harmattan.

Thévenot, L. (1984). Rules and implements: Investment in forms. *Social Science Information, 23*(1), 1–45. doi:10.1177/053901884023001001

Thompson, D. E. (2016). *The schematic state: Race, transnationalism, and the politics of the census.* Cambridge, England: Cambridge University Press.

Thorvaldsen, G. (2018). *Censuses and census takers: A global history.* London, England: Routledge.

Tilly, C. (1999). *Durable inequality*. Berkeley, CA: University of California Press.

Treviño Maruri, R., & Domingo, A. (2020). Goodbye to the Spanish census? Elements for consideration. *Revista Española de Investigaciones Sociológicas, 43* (171), 107–124. doi:10.5477/cis/reis.171.107

Udo, R. K. (1998). Geography and population censuses in Nigeria. In O. Areola (Ed.), *Fifty years of geography in Nigeria: The Ibadan story. Essays in commemoration of the golden jubilee of the University of Ibadan, 1948–1998* (pp. 348–372). Ibadan, Nigeria: Ibadan University Press.

UN. (2009). *Handbook on geospatial infrastructure in support of census activities* (Studies in Methods, Series F No. 103). New York, NY: United Nations. doi:10.18356/30560942-en

UN. (2016). *Handbook on the management of population and housing censuses: Revision 2* (No. ST/ESA/STAT/SER.F/83/Rev.2). New York, NY: United Nations.

UN. (2017). *Principles and recommendations for population and housing censuses: 2020 round*. New York, NY: United Nations.

UN. (2021). *Handbook on the management of population and housing censuses: Revision 2*. New York, NY: United Nations. Retrieved from https://unstats.un.org/unsd/publication/seriesF/Series_F83Rev2en.pdf

UNECE. (2007). *Register-based statistics in the Nordic countries*. Geneva, Switzerland: United Nations Economic Commission for Europe.

UNECE. (2018). *Guidelines on the use of registers and administrative data for population and housing censuses*. New York, NY: United Nations.

UNFPA. (2022). Global census tracker. Retrieved from https://experience.arcgis.com/experience/6c3954186a17429b84fe518af72aa674/page/Page/?views=Census-Timeline-%282020-Round%29

UNFPA Evaluation Office. (2016). *Evaluation of UNFPA support to population and housing census data to inform decision-making and policy formulation (2005–2014)*. New York, NY. Retrieved from United Nations Population Fund website: https://www.unfpa.org/admin-resource/evaluation-unfpa-support-population-and-housing-census-data-inform-decision-making

UNSC. (2012). *Report of the United States of America on the 2010 world programme on population and housing censuses* (No. E/CN.3/2012/2). New York, NY: United Nations.

UNSD. (2014). *Principles and recommendations for a vital statistics system* (Statistical Papers, Series M No. 19/Rev.3). New York, NY. Retrieved from United Nations Statistics Division website: https://unstats.un.org/unsd/demographic/standmeth/principles/M19Rev3en.pdf

UNSD. (2019). *Guidelines on the use of electronic data collection technologies in population and housing censuses*. New York, NY. Retrieved from United Nations Statistics Division website: https://unstats.un.org/unsd/demographic/standmeth/handbooks/guideline-edct-census-v1.pdf

UNSD. (2021). *2nd UNSD survey on impact of COVID-19 on censuses: [Responses 2021-4–19. Data provided by Srdjan Mrkic and Andrew Smith]*. New York, NY: United Nations Statistics Division.

UNSD. (2022). *Handbook on measuring international migration through population Censuses* (No. ST/ESA/STAT/SER.F/115). New York, NY. Retrieved from United Nations website: https://unstats.un.org/unsd/demographic-social/publication/SeriesF_115en.pdf

van der Brakel, J. (2022). New data sources and inference methods for official statistics. In A. L. Carriquiry, J. M. Tanur, & W. F. Eddy (Eds.), *Statistics in the public interest* (pp. 411–431). Cham, Switzerland: Springer.

Vega Valle, J. L., Argüeso Jiménez, A., & Pérez Julián, M. (2020). Moving towards a register based census in Spain. *Statistical Journal of the IAOS, 36*, 187–192. doi:10.3233/SJI-190516

Veira-Ramos, A., & Liubyva, T. (2020). Ukrainian identities in transformation. In A. Veira-Ramos, T. Liubyva, & E. Golovakha (Eds.), *Ukraine in transformation* (pp. 203–228). Cham, Switzerland: Springer.

Ventresca, M. J. (1995). *When states count: Institutional and political dynamics in modern census establishment, 1800–1993* (PhD thesis). Stanford University, CA.

Villaveces-Izquierdo, S. (2004). Internal diaspora and state imagination: Colombia's failure to envision a nation. In S. Szreter, H. Sholkamy, & A. Dharmalingam (Eds.), *Categories and contexts: Anthropological and historical studies in critical demography* (pp. 173–184). Oxford, England: Oxford University Press.

Vollmer, H. (2007). How to do more with numbers: Elementary stakes, framing, keying, and the three-dimensional character of numerical signs. *Accounting, Organizations and Society, 32*(6), 577–600.

Wallgren, B., & Wallgren, A. (2022). *Register-based statistics: Statistical methods for administrative data* (3rd ed.). Hoboken, NJ: John Wiley & Sons (Original work published 2014).

Walter, M., Kukutai, T., Carroll, S. R., & Rodriguez-Lonebear, D. (Eds.). (2020). *Indigenous data sovereignty and policy*. London, England: Routledge.

Watson, I. (2005). *Cognitive design: Creating the sets of categories and labels that structure our shared experience*. New Brunswick, NJ: Rutgers State University of New Jersey. Retrieved from http://ianwatson.org/cognitive_design_a4.pdf

Watson, I. (2018, March). *The Nordic national ID tradition: A guidepost for the future or a relic from the past?* Paper presented at FutureID3 Conference, Jesus College, Cambridge, England. Retrieved from http://www.ianwatson.org/nordic-national-id-tradition.pdf

Weiss, A. M. (1999). Much ado about counting: The conflict over holding a census in Pakistan. *Asian Survey, 39*(4), 679–693.

Will, A.-K. (2019). The German statistical category "migration background": Historical roots, revisions and shortcomings. *Ethnicities, 19*(3), 535–557.

Wright, J. D., & Devine, J. A. (1992). Counting the homeless: The census bureau's "S-night" in five US cities. *Evaluation Review, 16*(4), 355–364.

Yesufu, T. M. (1968). The politics and economics of Nigeria's population census. In J. C. Caldwell & C. Okonjo (Eds.), *The population of tropical Africa* (pp. 106–116). London, England: Longmans.

APPENDIX

Table 1.3 Types of census and modes of data collection in the 2020 census round.

Mode of data collection	Census type						
	Traditional	*Combined*	*Register*	*Rolling*	*No info*	*Total*	*%*
PAPI	19					19	8.7
CAPI	95	4				99	45.4
CAWI	2	5				7	3.2
Mixed mode	20	13				33	15.1
No info	22					22	10.1
Not applicable	0	0	26	2	10	38	17.4
Total	158	22	26	2	10	218	100
%	72.5	10.1	11.9	0.9	4.6	100	

Source: UNFPA (2022), own compilation.

Table I.4 2020 Global census round by type, method, country, and year.

Type	Method	Country (year)
Traditional	PAPI	Azerbaijan (2019), Bhutan (2017), Bolivia (Plurinational State of) (2024), Bosnia and Herzegovina (2023), Cambodia (2019), Chile (2024), Comoros (2017), Democratic People's Republic of Korea (2019), Ecuador (2022), Guatemala (2018), Madagascar (2018), Monaco (2016), Mozambique (2017), New Caledonia (2019), Pakistan (2017), Paraguay (2022), Peru (2017), Solomon Islands (2019), Tunisia (2024)
	CAPI	Albania (2022), Angola (2024), Antigua and Barbuda (2022), Aruba (2020), Bahamas (2022), Barbados (2022), Belize (2022), Benin (2023), Botswana (2022), Brazil (2022), Burkina Faso (2019), Burundi (2023), Cameroon (2023), Cape Verde (2021), Cook Islands (2022), Costa Rica (2022), Côte d'Ivoire (2021), Croatia (2021), Cuba (2023), Curaçao (2022), Cyprus (2021), Democratic Republic of the Congo (2023), Djibouti (2023), Dominica (2022), Dominican Republic (2022), Egypt (2017), El Salvador (2023), Equatorial Guinea (2015), Eswatini (2017), Ethiopia (2022), Fiji (2017), Gabon (2023), Gambia (2023), Ghana (2021), Grenada (2022), Guinea (2024), Guinea-Bissau (2023), Guyana (2022), Haiti (2022), India (NA), Jamaica (2022), Jordan (2015), Kenya (2019), Kiribati (2020), Lao People's Democratic Republic (2015), Lesotho (2016), Liberia (2022), Malawi (2018), Maldives (2022), Malta (2021), Marshall Islands (2021), Mauritania (2023), Mauritius (2022), Mexico (2020), Micronesia (Federated States of) (2022), Moldova (2024), Morocco (2024), Myanmar (2024), Namibia (2022), Niger (2022), Nigeria (2023), Panama (2023), Papua New Guinea (2024), Rwanda (2022), Saint Kitts and Nevis (2022), Saint Lucia (2022), Saint Vincent and the Grenadines (2023), Samoa (2022), Sao Tome and Principe (2023), Senegal (2023), Serbia (2022), Seychelles (2022), Sierra Leone (2021), Somalia (2023), South Sudan (NA), Sri Lanka (2023), Sudan (2023), Suriname (2022), Timor-Leste (2022), Togo (2022), Tonga (2021), Trinidad and Tobago (2022), Turkmenistan (2022), Turks and Caicos Islands (2022), Tuvalu (2022), Uganda (2023), United Republic of Tanzania (2022), Uruguay (2023), Vanuatu (2020), Viet Nam (2019), West Bank (2017), Zambia (2022), Zimbabwe (2022)
	CAWI	Canada (2021), Greece (2021)
	Mixed modes	Argentina (2022), Bangladesh (2022), Central African Republic (2023), Chad (2023), Colombia (2018), Georgia (2024), Kyrgyzstan (2022), Mali (2022), Nepal (2021), Philippines (2020), Portugal (2021), Puerto Rico (USA) (2020), Russian Federation (2021), South Africa (2022), Tadjikistan (2020), Ukraine (2023), United Kingdom of Great Britain & Northern Ireland (2021), United States of America (2020), United States Virgin Islands (USA) (2020), Uzbekistan (2023)
	No information on data collection methods	Algeria (2022), Anguilla (2021), Australia (2021), Bermuda (2016), British Virgin Islands (2022), Brunei Darussalam (2021), Bulgaria (2021), Cayman Islands (2021), French Polynesia (2017), Guam (USA) (NA), Honduras (2024), Japan (2020), Libya (2023), Malaysia (2020), Montserrat (2022), Nauru (2021), New Zealand (2018), Nicaragua (2022), Niue (2022), Palau (2021), Tokelau (2022), Venezuela (2022)

Combined	CAPI	Armenia (2022), Belarus (2019), China (2020), Iraq (2022)
	CAWI	Czechia (2021), Liechtenstein (2020), Luxembourg (2021), Switzerland (2020)
	Mixed modes	Germany (2022), Hungary (2022), Indonesia (2020), Iran (Islamic Republic of) (2016), Israel (2021), Italy (2021), Kazakhstan (2021), Mongolia (2020), North Macedonia (2021), Poland (2021), Romania (2022), Saudi Arabia (2022), Slovakia (2021), Thailand (2022)
Register-based		Andorra (2021), Austria (2021), Bahrain (2020), Belgium (2021), Canary Islands (2021), Denmark (2020), Estonia (2021), Finland (2021), Greenland (2021), Iceland (2021), Ireland (2022), Kuwait (2022), Latvia (2021), Lithuania (2021), Montenegro (2022), Netherlands (2021), Norway (2020), Oman (2020), Qatar (2021), Republic of Korea (2020), Singapore (2020), Slovenia (2015), Spain (2021), Sweden (2021), Turkey (2021), United Arab Emirates (2022)
Rolling census		France (2020), Martinique (NA)
No information on the type and year of census		Afghanistan, Eritrea, Holy See, Hong Kong, Lebanon, Macao, San Marino, Syrian Arab Republic, Taiwan, Yemen

Sources: UNFPA 2022, own compilation.

Notes: CAPI = Computer-assisted personal interview, CAWI = Computer-assisted web interview

I

The politics of ethnoracial categories

1 An avalanche of ethnoracial population data

On the productive politics of official ethnoracial statistics in 21st-century Latin America*

Mara Loveman

1.1 Introduction: an avalanche of ethnoracial population data

The first decades of the 21st century witnessed an unprecedented explosion of officially produced ethnic and racial demographic data in Latin America. This avalanche of ethnoracial statistics stems largely from the dramatic shift in how Latin American states classify and count their populations on national censuses.[1] Since the 2000s, almost every country in the region modified its national census to collect new types of data about individuals' ethnoracial identification. Across Latin America, states that had long refrained from collecting ethnoracial statistics in national censuses shifted course, adopting new questions that recognize ethnic or racial differences within their populations.

The rather sudden regional embrace of ethnic and racial data collection on censuses in the first decades of the 21st century is summarized in Figure 1.1. The shaded cells in Figure 1.1 indicate that a country took a national census in that decade. A white circle indicates that the census included a question that made indigenous populations statistically visible. A black circle indicates that the census included a question that made black or Afro-descendant populations statistically visible.

In the first decades of the 21st century, Latin America's national censuses increasingly made visible individuals who identify as indigenous or Afro-descendant. In the 1980s, approximately half of Latin American countries identified indigenous populations in the context of national censuses. By 2010, almost all of them had done so or planned to do so in the next census round. With respect to Afro-descendant populations, in the 1980s, only two countries – Brazil and Cuba – included census questions that differentiated them from others in the population. By 2010, nearly every Latin American country included a census question to count black or Afro-descendant identifying individuals, or planned to include such a question in their next census.[2] Taken together, the changes to Latin American censuses in this period marked a dramatic transformation of the datascape for the collection and production of ethnoracial demographic statistics.

This chapter argues that the transformation of the datascape for the production of ethnoracial population data in Latin America is both a *product* of politics and

DOI: 10.4324/9781003259749-3

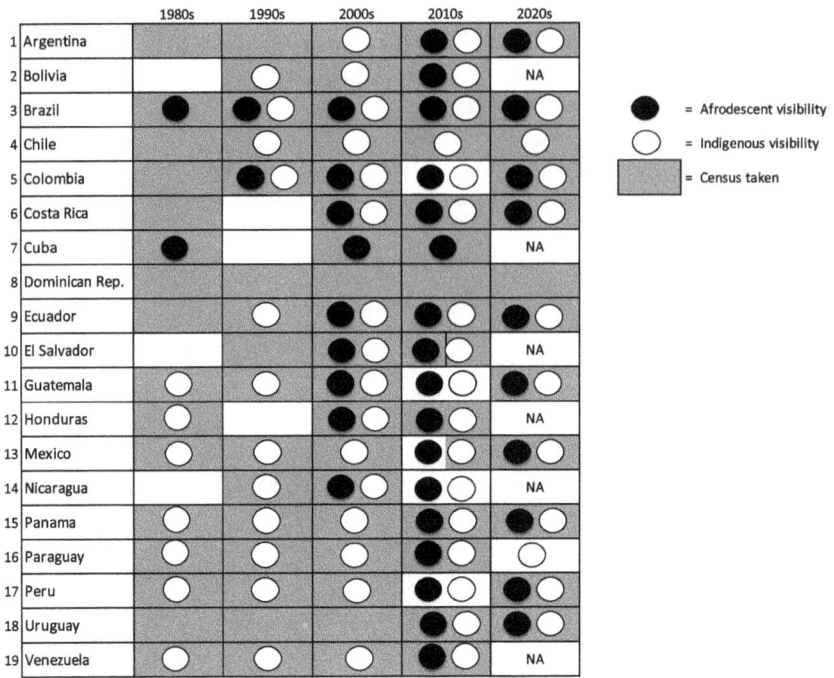

Figure 1.1 Visibility of Afro-descendant and indigenous peoples in Latin American censuses, 1980–2020s

Source: Loveman, 2014, p. 253 (updated with available information for the 2020 census round as of March 2023, with gratitude for research assistance by Byron Villacis)

productive of politics. The transformation breaks with several decades of official color-blindness in census enumeration in much of the region, a break that is attributable first and foremost to political struggle. Once in place, meanwhile, the transformed datascape contributes to shaping the ethnoracial realities it claims to merely describe, affecting the contours of the political terrain in which official ethnoracial statistics are subsequently produced and deployed. Placing the Brazilian experience in a regional comparative perspective illuminates how official ethnoracial statistics can produce outcomes that are simultaneously productive and counter-productive to the aims of those who struggle for their production in the first place.

1.2   Historical background: national censuses and the idea of progress as demographic whitening

To appreciate how politically momentous the late 20th and early 21st-century transformation of ethnoracial data production in national censuses is for Latin America, it is helpful to recall the longer history of ethno-demographic statistical production in the region. The recent shift in Latin America's national censuses to include questions about race and ethnicity marks a radical departure from the dominant practice in most of the region since mid-century. Prior to the 1960 census round, however, the collection of ethnic and racial statistics in Latin American national censuses has substantial historical precedent.

Throughout the 19th century and the first half of the 20th, most Latin American countries collected ethnoracial statistics on at least one national census if not more.[3] The ethnoracial demographic statistics produced in these decades were generated and deployed in heterogeneous ways. However, there were some shared themes that are evident in official census reports from across the region.

Prominent among these shared themes, many official census reports from the 19th to early 20th centuries use demographic data to substantiate arguments that national populations were becoming gradually more homogenous, and gradually whiter. Many authors of census reports noted declines in black and indigenous populations, whether through differential mortality and fertility or through *mestizaje*. They pointed to these supposedly objective demographic trends as positive signs of national progress.[4] In the context of published national census reports, statistics were used to support nationalist, and nationalizing, narratives that emphasized the gradual dissolution or disappearance of racially distinct indigenous or Afro-descendant groups within the population. With variations on the theme across contexts, Latin American political elites and scientists pointed to official statistics published in census results to craft and support stories of national progress conceived as ethnoracial homogenization in the direction of a whiter demographic future.

Official statistics were arranged in misleadingly clear comparative tables, and presented as irrefutable evidence of populations trending "naturally" in a whiter direction. Most of the time, such statistical tables were presented matter-of-factly and without much analysis, suggesting that the numbers were believed to speak for themselves. In some contexts, however, the accompanying narrative was decidedly triumphant. An infamous example of such racist triumphalism comes from the lengthy introduction to the Brazilian 1920 national census, penned by Oliveira Vianna. Discussing changes in Brazil's ethnoracial composition from 1872 to 1890, and ignoring the myriad known problems with census data he pointed to as evidence, Vianna wrote: "The delicate and complex mechanism of ethnic selection has been explained in the previous paragraphs; however, the demonstration of the excellence of its effects is this statistical table" (Table 1.1).

Vianna celebrated (his reading of) statistical trends, which he presented to readers as clear evidence of the racial progress of the nation. At one point, he enthused:

Table 1.1 Use of statistics to purportedly show the demographic "whitening" of Brazil in the 1920 census

ANNOS	Brancos %	Negros %	Indios %	Mestiços %
1872	38.1	19.7	3.9	38.3
1890	44.0	14.6	9.0	32.4

Source: Brazil. Directoria Geral de Estatistica, 1922, p. 344.

"See how fast the destruction of the black population is in the extreme south ... In contrast to the descendent evolution of the two inferior types [we can see] the magnificent ascendant movement of the aryan type" (Brazil. Directoria Geral de Estatistica, 1922, p. 344).

The introduction to Brazil's 1920 census stands out from other national censuses in the region in this period in the celebratory tone of the reading of statistics that purported to show the demise of distinct black and indigenous groups. It also stands out in its eccentric and problematic reliance on already-at-the-time anachronist strands of "race science." But the main argument in Vianna's essay is one that is echoed, in somewhat more neutral language and with context-specific variation, across many national censuses from this period. A common theme running through national census reports was to emphasize the gradual blending and demographic disappearance of distinct ethnic and racial groups as national populations became "mixed" and, ideally, "whiter."

Notably, up until the 1960s, these types of narratives were presented in official censuses results when data on indigenous and Afro-descent populations was collected, and also, often, when it was not. Thus, the presence or absence of questions about race or ethnicity in the first century or so of national census-taking in Latin America is not in itself a direct or clear signal of when, whether, or how much the producers of official demographic statistics were focused on the racial demographic trajectories of their populations. Brazil is again exemplary. After including a race/color question in the national census of 1872 and 1890, the question was omitted in 1900 and 1920. Yet the official report on the 1920 census focused much more intently on demographic trends in the racial composition of the population than in any of the decades prior (de Paiva Rio Camargo, 2010; Loveman, 2009; Nobles, 2000; Piza & Rosemberg, 1999).

As in Brazil, so in the rest of the region, national censuses were a strategic site, and official demographic statistics were an authoritative scientific language, for constructing and disseminating narratives of national progress via "racial progress." These narratives were anchored in and through the putative objectivity and authority of demographic science and stamped with the imprimatur of the state (Desrosières, 1998; Loveman, 2014). Scientific-demographic national-racial fictions complemented stories of national progress through whitening being told through other mediums. Beyond the world of official demographic data production, in literature and art, for example, the idea of national progress as and through racial progress became a pervasive theme. Again, this broad-brush generalization overlays enormous variations in the specifics of these narratives across and within countries over time. Still, as we know from a large and rich historiography, narrative and artistic constructions of national mythologies that celebrated the dissolution or disappearance of ethnoracial or ethnocultural difference became a common genre across the region, beginning especially in the latter decades of the 19th century and well into the 20th.[5]

The most well-known Brazilian version of this genre crystallized as a narrative of whitening through racial mixture, conceived as a pacific process through which the nation would ultimately develop into a more homogenous, and whiter, national

Figure 1.2 Modesto y Brocco, "The Redemption of Cain" 1895.

type. The core demographic notion of whitening through mixture is captured in the famous and frequently cited painting, Modesto y Brocco's "The Redemption of Cain" (1895) (Figure 1.2), which depicts racial mixture as an intergenerational process through which the Brazilian nation formed, and also as a process through which the population was and would be, continuously and miraculously, whitened.

There were of course dissenting voices, scientific and artistic both.[6] But the national narrative of Brazil as a racially-mixed and gradually whitening nation, curated from the mid-to-late 19th century, became an anchor for subsequent variations on the theme. In the lead-up to the 1920 census, for example, a political cartoon that

Figure 1.3 Representation of Brazil in publicity for the 1920 national census.

Source: A Careta, Ano XIII, n.620, May 8, 1920. Translation: "US!" – The census: "How many are you?" Reply – "One! Indivisible!"

publicized the event depicted Brazil as a singular, white, (male) national type – as if it were the baby from Modesto y Brocco's painting, all grown up (Figure 1.3).

Well into the 20th century, political, scientific, and artistic elites in Brazil and throughout much of Latin America advanced versions of national narratives that explicitly or implicitly equated national progress with the disappearance of distinct indigenous or Afro-descendant populations within the nation. National ideologies that championed race mixture as an alchemical ingredient of "racial democracy" (Gilberto Freyre) or "a nation for all" (José Martí) or a "cosmic race" (José

Vasconcelos) recognized and even valued some aspects of racial difference, but only in the context of assuming its eventual demographic disappearance and cultural dilution – processes conceived almost always with a directionality toward whiteness.

From the 1930s, and especially in the wake of World War II, the political, scientific, and moral legitimacy of the race concept became internationally suspect. This shifting international normative and scientific context unsettled and undercut the explicit equation of national progress with racial progress – and the definition of racial progress as demographic movement toward whiteness. In this altered context, Latin American census officials sought ways to reconcile their commitments to the continued pursuit of national progress with the global delegitimization of "race" as a scientific concept and of "racial improvement" as a political project (Loveman, 2014, Chapter 6). In many countries, this reconciliation involved the omission of race questions from national censuses. Countries that continued to enumerate ethnic difference in the context of the national census shifted increasingly toward questions about cultural traits and behaviors; these questions allow for the continued statistical visibility of indigenous populations in several countries, while contributing to the statistical invisibilization of Afro-descendants throughout most of the region. Brazil and Cuba became outliers in the region for the continuity of questions about "race" or "color" on the national census (with the exception of Brazil's 1970 census). Throughout almost all of Latin America, the second half of the 20th century became an era of official statistical color-blindness.

By the 2000 census round, Brazil and Cuba were outliers no more. Almost every Latin American country had introduced, or reintroduced, a census question to make Afro-descendant populations visible. And the same was true for the statistical visibility of indigenous populations throughout the region. Viewed against Latin America's long history of using official demographic data to document the blurring of ethnoracial distinctions and to mark, and even celebrate, the disappearance of differentiated Afro-descendant or indigenous populations within the nation, the regional embrace of ethnoracial data collection on national censuses appeared as a tectonic cultural and political shift. Instead of insisting on the blending and disappearance of ethnoracial distinctions in their populations, Latin American state actors now officially recognize enduring ethnoracial distinctions in their populations. And they increasingly proclaim the pluri-ethnic composition of their nations.[7]

This regional shift in Latin America's censuses is the product of political struggle. At the same time, the transformed datascape is itself politically productive: producing new subjectivities, new organizational sites, and new stakes of politics. In the two next sections, I elaborate on this argument. First, I briefly describe how the transformation of Latin America's ethnoracial statistical datascape resulted from political struggle, focusing on the articulation between actors embedded in social movements, the social sciences, and government agencies at local, national, regional, and international levels. I then turn to an analysis of some of the things that the production of ethnoracial statistics has become productive of, underlining the ways new data are transforming the political terrain that led to their production, and not always in ways that those who fought for their production intend.

1.3   How politics transformed the datascape

Existing accounts of the political struggles that led to the regional shift in census-taking practices in Latin America point to the critical role of mobilization by Afro-descendant and indigenous activists, strategic collaboration with allied academics, and support and pressure from interested international organizations (de Popolo, 2008; Hooker, 2005; Htun, 2004; Loveman, 2014; Nobles, 2000; Paschel, 2010, 2016). In Brazil, where the continuity of racial classification on national censuses was disrupted by the removal of the question by the military government in 1970, targeted political advocacy by a small number of black movement activists, social scientists, and technocrats succeeded in getting the question reintroduced in the census of 1980.[8] Outside Brazil, advocates of census reform had to persuade national statistics agencies to introduce race or ethnicity questions for the first time in decades, or in some countries for the first time in their history. With Brazil's experience as a crucial example, the addition of questions to national censuses crystallized as one concrete objective of activists who mobilized for ethnoracial recognition, rights, and redress within the broader political struggles for democratization and human rights in Latin America from the 1980s and into the early 2000s.

From the perspective of activists working to build the constituencies for mass movements domestically, and to increase the visibility of their struggles internationally, it made good sense to focus on the national census as a strategic political target. Activists identified the national census as a high-profile political stake to advance both symbolic and material goals. New questions on national censuses would, in themselves, repudiate narratives that denied enduring ethnoracial difference within the nation – narratives that historical census-taking practices had themselves helped to naturalize. To be named and counted on the national census was to win official recognition of the social existence of a collective identity or community. Official recognition, in turn, fortified the social reality of named groups. The census became a prime target of activists because the official classification was rightly seen as key to undermine the symbolic violence of the past and open up new avenues for effective claims-making in the future.

Activists also targeted the national census because of the practical and political value of data that could speak to the material conditions of minoritized populations. Census data are not just any data; they are *politically authoritative* data, precisely because their production is supposed to be above and outside politics. Their objectivity is legitimated through their production by institutionalized expertise that combines the (supposedly neutral) authority of Science and the State. Vested with this dual legitimacy, activists correctly recognized that ethno-demographic statistics could be deployed as ideal "tools of the weak" (Porter, 1995); the same tool used in the construction and naturalization of the unjust social order could be deployed to expose the existence of injustices and to "furnish arguments" to contest them (Desrosières, 2014, p. 351; cf. Lorde, 2018). In most Latin American countries, national censuses remain the primary source of statistical information about the population. Even in countries that conduct regular household surveys, the national census provides authoritative population data that serves as a baseline comparison for other surveys. Activists recognized that the inclusion of new questions on racial or ethnic identity or group membership would generate official, scientifically authoritative data to document the existence of and inequalities between distinct ethnoracial groups within the nation.

Brazil provided a model that activists elsewhere in the region sought to emulate and adapt to their own national contexts. The Brazilian experience suggested a more or less linear political trajectory: from the inclusion of a question on the census, to the production of statistical studies documenting inequalities between groups, to the development and implementation of policies to target those inequalities. In reality, of course, the Brazilian story was much more complicated and conjunctural, and riddled with particularities and tensions that made it far from a simple template for activists elsewhere to follow.

Brazilian activists' starting point for this political trajectory was quite distinct from the position of activists elsewhere in the region. As noted above, Brazilian national censuses had long included a race question. Its removal in 1970 was a deviation, so advocates were arguing for a return to the status quo ante rather than for a radical innovation, as was the case for several other countries. Also unlike the situation in many other countries, in Brazil activists drew upon and built from a long history of black activism (Andrews, 1991; Hanchard, 1994; Mitchell, 1992). While the black movement in the 1970s was relatively small and geographically concentrated, and while one of its main challenges was to "raise consciousness" of blackness among broader sectors of the Afro-descendant Brazilian population, the social fact of blackness in Brazil was broadly recognized (no one would question that someone who identifies as Brazilian could also identify as, or be, black). This state of things was similar to Colombia, but contrasted with the situation faced by black activists in countries such as Argentina, Chile, the Dominican Republic, Mexico, or Peru, for example). Most consequently, by the 1970s there was already an established social scientific and state interest in investigating questions of racial inequality in Brazil, an intellectual and political environment very distinct from the terrain in most of the rest of the region in these years.

The existence of prominent social scientists invested in the reintroduction of a race question in the 1980 census made possible an articulation of interests that proved politically effective.[9] In the 1970s, sociologists Carlos Hasenbalg (1979) and Nelson do Valle Silva (1978) published landmark studies that showed racial inequality in social mobility in Brazil. Their analyses were based on nationally representative statistics collected in the 1976 Brazil National Household Sample Survey (PNAD) and used statistical methods to document that Brazilians who self-identified as "white" experienced more intergenerational social mobility than "não-branco" [non-white] Brazilians.[10] The discovery of "statistically significant" racial inequality in Brazil was not new. In the 1940s and 1950s, researchers sponsored by United Nation's Educational, Scientific and Cultural Organization (UNESCO) analyzed Brazilian racial dynamics as part of the organization's ill-fated post-WWII search for examples of "harmonious race relations" to serve as a model for the world (Fernandes, 1972; Ianni, 1987; Maio, 1999; Wagley, 1952). These studies also documented severe racial inequality, but they interpreted it as a legacy of slavery that would be resolved with (and by) Brazil's modernization. The stratification models used by Hasenbalg and Silva disproved this interpretation by "controlling for" social origins to isolate contemporary societal dynamics and reveal the pervasive, present-day existence of racial discrimination.

These studies' core findings aligned with the diagnostic and political vision of black movement leaders (Campos, 2013). In addition to providing scientific proof of contemporary racial prejudice in Brazil, they used a dichotomous categorical comparison to undertake and interpret the findings. Hasenbalg and Silva chose to "lump" *pardos* [brown or mixed] and *pretos* [black] into one "*não-branco*" [non-white] category for the purpose of their analysis. They did so in part because the number of "*pretos*" was small for their statistical methods to productively analyze as a separate group, and in part because their models worked best with dichotomous data.[11] The vision of a Brazilian population divided into those who are white and those who are not corresponded to arguments from within the black movement at the time that all Brazilians of any African descent ought to recognize themselves/each other as "*negro*" [another word for "black", used in this context as a term to encompass *pardos* and *pretos*]. Though the preferred label for the category differed, the demarcation of the racial boundary was perfectly aligned. As Campos (2013) argues, this fundamental epistemological alignment smoothed the way to political alignment. That alignment secured the reintroduction of a race question on the census and continued to prove important when it came time to put racial statistics to use in the realm of public policy. Needless to say, this pivotal (and temporally circumscribed) epistemological and political alignment between activists and social scientists was not easily replicated in other countries in the region. (Nor, as we will see later, did it remain stable over time within Brazil).

Notwithstanding their very different situations, activists in other countries drew examples from the Brazilian experience as they formulated political projects to combat racial inequality. Inspired in part by the success of Afro-Brazilian activists, beginning in the early 1990s, Afro-Colombian and indigenous organizations lobbied the national statistics agency (DANE) to add new queries to Colombia's national census (Buvinic & Mazza, 2004). Activists called for new census queries on ethical grounds. They argued that the statistical invisibility of Colombia's indigenous and Afro-descendant populations in the national census perpetuated cultural violence against these segments of the Colombian population. The addition of new census questions was required, activists explained, to ensure that indigenous and Afro-descendant populations garnered official recognition of their existence, protection of their rights, and redress for their historical marginalization (Paschel, 2010).

Activists' demands for census reform were bolstered by the shifting stance of the Colombian state on the issue of recognition of domestic minority populations. Colombia became a signatory to ILO Convention 169 in 1991 and new data were required for compliance. Signatories to ILO Convention 169 are expected to track and report numerous indicators of indigenous peoples' well-being in comparison to the general population. States require data on socioeconomic conditions, education, health, land, working conditions, employment, impacts of large development projects, infrastructure, and human rights violations (Rodriguez-Piñero, 2005). The Colombian state also needed information about the size of indigenous reserves (*resguardos*) to carry out planned decentralization. Constitutional reforms granting special rights to Afro-Colombian communities, in turn, made it desirable to collect new information about the size and situation of these communities. Thus, momentum toward democratization domestically, growing pressure for official recognition of diversity internationally, and the specifically targeted demands of indigenous and Afro-descendant organizations converged to usher in a historic change to Colombia's national census.

Building from their success, Afro-Colombian activists and their allies within DANE set their sights on empowering their counterparts in other Latin American countries to introduce analogous reforms. Bottom-up activism targeting national statistics agencies pioneered in Brazil and then Colombia put the issue of statistical visibility of minoritized populations on political agendas in a growing number of Latin American countries. By the early 2000s, pressure to adopt new census queries had been disseminated across Latin America through the deliberate and coordinated efforts of domestic activists, their census agency allies, and regional and international organizations. In most cases, the appeals of domestic activists did not suffice to convince census agencies to introduce new questions on national censuses. Beyond Brazil and Colombia, it was the convergence of such appeals with new initiatives and demands from regional and international organizations that ushered in census reforms across much of the region.[12]

Activists in Brazil and Colombia took the lead in making census questions and categories a pivotal stake in broader political struggles for recognition, rights, and redress. In countries where national political elites resisted activists' calls to introduce racial or ethnic data collection, regional activist networks and international organizations played critical roles in pressuring national statistics agencies to introduce reforms. The pressure on national statistics agencies to add new ethnic or racial questions to censuses took varied forms. In some countries, pressure took the form of indirect but institutionally supported encouragement to voluntarily adopt new questions. International conferences and workshops brought together leaders of community groups, experts in the development of comparative economic and social indicators from regional organizations such as the UN Economic Commission for Latin America and the Caribbean (ECLAC), and academics with international comparative expertise to advise national statistics agencies on how to design and implement the new questions. In other contexts, pressure was more direct and coercive, such as understood conditionalities attached to loans from multilateral lending institutions for funding ongoing census operations.[13]

In sum: National statistics agencies became increasingly invested in ethnoracial data collection in response to targeted political pressure from coalitions of grass-roots activists, social scientists, and international development organizations, with the combinatory weight of these actors, and the means and modes of their alliances, varying substantially across different national contexts in the region. The particular political coalitions that managed to add new race and ethnicity questions to Latin American censuses differed in different Latin American countries, reflecting distinct histories of black and indigenous activism, relationships of activists to academics, technocrats, and political parties in power at the national level, and the relative status of national governments in the regional and international system of states.

Yet by the 2010s, with very few exceptions, the outcome of these political fights across almost the entire region was in one key sense the same: Afro-descendant and indigenous peoples were enumerated *as such* in national censuses. For the majority of Latin American countries, the statistical visibility of race, color, and ethnic identity in the national census departs from decades of *de facto* and *de jure* insistence on the absence or inconsequence of ethnic or racial distinctions within national populations. In the first decades of the 21st century, grass-roots political claims and their successful articulation to social science research agendas and to national and international political projects brought a prolonged era of official color-blindness in Latin America to an end.

1.4   How the datascape transformed politics

The fact that almost every country in Latin America now produces ethnoracial population data from the national census represents a momentous political accomplishment. In several countries, the inclusion of new questions and categories on national censuses has made indigenous and Afro-descendant individuals statistically "visible" for the first time in decades, or in some contexts, for the first time ever. This statistical visibility, in itself, marks a victory for communities that have long struggled to gain official recognition. It also helps to produce the conditions for rewriting Latin American narratives of nationhood, resetting discursive parameters for subsequent political struggles. On this altered political terrain, the processes of producing and using official ethnoracial statistics create new fronts and facets of political contestation within ongoing efforts to address contemporary inequalities and historical injustice. Ethnoracial statistics are not merely tools for engaging in politics, they contribute to constituting new sites, subjects, and stakes of political struggle.

1.4.1   *Producing political sites*

One striking feature of the transformed political terrain for the production of official ethnoracial statistics in some parts of Latin America is that it has opened novel sites for political participation. In a break from historical precedent, in some countries in the region, the enumerated have won a seat at the table to discuss the questions and categories that the census uses for their enumeration. In some contexts, the enumerated have also been invited to acquire the training needed to analyze the data that is produced.

For example, as part of the regional "Todos Contamos" ["We all count"] workshops, government statistics offices were encouraged to facilitate the participation of representatives from indigenous and Afro-descendant organizations in the process of ethnoracial data production. ECLAC, the World Bank, and other agencies supported the creation of training programs to provide indigenous and Afro-descendant individuals with access to the methods and software used by statisticians, policy analysts, and government officials to analyze census data. As one illustration: a technical cooperation agreement between the Ministry of Health of Chile (MINSAL) and ECLAC to produce a socio-demographic report on the metropolitan indigenous population using census data included "training workshops for indigenous technical personnel on the REDATAM software and the use of censuses." Financial and logistical support for such workshops demonstrated that calls by development agencies for greater inclusivity in the process of data production were not merely rhetorical.

Some agencies called for projects to go beyond the provision of technical training for data analysis, to incorporate input from community groups on the design of questionnaires. The UN's Permanent Forum on Indigenous Issues, for example, flagged the need for more information on whether survey questions adequately addressed the needs of indigenous communities: ".... indigenous peoples' understanding of poverty, or land rights, often differs considerably from that of dominant or mainstream populations. This is rarely taken into account in the collection of relevant data." In the lead-up to the 2010 census round, the UN/ECLAC sponsored conferences and seminars to ensure participation by representatives of indigenous and Afro-descendant groups. This was a primary objective of the 2008 ECLAC conference held in Santiago, Chile, with funding from UNICEF, the United Nations Population Fund (UNFPA), the United Nations Development Fund for Women (UNIFEM), and the World Health Organization (WHO). The workshop brought together "more than 100 experts from more than 20 countries, including from governmental and non-governmental organizations, representatives of indigenous and Afro-descendent organizations, academics and technical experts from international agencies."

Significantly, the meeting emphasized the importance of generating statistics that would be recognized as valid not only by governments and development agencies but also by indigenous and Afro-descendant communities. Toward that end, a conference report recommended, "... we should aim to obtain this information with the participation of the peoples (*pueblos*) and communities, which is what will make it appear legitimate in the eyes of the entire population." The conference participants underscored that participation should be construed in the broadest possible terms. The final recommendation of the conference report was that "participation of [indigenous and Afro-descendent] *pueblos* in the design of questions and the collection and analysis of data that refer to them should be institutionalized." The ECLAC conference report denotes a critical shift in how international organizations publicly construe the relationship between the producers of demographic data, the individuals from whom data are collected, and the uses to which the data are put.

In practice, efforts to ensure that the voices of those who will be enumerated actually get taken into account in designing instruments for enumeration remain the exception rather than the rule. Over time, it is conceivable that the continued legitimacy of the national census in the eyes of citizens will come to depend on the existence of formal venues for public input into census operations. In theory, such forums could become arenas of substantive democratization in action – where the historical objects of statistical inquiry become subjects empowered with the authority, expertise, and resources to investigate themselves. A much more likely scenario, however, is that national statistics officials, in collaboration with experts working for international development agencies, maintain *de facto* control over the design and execution of national surveys, while encouraging popular participation on carefully delimited issues with a range of acceptable outcomes more or less predetermined.

The inclusion of the enumerated in the design of tools of enumeration has opened up new sites of political struggle that used to be entirely *internal* to the administrative-technocratic census bureaucracies. These are discussions where demands for recognition – which categories and lines of ethnoracial distinction will get official sanction and which will remain statistically invisible – are tied to fights over representation – who gets to speak on behalf of whom, and who is not invited into the conversation at all? While these facets of official statistical production have always been political, this reality has been obscured historically through control of the means of enumeration in the hands of an elite few. Indeed, historically, the symbolic authority of the census as a source of objective demographic knowledge hinged on the successful obfuscation of the politics of production of that knowledge. Recent developments in the region have brought politics that inform every stage of data production into open view.

To the extent that the inherently political nature of producing official ethnoracial statistics becomes widely recognized, the authority of the census as a source of objective information is being eroded. This erosion of legitimacy, in turn, may undermine what remains of states' capacities to pass off official census categories as mere (non-political) descriptions of demographic realities. The democratization of the process of data production, however limited and circumscribed, is likely to invite growing critiques of the legitimacy of national census operations per se.

1.4.2   *Producing political subjectivities*

The politics that produced the transformed datascape of ethnoracial population data have also been productive of shifts in peoples' subjectivities. To be clear, this is not an argument that census categories create new identities from scratch or out of thin air. The constitutive power of census categories on identities is not nearly as direct or automatic as their most ardent critics pretend. To the contrary, census categories are derived from and interact with pre-existing subjectivities and understandings of categorical distinctions in society.[14] Their specific effects on self-understandings are contingent on their interactions with many other factors at play. The influence of census categories hinges, especially, on the particular ways they are *used* – in media, in research, in administrative governance, in public policy, etc. Thus, the influence of official census categories on individuals' identities is often indirect and diffuse, rather than direct and instrumental. And their productive effects on collective self-understandings are not always neatly aligned with what their producers may intend.

One prominent way that the introduction of new race and ethnicity questions on Latin American censuses became productive of shifts in ethnoracial subjectivities was through organized media campaigns targeted to this goal. In several countries, activists collaborated with census agency staff and international agencies to wage media campaigns that aimed to persuade would-be indigenous and Afro-descendant peoples to identify as such on national censuses. Testifying to the earlier successes of states' ideological and political assimilationist projects, many Latin Americans who, by the varied criteria used by activists, social scientists, and others, "qualify" as indigenous or Afro-descendant, nonetheless do not choose to identify as such when given a choice. For example, while many observers suggest that approximately one-quarter of Colombian's population is of African descent, in the 2005 census, only 11% of the population self-identified as Afro-Colombian. Thus, in preparation for the census enumeration, organized campaigns sought to persuade those who would be enumerated to embrace new classifications.

Activists in several countries worked to strengthen domestic demand for government action on behalf of ethnoracially defined populations by increasing the number of their fellow citizens who identify as indigenous or Afro-descendant. In one of the first such campaigns, in the lead-up to the 1990 census in Brazil, black movement activists launched a publicity campaign with support from the Ford Foundation to persuade Brazilians with any African ancestry to mark *preto* [black] instead of *pardo* [brown or mixed] or *branco* [white] on the census (Nobles, 2000). The campaign slogan admonished: "Não deixe sua cor passar em branco" ["Don't let your color pass into white"]. Prior to the 2005 census in Colombia, to take another example, Afro-Colombian organizations launched a campaign to encourage Colombians to recognize the "beautiful faces of my black people" ("Las caras lindas de mi gente negra") (Estupiñón, nd; Paschel, 2013). The organizations produced a television commercial featuring individuals who identified themselves as "morena," "negra," "mulata," "zamba," or "raizal," and concluded with the slogan: "In this census, make yourself counted. Proudly Afro-descendant." The advertisement was aired through an agreement with the national statistics agency (DANE) and also disseminated via social media. In Panama prior to the 2010 census, a campaign with the same slogan did double duty as a commemoration of Afro-descendant women on the International Day of Women. The 30-second spot featured Afro-descendant men engaged in traditional women's work (ironing, cooking, hanging laundry) and appealing to viewers to commemorate International Women's Day and to show gratitude to Afro-descendant Panamanian women in particular by proudly reporting their Afro-descent on the census.

Campaigns to educate Latin American populations about the importance of responding to questions about ethnic or racial heritage, and how to respond, escalated significantly in preparation for the 2010 census round. Domestic activists, census officials, national media outlets, and regional and international organizations worked together to launch major publicity campaigns to inform Latin Americans of African descent about the new census queries and to encourage them to acknowledge their African heritage in selecting a response. For example, a regional census 2010 working group comprised of Afro-descendant leaders partnered with the UN Development Fund for Women as well as a Brazilian communications firm to produce a four-part television series called "The Americas have color: Afro-descendants in 21st-century censuses." According to a press release, the series was "created to inform the population of the Americas about the 2010–12 census round" and covered "the conditions of life of black men and women, black resistance throughout history and a panorama of public policies to confront racism." The series described the living conditions of Afro-descendant populations in Brazil, Ecuador, Panama, and Uruguay and underscored the importance of self-identification as Afro-descendant in the census.

Publicity for new ethnic and racial ancestry questions often made explicit mention of racial prejudice, discrimination, and inequality as social ills that the new statistics would help to combat. In Brazil, for example, the "Americas have color" documentary series opened with a special episode of a weekly news program, *Cenas do Brazil*, in which journalist Lúcia Abreu discussed "the importance of declaring one's color on the 2010 census, the evolution of demographic data that refer to color or race, and their contribution to the design of public policies." Public campaigns encouraging Latin Americans to acknowledge and embrace the African part of their heritage, coupled with increasing mainstream media coverage of racial discrimination and prejudice, represents a significant break from decades of public silence around this issue.

While the effect of public information campaigns on census results is difficult to isolate from other sources of influence on how individuals identify themselves (and sometimes their household members) on census questionnaires, Brazil and Colombia saw growth in the relative size of self-identified Afro-descendant populations over the past two decades. Demographers and other social scientists have estimated how a change in racial self-identification between censuses is shaping these shifts (e.g., Carvalho, Wood, & Andrade, 2004). Some observers attribute these trends to census outreach campaigns in the context of broader cultural shifts in the valuation of blackness, especially among younger cohorts; others note the likely influence of the growing use of official census categories in ethnoracially targeted public policies. As discussed further below, when ethnoracial census categories are used to design, implement, or monitor affirmative action programs, instrumental incentives may play a role in shifting self-identifications as well (Bailey, Fialho, & Loveman, 2018; Canessa, 2007; Muniz, 2010).

When states collect and use ethnic and racial data, this shapes subjectivities partly by triggering public discussions and renegotiations of what particular racial categories signify, who belongs within them, and the criteria for belonging (Mora, 2014). To the present in Brazil, there is an active debate among and between activists, scholars, technocrats, politicians, and increasingly many other public figures over the meaning of racial categories that itself contributes indirectly to the constitutive effects of those categories. University undergraduates interviewed in Rio de Janeiro in the early 2000s were not always sure of the "correct" way to answer a question about their race or color. They often referenced their appearance and their family background in discussing whether or not to identify themselves as *pardo* or *preto* (Schwartzman, 2009). In the years since then, there has been indication that responding based on understandings of origins or ancestry has gained ground (Guimarães, 2011). Additionally, there is evidence that between 1995 and 2008 (before the implementation of quota policies, and then some years after implementation) Brazilians became increasingly likely to choose one of the census race/color categories to self-identify in an open-format survey question (Bailey et al., 2018). This suggests the recursivity of questions and categories used in the census and public policy in shaping the social phenomenon they aim to describe.

Importantly, national media campaigns and public debates about ethnoracial census categories and their uses influence the subjective self-understandings of those who are *not* the principle targets of the campaigns or public policies, as well as those who are. In recent years in several Latin American countries, groups have emerged to defend official recognition of *mestizos*, using many of the same arguments invoked by Afro-descendant and indigenous groups to demand visibility on censuses. For these critics, official ethnoracial classification is not inherently problematic; it is the depreciation or negation of "mixed" categories in favor of "absolute" or "pure" categories that is cause for concern. In Brazil, for example, in the mid-2000s, an NGO based in Amazonia with the name "Nação Mestiça" sought to advance

> The valorization of the process of miscegenation (mixture) between the diverse ethnic groups that created the Brazilian nationality, the promotion and defense of *pardo-mestiça* identity and the recognition of *pardo-mestiços* as cultural and territorial inheritors of the people from which they are descended.

Arguing against campaigns to "unmix" Brazilians and echoing nationalists from the past, the group's slogan announced: "*A miscigenação une a nação*" ["miscegenation unites the nation"].

Such slogans foreshadowed Jair Bolsonaro's skillful deployment of the same nationalist trope to foment resentment against affirmative action policies as part of his presidential campaign. For instance, Bolsonaro's response to questions about racial or color identification for affirmative action was to insist on the preeminence of national identity, as signaled in slogans such as "My color is Brazil", as seen in Figure 1.4.

As is well known, race-targeted social policies in Brazil produced fierce scholarly, public, and political debate (On these debates see: Bailey & Peria, 2010; Feres Júnior, Campos, Daflon, 2011; Fry, Maggie, Maio, Monteiro, & Santos, 2007; Guimarães, 1999; Heringer & Johnson, 2015; Teixeira, 2003). Above and beyond the immediate stakes and specific arguments of these debates, the existence and tenor of the debates themselves have arguably had productive effects on Brazilians' subjectivities. Of particular importance, the public controversies helped catalyze the salience and broad popular resonance of a resurgent nationalist identity that insists on the (re)subordination of distinct ethnoracial identities to identification as Brazilian.

Thus, the production and use of official ethnoracial statistics may be productive of subjectivities that shift toward conformity with official categories and

Figure 1.4 Jair Bolsonoro wearing a t-shirt that states "My Color is Brazil."

Source: Screenshot of image shared on social media during Bolsonaro's presidential campaign

simultaneously productive of subjectivities of reaction, that take shape as alternatives to official categories or in opposition to official categorization per se. The recursivity of census categories – their ability to produce a "looping effect" in Ian Hacking's (1996) phrasing – is actively contested and thus politically contingent. The productive effects of new census questions on subjectivities may not resound within closed loops. They may instead reverberate through an open-ended spiral. In some ways, the growing use of official ethnoracial categories in Latin America is shaping subjectivities in patterned and (more or less) predictable ways. In other ways, the constitutive consequences of official ethnoracial classification are spilling out in unanticipated forms and directions.

1.4.3   *Producing political stakes*

The way individuals answer census questions about their racial or ethnic identification reflects, in part, their assessments of what the census is really asking based on prevailing societal understandings of the terms used, why it is being asked, and what they are expected to take into account in providing their response. Their answers reflect, as well, shifting understandings of what exactly is at stake.

By the latter part of the 2000s, political struggles to get Latin American states to collect ethnoracial statistics on their national censuses mostly gave way to increasing demands to *do* things with the new data. As the data were analyzed and "put to use" to motivate, develop, implement, and monitor public policies, new political stakes emerged – stakes that were directly tied to official ethnoracial classification.

The existence of new stakes tied to ethnoracial categories reverberated through the political terrain, shifting identities, alliances, and alignments (epistemological and political) that had fueled the success of political struggles to transform the datascape in the first place.

Before the new data collected in Latin American censuses could become "tools of the weak" in Porter's sense, the avalanche of raw *numbers* needed to be transformed into useful/usable *knowledge*. As the authors of a report sponsored by the Inter-American Development Bank pointed out, "The inclusion of the ethnic variable in censuses and surveys is pointless if it isn't used in analyses." Thus ensued the production of a wave of census agency summary statistics, ECLAC and World Bank reports, and social scientific studies. These assembled descriptive statistics and statistical analyses document ethnoracial inequalities on various indicators of material well-being, including studies of disparities in life expectancy, educational attainment, access to electricity and sanitation, and income, among other "outcomes." International and regional development agencies also began to generate comprehensive reports on the condition of Afro-descendant or indigenous communities in particular Latin American countries and in comparative perspective. The ECLAC, for example, committed resources to the production of statistical reports on the status of indigenous and Afro-descendant youth in Latin America based on results from the 2000 census round. The World Bank and agencies of the United Nations also sponsored the production of reports on various aspects of health, education, and living conditions of minority populations in several parts of Latin America.

Armed with a growing body of statistical evidence of ethnoracial inequality, activists and their allies within state governments began to push for targeted social policies. Statistical documentation of ethnoracial inequality supported calls for states to supplement legal promises of protection from discrimination with proactive, corrective, and reparative measures to reduce existing disparities of condition and opportunity. In contexts where social movement actors or their allies occupied positions within the state, policies that directly address ethnoracial inequality became a visible priority.

Brazil pioneered such policies in the region, and to date, Brazil has gone farthest with their use, with a broad range of affirmative action initiatives in government agencies, university admissions, and industries such as fashion and television.[15] In Brazil, affirmative action policies came to rely upon official ethnoracial statistics to justify their introduction, to guide their implementation, and to monitor the results. The implementation of affirmative action programs created new stakes of racial classification, and thus drew sharp attention to the categories used in the census, in social science, and in social policy.

The production of new political stakes introduced fissures into the political terrain that stressed, and in some contexts fractured, alliances that had been strong in the political struggle to get affirmative action legislation passed in the first place. In different ways, these fissures ran through the social sciences, social movements, universities as institutions, and the political domain (Bailey & Peria, 2010; Carvalho, 2005). As just one example of such a fissure, it is instructive to consider how the politics and practice of drawing boundaries around and within the "não branco" [non-white] population shifted from the period of mobilization for affirmative action policies to the period of their implementation. As discussed above, social science research in the 1970s and 1980s favored the use of a dichotomous analytic comparison between whites and non-whites, with the latter grouping together census *pardos* and *pretos*. This scheme was easily translated into comparisons between whites and "*negros*," using black movement activists' preferred framing at that time, which advocated for an expansive definition of *negro* that included census *pretos*, *pardos*, and anyone who identified as Afro-descendant (Campos, 2013). In scholarship on racial inequality in Brazil written in English, this translation was even more straightforward since the same word – black – is used to translate both *preto* and *negro*.

In the period of political mobilization leading up to the passage of racial quota policies, methods that downplayed distinctions within the non-white population were favored by social scientists, just as they were important to activists' vision of building the broadest possible constituency to press for social policies to combat racial inequality. Fast forward some years, and there is a much more active debate among social scientists over whether it is appropriate or misleading to lump *pretos* and *pardos* together in statistical analyses.[16] Citing evidence of socioeconomic heterogeneity within the *pardo* category, variation in perceptions of discrimination by class status, and regional variation in its meaning, some researchers advocate avoiding binary models [white/non-white] because they obscure class and regional inequality within the *pardo* category, and the severity of disadvantage suffered by *pretos* (Bailey, Loveman & Muniz, 2013; Bailey & Telles, 2006; Daflon, Carvalhães, & Júnior, 2017; Muniz, 2010). Alongside these emerging strands of research, there are political realignments within the coalitions of activists and academics supportive of affirmative action policies. Programs that tied opportunities for university admission or government employment to categorical ethnoracial identification raised thorny questions about who qualifies for such programs and who does not. This question opened the door to debates over *who decides* who qualifies, and on what basis. The distributional stakes of affirmative action programs produced reassessment and revision to earlier views of how racial boundaries in Brazil should be drawn.

Whereas in a mobilizational moment of politics, the favored approach of activists and allied scholars alike was to group *pardos* and *pretos* together, in a distributional moment of politics, the favored approach has shifted to one that differentiates *pardos* from *pretos*, "brightening" a boundary they had previously blurred. This is evident, for example, in some of the discourses and practices of "anti-racial fraud" campaigns (Figure 1.5).

Figure 1.5 is a screenshot of a poster taken from a website of a black student group at a Brazilian university in 2016. The poster asks "Who are the new 'black' quota students at UFES [the Federal University of Espírito Santo]?" The text superimposed on the foreheads states that "51% of the Brazilian population is black," and that "the other half has double the opportunities." To arrive at 51%, they must include census *pretos* (black) and census *pardos* (brown or mixed), since

QUEM SÃO OS NOVOS COTISTAS "NEGROS" DA UFES?

Publicado em 11 de fevereiro de 2016 por mirtesants

51% DA POPULAÇÃO DO BRASIL É NEGRA.

E A OUTRA METADE TEM O DOBRO DE OPORTUNIDADES.

AUTODECLARAÇÃO FALSA É CRIME! - ART. 299 CP

Figure 1.5 A poster denounces fraudulent racial identification for affirmative action at a Federal University in Brazil.

Source: Screenshot from a website of a student group at the Federal University of Espirito Santo in 2016.

self-identified censos *pretos* alone would be less than 8% (according to the 2010 census). Nonetheless, the picture is clearly one that represents Brazil as a population made up of black and white. At the bottom, the poster notes that "False Self Identification is a Crime." Most media coverage of this campaign and legislation focused on people accused of falsely claiming to be non-white when they are "actually" (socially) white. However, racial fraud charges were also levied against self-identified *pardos* seen as not being black enough. An article that appeared in English language press in 2017, in the journal *Foreign Policy*, told the story of a woman who stood accused of racial fraud for identifying herself as "*Parda*" in the application process for a public prosecutor job. Some activists criticize self-identified *pardas* for using programs that were in fact set up for *pardos* and *pretos*, arguing that really these spaces were not meant for "people with black grandmothers"; rather they were meant for those with "very dark skin" (de Oliveira, 2017). Criteria used to draw distinctions within the *negro* [black] category – distinctions that were downplayed or even denied in a mobilizational moment of politics -- were being resurrected and redrawn to more narrowly delimit beneficiaries in the context of determining the distribution of scarce resources. An unintended consequence of the ways that Brazil's racial quota policies have been implemented, and the varied reactions to these policies, has been to (re)introduce political and sociological fissures through the "não branco" [non-white] population, sharpening the boundary between *pretos* and *pardos* that was previously deliberately blurred.

Ethnoracially targeted affirmative action programs of various sorts are in place or on political agendas in several other Latin American countries. These initiatives range from policies focused on health services, nutrition, housing, poverty alleviation, and land titling, to educational benefits and guaranteed political representation. As ethnoracially targeted social programs spread in Latin America, controversies over why and how states classify their populations by race or ethnicity escalate. As occurred in Brazil, critics of these programs point to opportunistic self-identification; they note how official categories will create or grow constituencies along those categorical lines; and they warn of unintended consequences of well-meaning policies, including the consolidation of more rigid group boundaries and attendant polarization or fragmentation of national societies.

Those who support the aims of targeted social policies are likely to raise concerns about official ethnoracial classification as well. Policymakers, social scientists, and activists will debate whether the categories used for implementation effectively funnel resources or opportunities to the most deserving and intended beneficiaries. The ethnic or racial categories used to implement targeted social policies will never correspond perfectly to all of the categorical distinctions that are operative in the lives of individuals and communities. In some contexts, official ethnoracial categories may be drawn narrowly and inadvertently exclude individuals who merit inclusion. In others, official categories will be deemed too broad, diffusing the impact of initiatives to address explicitly ethnoracial facets of poverty and inequality. There is an inherent tension between census question formats or social science analytic methods that aim to maximize the total number of those who count as indigenous or black, and the design of social policies that use such categories to target those who are most disadvantaged.

The transformed datascape for the collection, production, and use of official ethnoracial statistics in Latin America can be used both as information and as ammunition in political struggles over targeted social policies and over competing visions of social justice and social progress. In previous decades, the disaggregation of poverty and health statistics by sex revealed the gendered nature of inequality in Latin America. Such statistics helped support a shift toward development strategies focused on women's roles in the distribution of scarce resources within families and communities. The emergent avalanche of racial and ethnic statistics will likewise continue to expose severe ethnic and racial inequalities in some regions and provide support for government interventions that aim to reduce them. In some contexts, these projects may respond to real and pressing needs of historically neglected populations. In others, they may amount to what anthropologist Charles Hale terms "multicultural neoliberalism," serving primarily to *appear* to address inequalities while effectively dissipating bottom-up pressure for development strategies that would entail more substantive structural change. Either way, the production and use of official ethnoracial statistics produce new political stakes which are themselves productive: as they are put to use in government administration and public policy, they produce tremors, fissures, and fractures that alter the conditions of possibility for the emergence of new coalitional movements that may fundamentally alter the political terrain.

1.5 Conclusion: looping effects or spirals? Navigating a datascape transformed

The inclusion of new race, ethnicity, and color questions on national censuses and nationally representative household surveys in Latin America in recent decades is a momentous political accomplishment. This transformed datascape for the collection, production, and use of ethnic and racial statistics, in turn, has become productive of new sites, subjectivities, and stakes of politics.

The new politics of ethnoracial data production have seeded new struggles within the political field of grass-roots activism, social scientists, state actors, and international organizations. This is especially evident in contexts where the inclusion of new questions or the revision of question-wording or response options remains open for discussion. Struggles over the terms of official recognition – which ethnoracial categories and boundaries will be officially sanctioned and which will remain officially invisible – cede easily into struggles over representation – who gets to speak on behalf of whom? These latter struggles have shaped the field of social movement organization and ties between NGOs, activists, political parties, and national and international actors (Paschel, 2016).

The unprecedented availability of ethnic and racial population data has predictably unleashed an avalanche of new statistical descriptions and analyses of ethnoracial inequalities. In most countries in the region, it remains an open question whether or how the accumulation of quantitative studies documenting pervasive ethnoracial disparities in indicators of income, health, and education will translate into successful political struggles for either incremental or more transformative change. The new availability of ethnoracial population data has bolstered activists' demands for expanded benefits of social citizenship, including demands for ethnoracially targeted social benefits as a means of redress for historical marginalization and/or contemporary discrimination. As noted above, affirmative action policies for ethnoracially defined groups are already in place in several Latin American countries and there is pressure on states from both domestic activists and international organizations to introduce more initiatives of this kind. Such policies focus on targeted delivery of benefits ranging from health services to housing, and poverty alleviation to political representation. Among the most visible and contentious initiatives have been those focused on affirmative action in the field of higher education.

In Brazil, the proliferation of statistical social scientific studies of racial disparities in the late 1980s and 1990s supported claims made by the black movement and allies for targeted policy interventions; Brazil became a leader in the introduction of affirmative action in government employment and university admissions. More recently, however, Brazil has also become a leading example of organized backlash. Brazil had been seen as a regional model for how political mobilization could produce changes to the census, which could then produce ethnoracial statistics, which could then be used to produce social science research on racialized inequalities, which could in turn inform public policies that aim to address those inequalities. Even as this example remains, the more recent Brazilian experience of reactionary backlash to this political project – and its links between activists, social scientists, and state actors – is now weighed by its neighbors as a precautionary example as well.

As the Brazilian experience demonstrates all too well, social policies that explicitly aim to address ethnoracial inequalities through interventions that target ethnoracially defined beneficiaries often draw attention to the political processes that inform the production of ethnoracial data in the first place. Political battles fueled in part by the statistical documentation of ethnoracial inequalities tend to circle back around to political battles over the production of ethnoracial statistics *per se*. Thus, as ethnoracially targeted social programs spread, controversies over why and how states classify citizens by race or ethnicity will escalate. As in the United States, so in Brazil, Bolivia, Mexico, and a growing list of other countries in the Americas, we are likely to witness escalating opposition to the idea that states may legitimately produce ethnoracial statistics at all.

The future of ethnoracial demographic data production in Latin America remains uncertain. Desrosières (2014, p. 352) argues that "For a statistic to play its social role as a neutral reference, above the conflicts of social groups, it must be instituted, guaranteed by democratic procedures, themselves legitimate. It then contributes to making reality and not simply reflecting reality." While the analysis in this chapter could lend support for the first part of this claim, it questions the second: it is clear that the *official institutionalization* of ethnoracial statistics fuels and unleashes their productive potential; it is also evident, however, that official ethnoracial statistics may contribute to constituting the realities they aim to measure even when their neutrality is questioned and their legitimacy is contested. Indeed, this is so even when their neutrality is questioned and their legitimacy contested at every stage of their production, institutionalization, analysis, and use. It may even be the case that the productive power of ethno-demographic statistics is amplified, if in unintended ways, as a result of societal contestation of their neutrality and legitimacy. The productive power of a statistic and the democratic acceptance of its means of production and institutionalization are not so tightly coupled.

Whether ethnoracial statistics contribute to (re)making reality in the ways their advocates and producers intend is another matter altogether. Official ethnoracial categories, institutionalized in statistics and operationalized through public policy, may be partly constitutive of subjectivities but not in a controlled and contained way, such as we might imagine when invoking Ian Hacking's (1995) "looping effect" metaphor. Rather than a closed loop, the productive reverberations of official ethnoracial statistical categories may reverberate in an open-ended spiral, wound more or less tightly and with the ends pointing inward or outward in relation to other actors and identities in the political field, making possible/probable different articulations and coalitions of political actors and movements in the political terrain. As the metaphor of a "spiral effect" suggests, the analysis in this chapter and growing evidence from the region affirms that the productive effects of ethnoracial statistics are both predictable and unpredictable, constitutive of recursive but open-ended "spiral" effects rather than self-reinforcing and closed "loops." As a consequence, the creation and use of official ethnoracial statistics are apt to be politically productive in ways at once productive and counterproductive to the aims of those who produce them.

The transformed datascape is not only fueling the creation of new knowledge about ethnoracial inequalities in Latin America; it is also helping to define new sites, subjectivities, and stakes of political struggle over recognition, rights, and redress for historically marginalized communities and contemporary fights for racial justice – while also stoking reactionary nationalist opposition. Thus, while the early 21st-century boom in the production of ethnoracial statistics across most of Latin America is a historic political accomplishment, it is simultaneously a politically contentious and tenuous accomplishment that could well be short-lived.

New data have generated new knowledge about ethnic, racial, and color inequalities in Latin American societies; this new knowledge, in turn, together with the process and politics of its production, has stoked new sites, subjectivities, and stakes of politics. These have contributed to the political constitution of ethnoracial identities and boundaries, to the advancement of claims for greater equality and recognition for ethnoracially defined individuals and communities, and to stoking the rise of organized reactionary opposition. Thus, in the context of unfolding political struggles to challenge existing relations between states, scientists, and (non) citizens, as social scientists grapple with the avalanche of ethnoracial population data in Latin America with the aim to advance understanding of ethnoracial inequalities, they must also keep the politics of the production of such data, *and their productive effects on politics,* within the frame of analysis.

Notes

* This chapter includes material previously published in Portuguese in Sociologias 23(56), 2021: 110-153. https://doi.org/10.1590/15174522-109784, under CC licence 4.0.
1 The use of the term "avalanche" in this context is in reference to Ian Hacking's (1982) essay "Biopower and the Avalanche of Printed Numbers." While Hacking analyzes a very different context and moment in the history of demographic statistics, echoes of that earlier avalanche still reverberate in this one.

2 Table 1.1 provides a summary overview of the change in statistical visibility of indigenous and Afro-descendant populations, but it also obscures the considerable variation in the particularities of census question formats and response options across countries, which make visible distinctions within these categories in some countries. Table 1.1 also omits consideration of other lines of ethnoracial distinction enumerated in some national censuses in the region (for example, some censuses allow responses for "whites" or "Asian," while some deliberately omit such response options). For more on these variations and their justifications and consequences, see Loveman, 2014, pp. 250–300; De Popolo, 2008.

3 For an overview of ethnoracial classification on censuses in Latin America in this period, see Tables in Loveman, 2014, p. 233, 241.

4 These generalizations are developed at more length and with many examples in Loveman, 2014, pp. 121–300. There are also many excellent studies of these practices for specific countries and census years – too many to cite them all here, but see, for example, Otero, 2006; Camargo, 2010.

5 The historiography on this theme is much too long to cite comprehensively here. Important contributions to this large body of work include Vejo and Yankelevich (orgs), 2017; Sommer, 1991; Martinez-Echazábel, 1998, 1996).

6 On dissenting voices, see for example the views of a military doctor who used data collected from soldiers to argue against Vianna's thesis of evolution toward a singular, whitened, national Brazilian "type" (Loveman, 2009). See also Skidmore, 1993; Schwarcz, 1993; Borges, 1993.

7 An extended analysis of the recent shift in state practices of ethnic and racial classification in Latin America can be found in Loveman (2014).

8 Nobles 2000, pp. 98–110, 116–119. Brazil's census almost omitted the race question in 1940, but it was ultimately retained with the justification and expectation that it would allow Brazil to show the world its continued progress toward whitening (Camargo, 2010, pp. 255–266).

9 Nobles (2000, pp. 98–110, 116–119) argues that absent organized pressure from academics and activists, Brazil's 1980 census would have been fielded without a color question.

10 The question of whether mobility itself influenced self-classification as "white" could not be disentangled without longitudinal data. The debate over whether (or when, or in what sense) "money whitens" in Brazil, which dates back decades, continues to be debated by social scientists to this day.

11 They also argued later that the grouping of *"pardos"* and *"pretos"* made substantive sense given their proximity in social status in comparison to "whites." To this day, scholars debate whether it is best to use dichotomous, trichotomous, an analytic combination of both, or none of the above to analyze racial inequalities in Brazil.

12 For more on international organizations' role, see Loveman, 2014, pp. 250–300.

13 One example of the former type is the international conference in Lima, Peru 2016, which brought together social movement spokespersons, representatives of the national statistics agency, government officials, social scientists from independent research institutes and universities, UN representative, and "international academic experts" to discuss the adoption of new questions on the 2017 census. For an example of the latter, see account of Nicaragua's, 2005 census in Loveman (2014, pp. 290–293).

14 On problems with overly state-centric, top-down approaches to understanding census categories and their consequences, see Emigh, Riley, and Ahmed (2016).

15 For detailed description and analysis of these programs and their politics, see for example: Feres Júnior, 2007, 2008; Fry et al., 2007; Petruccelli, 2015.

16 There were dissenting voices on the use of dichotomous analytic scheme earlier, and some scholars continued to produce analyses that kept *pardo* and *preto* separate throughout these years. But the approach that was seen and presented as most in line with politics of racial justice was to adopt a dichotomous lens.

References

Andrews, G. (1991). *Blacks and whites in Sao Paulo, Brazil, 1888–1988*. Madison, WI: University of Wisconsin Press.

Bailey, S., Fialho, F. M., & Loveman, M. (2018). How states make race: New evidence from Brazil. *Sociological Science, 5*(31), 722–751.

Bailey, S., Loveman, M., & Muniz, J. (2013). Measures of 'race' and the analysis of racial inequality in Brazil. *Social Science Research, 42*(1), 106–119.

Bailey, S., & Peria, M. (2010). Racial quotas and the culture war in Brazilian academia. *Sociology Compass, 4*(8), 592–604.

Bailey, S., & Telles, E. (2006). Multiracial versus collective black categories: Examining census classification debates in Brazil. *Ethnicities, 6*(1), 74–101.

Borges, D. (1993). 'Puffy, ugly, slothful and inert': Degeneration in Brazilian social thought, 1880–1940. *Journal of Latin American Studies, 25*, 235–256.

Bourdieu, P. (1999). Rethinking the state: Genesis and structure of the bureaucratic field. In G. Steinmetz (Ed.), *State/culture: State-formation after the cultural turn* (pp. 53–75). Ithaca, NY: Cornell University Press.Brazil. Directoria Geral de Estatística. (1922). *Recenseamento do Brazil realizado em 1 de setembro de 1920*. Rio de Janeiro, Brazil: Typografia da Estatistica.

Campos, L. A. (2013). O pardo como dilema político. *Insight Inteligência*, Out-Nov-Dez.

Canessa, A. (2007) Who is indigenous? Self-identification, indigeneity, and claims to justice in contemporary Bolivia. *Urban Anthropology, 36*(3), 195–237.

Carvalho, J. A. M., Wood, C., & Andrade, F. (2004). Estimating the stability of census-based racial/ethnic classifications: The case of Brazil. *Population Studies, 58*(3), 331–343.

Carvalho, J. J. (2005). Usos e abusos da antropologia em um contexto de tensão racial: o caso das cotas para negros na UnB. *Horizontes Antropológicos, 11*(23), 237–246.

Daflon, V. T., Carvalhães, F., & Júnior, J. F. (2017). Sentindo na pele: Percepções de discriminação cotidiana de pretos e pardos no Brasil. *Dados, 60*(2), 293–330.

de Oliveira, C. (2017, July 24). One woman's fight to claim her 'blackness' in Brazil. *Foreign Policy*.

de Paiva Rio Camargo, A. (2010). Classificações raciais e formação do campo estatístico no Brasil (1872–1940). In N. D. C. Senra & A. D. P. R. Camargo (Eds.), *Estatísticas nas Américas: por uma agenda de estudos históricos comparados* (pp. 229–264). Estudos & Analises. Rio de Janeiro, Brazil: IBGE.

de Popolo, F. (2008). *Los pueblos indígenas y afrodescendientes en las fuentes de datos: Experiencias en América Latina*. Santiago. Chile: Comisión Económica para América Latina y el Caribe, CEPAL.

De Popolo, F., & Schkolnik, S. (2012). Indigenous people and afro-descendants: The difficult art of counting. In F. A. Ferrandez & S. Kradolfer (Eds.), *Everlasting countdowns: Race, ethnicity, and national censuses in Latin American states* (pp.304–334). Cambridge, England: Cambridge Scholars.

Desrosières, A. (1998). *The politics of large numbers: A history of statistical reasoning*. Cambridge, MA: Harvard University Press.

Desrosières, A. (2014). Statistics and social critique. *Partecipazione e conflitto, 7*(2), 348–359.

Dos Santos, J. T. (1999). Dilemas nada atuais das políticas para os afro-Brasileiros: Ação afirmativa no Brasil dos Anos 60. In J. Bacelar & C. Caroso (Eds.), *Brasil: Um pais de negros?* (pp. 221–233). Rio de Janeiro, Brazil: Palias.

Economic Commission for Latin America and the Caribbean. (2009). *Progress report on the biennial programme of regional and international cooperation activities of the Statistical*

Conference of the Americas of the Economic Commission for Latin America and the Caribbean (ECLAC) 2007–2009. Santo Domingo, Dominican Republic: ECLAC.

Economic Commission on Latin America and the Caribbean, Population Division. (2005). *Pueblos indígenas y afrodescendientes de América Latina y el Caribe: relevancia y pertinencia de la información sociodemográfica para políticas y programas.* CEPAL, 27 de abril al 29 de abril de 2005.

Emigh, R. J., Riley, D., & Ahmed, P. (2016). *Changes in censuses from imperialist to welfare states: How societies and states count.* New York, NY: Palgrave Macmillan.

Estupiñón, J. P. (n.d.). Afrocolombianos y el Censo 2005. *Revista de Información básica, 1*(1), 7. Retrieved from https://sitios.dane.gov.co/revista_ib/html_r1/articulo7_r1.htm

Feres Júnior, J. (2007). Comparando justificações das políticas de ação afirmativa: Estados Unidos e Brazil. *Estudos Afro-Asiáticos, 29*(1–3), 63–84.

Feres Júnior, J. (2008). Ação afirmativa: política pública e opinião. *Sinais Sociais, 3*(8), 38–77.

Feres Júnior, J., Campos, L. A., & Daflon, V. T. (2011). Fora de quadro: a ação afirmativa nas pagina d'O Globo. *Contemporânea - Revista de Sociologia da UFSCar, 1*(2), 61–83.

Fernandes, F. (1972). *O negro no mundo dos brancos.* São Paulo, Brazil: Difusão Europeia do Livro.

Fontaine, P. M. (1981). Transnational relations and racial mobilization: Emerging black movements in Brazil. In J. E. Stack Jr. (Ed.), *Ethnic identities in a transnational world* (pp. 141–162). Westport, CT: Greenwood Press.

Fontaine, P. M. (1985). Blacks and the search for power in Brazil. In P. Fontaine (Ed.), *Race, class and power in Brazil* (pp. 56–72). Los Angeles, CA: University of California, Center for Afro-American Studies.

Fry, P., Maggie, Y., Maio, M. C., Monteiro, S., & Santos, R. V. (Eds.). (2007). *Divisões perigosas: políticas raciais no Brazil contemporâneo.* Rio de Janeiro, Brazil: Editora Civilização Brasileira.

Fundacao Perseu Abramo. (2003). *Discriminação racial e preconceito de cor no Brasil.* São Paulo, Brazil: Fundação Perseu Abramo.

Guimarães, A. S. (1999). *Racismo e anti-racismo no Brazil.* São Paulo, Brazil: Editora 34.

Guimarães, A. S. (2011). Raça, cor da pele e etnia. *Revista Cadernos de Campo, 20*(20), 265–271.

Hacking, I. (1982). Biopower and the avalanche of printed numbers. *Humanities in Society, 5,* 279–295.Hacking, I. (1986). Making up people. In T. Heller, M. Sosna & D. Wellbery (Eds.), *Reconstructing individualism: Autonomy, individuality, and the self in Western thought* (pp. 222–236). Stanford, CA: Stanford University Press.

Hacking, I. (1996). The looping effects of human kinds. In D. Sperber, D. Premack, & A. J. Premack (Eds.), *Causal cognition: A multidisciplinary debate* (pp. 351–394). Oxford, England: Clarendon Press/Oxford University Press.

Hale, C. (2002). Does multiculturalism menace? Governance, cultural rights, and the politics of identity in Guatemala. *Journal of Latin American Studies, 24,* 485–524.

Hanchard, M. (1994). *Orpheus and power: The Movimento Negro of Rio de Janeiro and São Paulo, Brazil, 1945–1988.* Princeton, NJ: Princeton University Press.

Hasenbalg, C. (1979). *Discriminação e desigualdades raciais no Brasil contemporâneo.* Rio de Janeiro, Brazil: Graal.

Hasenbalg, C. (1985). Race and socioeconomic inequalities in Brazil. In P. M. Fontaine (Ed.), *Race, class, and power in Brazil* (pp. 25–41). Los Angeles, CA: University of California, Center for Afro-American Studies.

Hasenbalg, C. (1990). Raça e oportunidades educacionais no Brasil. *Cadernos de Pesquisa, 73,* 5–12.

Hasenbalg, C. (1996). Entre o mito e os fatos: Racismo e relações raciais no Brasil. In M. C. Maio & R. V. Santos (Eds.), *Raça, ciência e sociedade* (pp. 235–249). Rio de Janeiro, Brazil: Fiocruz/CCBB.

Hasenbalg, C., & do Valle Silva, N. (1988). *Estrutura social, mobilidade e raça.* Rio de Janeiro, Brazil: IUPERJ.

Hasenbalg, C., & do Valle Silva, N. (1999). Notes on racial and political inequality in Brazil. In M. Hanchard (Ed.), *Racial politics in contemporary Brazil* (pp. 154–178). Durham, NC: Duke University Press.

Heringer, R., & Johnson, O. (Eds.). (2015). *Race, politics and education in Brazil: Affirmative action in higher education.* New York, NY: Palgrave Macmillan.

Hooker, J. (2005). Indigenous inclusion/black exclusion: Race, ethnicity and multicultural citizenship in Latin America. *Journal of Latin American Studies, 37,* 1–26.

Htun, M. (2004). From 'racial democracy' to affirmative action: Changing state policy on race in Brazil. *Latin American Research Review, 39*(1), 60–89.

Ianni, O. (1987). *Raças e classes sociais no Brasil* (3rd ed.) São Paulo, Brazil: Brasiliense.

Interamerican Development Bank. (2000, November 7). *BID, Banco Mundial y Colombia auspician seminario internacional sobre factores raciales y étnicos en censos de América Latina.* Press release. Retrieved from http://www.iadb.org/es/noticias/comunicados-de-prensa/2000-11-07/bid-banco-mundial-y-colombia-auspician-seminario-internacional-sobre-factores-raciales-y-etnicos-en-censos-de-america-latina,799.htmlLorde, A. (2018). *The master's tools will never dismantle the master's house.* London, England: Penguin Classics.

Lovell, P. (Ed.). (1991). *Desigualdade racial no Brasil contemporâneo.* Belo Horizonte, Brazil: MGSP Editores.

Lovell, M. (2009). The race to progress: Census-taking and nation-making in Brazil, 1870–1920. *Hispanic American Historical Review, 89*(3), 435–470.

Loveman, M. (2014). *National colors: Racial classification and the state in Latin America.* Oxford, England: Oxford University Press.

Loveman, M., Muniz, J., & Bailey, S. (2011). Brazil .in black and white? Race categories, the census, and the study of inequality. *Ethnic and Racial Studies, 1*(18), 1466–1483.

Magnoli, D. (2007). Ministério de classificação racial. In P. Fry, Y. Maggie, M. C. Maio, S. Monteiro, R. V. Santos (Eds.), *Divisões perigosas: Políticas raciais no Brasil contemporâneo* (pp. 89–94). Rio de Janeiro, Brazil: Civilização Brasileira.

Maio, M. C. 1999. O Projeto Unesco e a agenda das ciências sociais no Brasil dos anos 40 e 50. *Revista Brasileira de Ciencias Sociais, 14*(41), 141–158.

Maio, M. C., & Santos, R. V. (Eds.). (1996). *Raça, ciência e sociedade.* Rio de Janeiro, Brazil: Editora Fiocruz.

Martinez-Echazábal, L. (1996). O culturalismo dos anos 30 no Brazil e na America Latina: Deslocamento retórico ou mudança conceitual? In M. C. Maio & R. V. Santos (Eds.), *Raça, ciência e sociedade* (pp. 107–24). Rio de Janeiro, Brazil: Editora FioCruz, Centro Cultural do Brasil.

Martinez-Echazábal, L. (1998). Mestizaje and the discourse of national/cultural identity in Latin American 1845–1959. *Latin American Perspectives, 25*(100), 21–42.

Mitchell, M. (1992). Racial identity and political vision in the Black press of São Paulo, Brazil, 1930–1947. *Contributions in Black Studies, 9*(3), 1–13.

Mora, G. C. (2014). *Making Hispanics: How activists, bureaucrats, and media constructed a new American.* Chicago, IL: University of Chicago Press.

Muniz, J. (2010). Sobre o uso da variável raça-cor em estudos quantitativos. *Revista de Sociologia e Política, 18*(36), 277–291.

Nascimento, A., & Nascimento, E. L. (2000). Reflexões sobre o movimento negro no Brasil (1938–1997). In A. S. Guimarães & L. Huntley (Eds.), *Tirando a máscara: ensaios sobre o racismo no Brasil* (pp. 203–234). São Paulo, Brazil: Paz e Terra.

Nobles, M. (2000). *Shades of citizenship: Race and the census in modern politics.* Stanford, CA: Stanford University Press.

Oliveira, L. E. G., Porcaro, R. M., & Araujo Costa, T. C. N. (1983). *O lugar do negro na força do trabalho.* Rio de Janeiro, Brazil: IBGE.

Otero, H. (2006). *Estadística y nación: Una historia conceptual del pensamiento censal de la Argentina moderna, 1896–1914.* Ciudad de Buenos Aires, Argentina: Promoteo Libros.

Paschel, T. (2010). The right to difference: Explaining Colombia's shift from color-blindness to the law of black communities. *American Journal of Sociology, 116*(3), 729–769.

Paschel, T. (2013) 'The beautiful faces of my Black people': Race, ethnicity and the politics of Colombia's 2005 census. *Ethnic and Racial Studies, 36,* 1–20.

Paschel, T. (2016). *Becoming black political subjects.* Princeton, NJ: Princeton University Press.

Petruccelli, J. L. (2015). Brazilian ethnoracial classification and affirmative action policies: Where are we and where do we go? In P. Simon, V. Piche, & A. A. Gagnon (Eds.), *Social statistics and ethnic diversity: Cross-national perspectives in classifications and identity politics* (pp. 101–109). Cham, Switzerland: Springer.

Piza, E., & Rosemberg, F. (1999). Color in the Brazilian census. In R. Reichmann (Ed.), *Race in contemporary Brazil* (pp. 37–52). University Park, PA: Pennsylvania State University Press.

Porter, T. R. (1986). *The rise of statistical thinking, 1820–1900.* Princeton, NJ: Princeton University Press.

Porter, T. R. (1995). *Trust in numbers: The pursuit of objectivity in science and public life.* Princeton, NJ: Princeton University Press.

Rodriguez-Piñero, L. (2005). *Indigenous peoples, postcolonialism and international aw: The ILO regime (1919–1989).* Oxford, England: Oxford University Press.

Schwarcz, L. M. (1993). *O espetáculo das raças: Cientistas, instituições, e questão racial no Brasil 1870–1930.* São Paulo, Brazil: Companhia das Letras.

Schwartzman, L. F. (2009). Seeing like citizens: Unofficial understandings of official racial categories in a Brazilian university. *Journal of Latin American Studies, 41*(2), 221–250.

Senra, N., & Camargo, A. D. P. R. (Eds.). (2010). *Estatísticas nas Américas: Por uma agenda de estudos históricos comparados.* Rio de Janeiro, Brazil: IBGE, Centro de Documentação de Informações.

Silva, G. M., & Leão, L. S. (2012). O paradoxo da mistura: Identidades, desigualdades e percepção de discriminação entre brasileiros pardos. *Revista Brasileira de Ciências Sociais, 27,* 117–133.

Silva, G. M., & Paixão, M. (2014). Mixed and unequal: New perspectives on Brazilian ethnoracial relations. In E. Telles & PERLA (Eds.), *Pigmentocracies: Ethnicity, race, and color in Latin America* (pp. 172–217). Chapel Hill, NC: University of North Carolina Press.

Silva, N. do V. (1978). *Black-white income differentials in Brazil* (Doctoral thesis in Sociology). University of Michigan, Ann Arbor.

Silva, N. do V. (1985). Updating the cost of not being white in Brazil. In P. M. Fontaine (Ed.), *Race, class, and power in Brazil* (pp. 42–55). Los Angeles, CA: University of California, Center for Afro-American Studies.

Silva, N. do V. (1995). Morenidade: Modos de usar. *Estudos Afro-Asiáticos, 30,* 79–95.

Silva, N. do V. (1981). Cor e o processo de ealização socio-economica. *Dados, 24,* 391–409.

Skidmore, T. (1993). *Black into white: Race and nationality in Brazilian thought.* Oxford, England: Oxford University Press.

Sommer, D. (1991). *Foundational fictions: The national romances of Latin America.* Berkeley, CA: University of California Press.

Teixeira, M. de P. (2003). *Negros na universidade: Identidade e trajetórias de ascensão social no Rio de Janeiro.* Rio de Janeiro, Brazil: Pallas.

Telles, E. (2004). *Race in another America: The significance of skin color in Brazil.* Princeton, NJ: Princeton University Press.United Nations. (2009). Censos 2010 y la inclusión del enfoque étnico: hacia una construcción participativa con pueblos indígenas y afrodescendientes de América Latina. Serie seminarios y conferencias, n.57. Santiago de Chile.

United Nations Development Programme. (2004). *Human development report 2004: Cultural liberty in today's diverse world.* New York, NY: United Nations Development Programme.

United Nations. Secretariat of the Permanent Forum on Indigenous Issues. (2004). *Workshop on data collection and disaggregation for indigenous peoples. Department of Economic and Social Affairs.* New York, NY: Division for Social Policy Development, 19–21 January.

Vejo, T. P., & Yankelevich, P. (Eds.). (2017). *Raza y política en Hispanoamérica.* Diásporas. Madrid, Spain: Iberoamericana Vervuert; Mexico City: Bonilla Artigas Editores; Mexico City: El Colegio de México.

Wagley, C. (Ed.). (1952). *Race and class in rural Brazil: A UNESCO study.* Paris, France: UNESCO.

2 Census, politics and the construction of identities in India

Ram B. Bhagat

2.1 Introduction

Of late, identity has acquired centre stage in politics, a yardstick for evaluating the fulfilment of social and economic rights and a force of state social policy in many countries of the world. Based on the social and cultural characteristics of people, political parties lured by vote bank politics have used identity as an instrument of political mobilisation and consequently of shaping state policies. In this context, knowing the social and cultural characteristics of people and classifying them has been an essential state project executed through the institution of census, both, historically as well as in contemporary times.

Census enumeration is considered to be a scientific exercise in order to know the size, growth composition and characteristics of the population. The mandate of the census is derived from the state, an institution that embodies the relationship between a territory and its people. How the state views its people and their charac- teristics is very much a political phenomenon which changes according to the na- ture of the state and its strategy to maintain power. In this situation, the census turns out to be an instrument of the state; it converts people into the population and uses a classificatory principle to divide them into mutually exclusive ethnic, religious, racial and caste groups. This is being done under the assumption that the method- ology of natural sciences as such could be applied to the social reality as well. In the context of India, historical records show the interwoven and inclusive nature of social identities in many spheres of life. The Census Commissioners of Brit- ish India presented a plethora of such examples, highlighting many religious and cultural practices in India as overlapping, which could not be classified in a form that is mutually exclusive (Bhagat, 2003, 2013). However, such identities could be better represented if multiple choices were provided in the census schedule. The present chapter argues that mutually exclusive classification has a deep impact on creating a contrived social reality and the unfolding of identity politics suitable for statecraft. In India, this process began during the colonial regime and has not changed after independence; it was rather reinforced and re-emerged in various new forms. Along with ethnic categorisation, size and growth have also been added to the construction of identities in the era of democratic politics influenced by the emerging demographic reality.

DOI: 10.4324/9781003259749-4

According to Anderson (1991), the census is one of the institutions of power along with the map and the museum. Although they were invented before the mid-19th century, they turned out to be very powerful instruments of domination during colonial rule (Anderson, 1991, p. 163). These three institutions profoundly shaped the way in which the colonial state imagined its dominion. The census created mutually exclusive identities imagined by the classifying mind of the co-lonial state. Similarly, the map worked as an exhaustive territorial classification, showing definite control, which also served as a logo, i.e. an instantly recognisable symbol of power. The museum allowed the colonial state to appear as the guardian of local tradition. However, all three institutions have changed their role with the disappearance of colonialism and the emergence of nationalism. In the sovereign and republican states, the census mirrored the new demographic imaginations of the modern state, deeply influenced by their political imaginations of the demos. While at times the census reflected many characteristics of the modernising state, sometimes it did not completely abjure the colonial legacies. Scholars also argue that various forms of governmentality targeting the population were novel tactics of colonialism; unlike such instruments also existed during pre-colonial times in various forms of statistical and record-keeping practices, the state hardly assumed an interventionist role (Kalpagam, 2014, p. 68). In the colonial state, the process of intervention was not only confined to administrative, record-keeping, statistical and knowledge production but also in the entire social, cultural and political sphere reshaping the institutions and practices in the mirror image of the west. The orient was almost a European invention, which helped to redefine the orient in contrast to the occidental image, ideas, practices and experience. The orient was portrayed as irrational, unscientific, uncivilized and primitive. In other words, orientalism was a Western style of knowledge for dominating, restructuring and having authority over the orient (Said, 1978). In this context, census, record-keeping, mapping and vital statistics were the new state apparatuses introduced to modernise, civilise and govern Indians during the colonial times. The British mode of thought, imagination and classification of Indian society constructed exclusive social and religious iden-tities in India that undermined the pluralistic understanding of Indian society (Sen, 2005). As far as the census is concerned, there are at least three pathways through which it constructs ethnic, racial, caste and religious identities driven by politics. Firstly, the enumeration of people through the census requires a clear setting of the boundaries of ethnic and religious communities to know their size, growth and various characteristics. Sometimes, boundaries are not clearly defined in daily life, but the census makes it possible to draw a neat social and geographical boundary because enumeration demands mutually exclusive categories. In reality, however, this may not be the case due to the overlapping nature of social practices of many communities, as well as variations of social and cultural practices over space and time. However, censuses do not only fix ethnic and cultural boundaries over time and space but also create a homogeneous community. Thus, the key strategy of cen-sus taking has been to construct fixed and immutable social categories, like caste and religion, whose boundaries have been changing historically in India (Bhagat, 2006; Cohn, 1987). Second, through census enumeration, the size, growth and social and economic characteristics of communities are known. The knowledge of these attributes is not independent but enters into the community consciousness, shaping social relations and recreating communities (Bhagat, 2006). In fact, the census produces knowledge about the community which creates a consciousness of shared belongingness and affinity that is geographically discontinuous and spread over far-off places. In other words, census enumeration through demographic and geographical imagination reconstructs the ethnic and religious communities. Third, in a democratic setup, where number means power, where numbers determine access to social and economic resources, various types of fission and fusion among communities take place, being unfolded by competitive communitarian politics (Bhagat, 2001, 2013). Communitarian politics sometimes demand the inclusion of a separate community in the subsequent census, delinked from the larger community. However, the tendency of the census is essentially to create few but large homogeneous communities, ironing out differences. This principle of enumeration is mirrored in the political process. Thus, the census cannot be separated from social and political processes, but it abets and is embedded in the same processes. This chapter describes the practices of various censuses in India constructing social identities with reference to religious and caste categories resulting from the colonial and post-colonial imagination and politics.

2.2   Census in India – continuity and change

In India, as early as 1856, the British Government decided to hold a census in 1861, which eventually could not be held due to a mutiny in 1857. After the great mutiny, the colonial government reorganised the army on caste and ethnic lines and also notified certain castes and tribes as criminals (Kumar, 2004, p. 1087). Against this backdrop, it was decided in 1865 that a general population census would be taken in 1871. But, the years 1867–1872 were actually spent in census taking. This series of census activities was in fact known as the census of 1872, which was neither a synchronous census nor covered the entire territory controlled by the British (Srivastava, 1972, p. 9). In fact, the first synchronous census may be called to have begun from the census of 1881.

There were several reasons why the British Government started census taking in India. From the administrative point of view, a clear knowledge of the composition of Indian society was necessary to exert control and extract revenue. Colonial rulers understood the administrative exigencies of the role of knowledge in maintaining power and embarked upon the codification of Indian society. They were also amazed to see the complexities of Indian society. At this juncture, a new classificatory trend in the European intellectual tradition was also emerging which motivated the colonial rulers to develop a taxonomy based on their perception of Indian society primarily belonging to primordial categories (Bandyopadhyay, 1992, p. 26). A distinguished historian of India, R.S. Sharma, argues that the 1857 revolt caused Britain to realise that it badly needed a deeper knowledge of manners and social systems of the Indian people for a continued ruling. As a consequence, not only census was undertaken but also ancient scriptures were translated on a massive scale under the editorship of Max Mueller (Sharma, 2005). Britain thus embarked upon consolidating the regime by the creation of a body of appropriate knowledge not only to understand Indian society and culture better but also to offer legitimacy to the administrative concerns and decisions of the colonial rule (Kumar, 2004, p. 1086). On the other hand, Guha (2003) argues that census enumeration was not a novel practice adopted by the British, but was equally practiced in earlier times, particularly in the Moghul period. But the fact remains that the earlier enumerations were very much confined to the purposes of land revenue and taxation and also geographically more limited. In contrast to this, the colonial census was interested in anthropological knowledge and was carried out covering most parts of the country.

Thus, the census at this juncture was an instrument for creating the requisite anthropological knowledge about Indian society. It was not a coincidence that most of the census commissioners were trained Anthropologists. Noted among them were Herbert Risley and J.H. Hutton who looked after the census operations of 1901 and 1931 respectively, and contributed immensely through their writings and publications in producing anthropological knowledge about India. Risely wrote a monumental book entitled *The People of India* first published in 1908, and Hutton wrote the famous book, *Caste in India: Its Nature, Function and Origins* first published in 1946. In the mind of the colonial power, there was a desire to know the differences between the Indian people which could be helpful in matters of governance and continuance of power; and there was also a curiosity to know whether any social group in India constituted a nation in the European sense of the term. In fact, Indian identities do not exclusively belong to race, language and religion, but all of them are overlapping. As such, they do not constitute a nation in the European sense. For example, Hindus and Muslims speak the same language in different parts of India and possess similar racial characteristics and caste structures. This is true for other religious groups as well (Ahmad, 1978, 1999; Hutton, 1931).

Since the very first census of 1872 the colonial state incorporated caste and religious categories in the enumeration of the Indian population. In the beginning, caste was equated with class (1872 census); in 1881 sect was also introduced, followed by race in 1891, and tribe was included since the 1901 census which continued until 1941 (Census of India n.d.). However, religion was included since the beginning of the census. This is in marked contrast to the British census which introduced religion only in 2001 on a voluntary basis. It also contrasts with the USA, where asking a question on religion in the census is prevented by the American constitution. In independent India, asking about religion in the census continued, but caste was dropped, except for Scheduled Castes (SCs) and Scheduled Tribes (STs). The SCs and STs are the official social categories identified under the provision of Article 341 of the Indian Constitution. According to this article, the President of India, after consultation with the Governor of a State or Union Territory (UT), may declare castes, races or tribes or groups within castes, races of tribes as SC or ST belonging to that state or UT based on their historical deprivations and exclusions. It is a state strategy of social protection which accelerated the development of the identified groups known as SCs and STs in independent India (cf. Piketty, 2022, pp. 175–202).

While at times the census reflected many characteristics of the modernising state, sometimes it did not completely abjure the colonial legacies. This is evident in the fact that the census in independent India is a part of the Ministry of Home Affairs unlike the British census being controlled by an independent statistical authority, answerable to the British Parliament directly. In the USA, the census is a part of the Department of Commerce. While there is nothing wrong with the Census of India being part of the Ministry of Home Affairs, it reflects the perception that the census is rooted in the colonial legacy of governance and politics.

2.2.1   Construction of caste identities

Since the 1970s, with increased globalisation, identities have acquired centre stage in national politics in several developing countries (Kukutai & Thompson, 2015), including India. India has been a multi-ethnic and multi-cultural society historically. However, colonial interaction and a new mode of knowledge production reconstructed the existing social and religious relations in newer forms. This is in contrast to the evolution of social groups, a process in which primary groups (Gemeinschaft) changed into secondary groups (Gesellschaft) alongside with the agricultural-industrial economic transition in European countries (Tönnies, 1957). As India did not experience the same type of industrialisation and the historic trajectory of social evolution was disrupted by colonialism, social identities were reconstructed through colonial powerholders armed with the means of knowledge production that served their immediate needs but had significant implications for social structure. This is not to say that what existed was good but to highlight that the social evolution as a project of construction of social categories during colonial times shaped India's present social and political structure and identities.

Bernard Cohn (1987) wrote a seminal article advancing an argument that the colonial census played an important role in identity formation in South Asian countries. The religious and caste identities were two important forms recreated during colonial rule. Both these identities are intertwined, competitive and reinforcing each other. The census played an important role as an administrative apparatus through which communities were made to view themselves in relation to others in definite terms, and a gradual realisation of demographic strength cropped up to promote self-interested collective action. Several scholars later believed that modern politicised identities in India found their definite geographical and social boundaries through census enumeration initiated during British rule (Anderson, 1991; Appadurai, 1993; Dirks, 2001; Kaviraj, 1993). Further, the census practice in colonial India was in sharp contrast to census taking in Britain where the question on ethnicity was kept aside from enumeration until the 1991 census and religion was reintroduced in the 2001 census (Bhagat, 2001). Categorisation and enumeration were two strategies of the census exercise. Categories were identified as distinct, homogeneous and mutually exclusive which is essential to the scientific understanding. According to Appadurai (1993) enumeration played a more important role than categorisation in the formation of identities. However, this is a debatable issue because enumeration requires a neat categorisation. It is through categorisation that the social boundaries of ethnic and social groups are delineated and fixed, boundaries which gradually become immutable. This process of quantification established the predominance of group size in democratic politics, converting people into populations and their size becoming a marker of social identities.

Historically numerous communities existed in India but they have existed as fuzzy communities from time immemorial, but their congealing into distinct, discrete and mutually antagonistic communities was certainly aided to a great extent by the counting of heads. Fuzzy communities are defined as indistinct groups with neither internal cohesion nor well-known external boundaries and as such, are communities without overt communication. The group did not know how far it extended and what was its strength in numbers, therefore, had less accurate and less aggressive self-awareness. In other words, the geographical and social boundaries were not clear. The fuzzy communities also did not require any developed theory of otherness (Das 1994; Kaviraj, 1993).

India's social structure was also interwoven in its nature. Caste and religion – two predominant identities have been overlapping in the past and their interwoven nature is still visible in many parts of the country in spite of political attempts to separate them. Religions like Islam and Christianity, which came from outside, have also a strong presence of caste within them (Ahmad 1978; Ansari 1959; Wyatt 1998).

However, there have been attempts to define, categorise and understand caste, tribe, race and religion in India through censuses. In the beginning, caste and tribe were loosely used in the censuses and hardly any distinction was made (Baines, 1893). Among the categories of caste, tribe and race, caste has been most pervasive and distinct in Indian society, but people perceived and replied about caste status, not in a uniform manner.

For some, caste meant a status in the caste hierarchy, for others, it was a sense of belonging to an occupational group, a linguistic identity like Gujarati, Bengali and for still others a place or regional identity like Madrasi (Samarendra, 2011). Herbert Risly, the Census Commissioner of the 1901 census tried to develop a theory of the origin of caste based on racial lines. He attempted to classify castes based on colour and physical characteristics (Risley, 1908). However, due to the enormous diversity in the physical characteristics and caste traits of the population, many questions on the possible link between caste and race were raised, and in the end, this exercise was not conclusive (Gait, 1913). In fact, various studies point out that caste was historically a localised status group arranged hierarchically from higher to lower status with little sense of solidarity with similar groups outside of their local habitat. Furthermore, castes as social groups were mobile and mutable in the past, but census enumeration and categorisation rendered them fixed and immutable (Bhagat, 2006).

During colonial rule, the concept of race was also used to differentiate between the natives of Indian origin and those coming from the West or European countries. The population was divided into races such as Indo-Aryan, Iranian, Semitic and primitive races followed by categories of religious groups within each of them. Also, an effort was made to categorise population into British, Europeans other than British a hybrid category like Anglo-Indians resulting from mixed marriages. It represented physical characteristics in earlier censuses but later it represented nationality. The Census Commissioner of 1931 J.H. Hutton (1931, p. 425) believed that race is so loosely used as to defy any definition, but generally, it represented nationality or place of birth of foreign origin. Race is no longer an important issue either in politics or the census of independent India. The term tribe was understood as a category to the extent it is not converted into caste and also represented the residual category of a race.

There were more than 4,000 castes and sub-castes identified in the censuses during British rule. The number of castes gradually increased after every census. For example, the number of castes identified in the 1901 census was 1,646, which increased to 4,147 in the 1931 census, the last census on caste in India. The total number of castes also included over 300 castes whose religion was recorded as Christianity and over 500 who were recorded as Muslims. The Anthropological Survey of India launched the Peoples of India Project under the leadership of K. S. Singh, which was completed in the 1990s. According to this study, there were 4,635 castes or communities in India (Bhagat, 2006). It was noted by the Census Commissioner of the 1921 census, E.A. Gait that, historically, caste had not been a fixed and immutable category but new castes emerged through fission and fusion. Sometimes a group of people developed a different occupation or a religious cult or social practice and regarded themselves as distinct from the others, leading to the emergence of a new caste. In other cases, several castes of the same occupation or followers of a cult or social practice merged creating a new caste. This resulted in the fixing or re-fixing of social status in the local context (Gait, 1913, p. 371). However, due to the institutionalisation of the census, the historical characteristics of caste changed from a mobile and mutable category to a fixed and immutable category. At the same time, the category of caste was uprooted from the local context of status hierarchy and was raised to all-India level. Castes positioned themselves face to face with the government as an identity group competing for resources and privileges with other castes. The demographic size of castes added an objective sense of strength to their negotiations and claims for access to resources and power. Some of the powerful castes of contemporary India, namely Yadavas, Reddys, Kamma, Marathas and Patels/Patidars, to mention just a few, have re-emerged against this backdrop by the amalgamation of several castes and sub-castes (Hutton, 1931; Upadhya, 1997).

The decennial censuses not only updated the population figures but also gave them specific labels and ranks. The British understanding of caste was based on their reading of the oriental literature as well as their reliance on the local scholars of Hindu scriptures. The sources of knowledge about caste based on religious texts presented a model of caste systems primarily based on *Varna,* which literally means colour in Sanskrit. Varna is a hierarchical system of caste status based on social origin in the Hindu society, ranging from Brahmins (priestly castes) at the top, followed by Kshatriya (warrior castes), Vaishyas (trading castes) and Shudras (labouring castes) at the bottom of the social ladder. However, in reality, within each of the Varna categories, there existed numerous castes and sub-castes whose names were not similar in the different parts of India and their hierarchical social status was also not clear. Varna, in fact, provides a principle of caste ranking based on Hindu scriptures, among them *Manusmriti* is often cited for its enunciation. However, many times, the Varna hierarchy did not represent the way caste was actually practised in different parts of India and hence represented a distorted view of the caste system (Bandyopadhyay, 1992; Srinivas, 1962). In actual census practice, many people perceived that the object of the census was to fix the relative social position of the different social classes and to deal with questions of social superiority. As a result, the actual reporting of caste was also distorted, as many lower castes returned to higher castes in order to raise their social status. In the census, the underprivileged found an opportunity to express their aspiration and if possible to acquire a new identity through enumeration (Ahmed, 1981, p. 116). In many places, several caste groups petitioned the census officials and claimed to be included in the census different from what was entered in the census record. In several cases, their claim was accepted. On the other hand, a number of castes were not happy with the way they were entered and grouped in the census, and there was widespread resentment over this (Cohn, 1987).

The census of 1921 was conducted in a changed political scenario when the growing force of a national movement led by Mahatma Gandhi challenged the edifice of British rule. The census as a tool of the empire responded to the growing unity of Indian people by introducing a category of "depressed classes," which for the first time reflected the state's gesture to provide protection to this category of people. However, the term depressed classes was not favoured in the 1931 census and was replaced by "exterior castes" (Hutton, 1931). The tabulation of figures for individual castes in the 1931 census was based on four criteria namely (i) those who belong to the lower caste hierarchy, (ii) primitive tribes and (iii) all other castes with the exception of (iv) those short of four per 1,000 of population and those for which separate figure were deemed to be unnecessary by local government (Census of India, 1961). Thus, group size and social backwardness[1] were the yardsticks adopted for a new classification of castes for administrative and governance purposes. A remarkable political development following this categorisation was the promulgation of the Scheduled Order of 1936 that officially recognised the listing of castes in every province of India for providing reservation in political representation and protection from the British Government (Bandyopadhyay, 1997, p. 33). This practice continued in independent India, and the castes identified were known as SCs and tribes as STs, declared under constitutional provisions to provide the reservation and social protection to them. In the censuses of independent India, the status of SCs and STs was crucially linked with religion. Until the 1981 census, SCs could only belong to Hindus and Sikhs, while STs could belong to any religion. However, since the 1991 census SCs could also belong to the Buddhist religion (Census of India, 1991).

In the early 1990s, the Government of India, based on the recommendations of the Mandal Commission, created a group of castes known as Other Backward Classes (OBCs) and extended to them quotas for the reservation of government jobs and also for the admission to the colleges and universities. The creation of

OBCs was based on the caste data of the 1931 census and the criteria adopted to identify OBCs were based on educational and social backwardness. This raised a demand for a renewed inclusion of caste in the 2001 census.

Although there was a strong demand to include caste in order to identify the demographic size of the OBC group in 2001 it was rejected on two grounds. First, the task of enumerating 4,000 odd castes in the census is fraught with many difficulties and second, it will strengthen the caste sentiments and caste identities which the Indian state aimed to eradicate with independence. However, the demand was again raised in the run-up to the 2011 census. The Central Government finally decided that the caste census would be held as a separate exercise, but not as a part of the 2011 census; the task of collecting data on caste was entrusted to the Ministry of Rural Development for rural areas and to the Ministry of Housing and Urban Poverty Alleviation for the urban areas. The rural areas also included urban out-growth and urban was confined to statutory urban centers only. This census was called "Socio-economic and Caste Census 2011," but data was neither made available for individual castes nor for the OBCs (Socio-Economic Caste Census, 2011). However, it did not serve the political demand for demographic data on the OBCs. The re-inclusion of caste continues to remain a contentious issue with scholars arguing for and against it. However, the complexities of caste show that a particular caste has several sub-castes and there are various nomenclatures used for the same caste in different linguistic set-ups. The same caste can also have varied social organisation in different parts of the country. It may also not hold the same social ranking, and therefore aggregation at the all-India level is problematic (Bhagat, 2007; Deshpande & John, 2010). While it is an open question if caste will be included or not in future censuses, the genie is already out of the bottle as censuses of the past have already played an important role in shaping its form and content. Presently, caste hardly represents the hierarchical status in a traditional sense, but forms an identity and interest group driven by the politics of kinship.

2.2.2   *Construction of religious communities*

The demographic data published by the census on religion and caste not only re-shaped Hindu-Muslim and inter-caste relationships but also shaped the political imagination of India. Like caste, the question of religion was asked since the beginning of the census in 1872. Most of the demographic data were also presented cross-classified by religion. Census officials took great pains to classify the Indian population in terms of homogeneous and mutually exclusive religious communities in spite of several difficulties. The difficulties encountered by census officials were mainly related to the overlapping of beliefs, faiths and practices, sects, indistinct boundaries among religious communities in many places and local perception of a religious community based on the contextual status hierarchy when a question on religion was asked (Bhagat, 2013). While the caste known as Jati is a localised status hierarchy based on birth and maintained among endogamous social groups in a region, religion in India has been defined traditionally and practised as *Dharma* which means a moral duty. According to one of the Census Commissioners,

The Hindu word 'dharma' which corresponds most closely to our word 'religion' connotes conduct more than creed. In India the line of cleavage is
social rather than religious, and the tendency of the people themselves is to
classify their neighbours, not according to their beliefs, but according to their
social status and manner of living.

(Census of India, 1911a, p. 113)

However, contrary to this notion of Indian religious tradition, Indian people were
divided based on their membership (creed) in religious groups, leading to the creation of religious identities based on demography, shaping the national politics of
the country. While census officials were very clear about what constituted Christianity and Islam, they came across enormous difficulty in defining and understanding who Hindus were. Thus, we find in various reports of Census Commissioners
of British India descriptions of how they deliberated, debated and wavered as
to whether the term 'Hindu' implied a 'race', nationality', a 'religion' or a 'set of
practices'.

In one census report, for example, Hindus were defined as,

A native of India who is not of European, Armenian, Moghul, Persian or
other foreign descent, who is a member of a recognised caste, who acknowl-
edges the spiritual authority of *Brahmans* (priestly caste), who venerates or
at least refuses to kill or harm kine, and does not profess any creed or religion
which the Brahman forbids him to profess.

(Census of India, 1911, p. 119)

Census officials also tried to distinguish animism from Hinduism. Animism was
considered as a pre-Hindu religion, but it was not clear at what stage animism
ceased and an animist became a Hindu (Census of India, 1911a, p. 129). The category of animism was so vague that its name was changed to Tribal Religion in
the1921 census (Census of India 1921). It was also pointed out by the census officials that many social and cultural practices among Hindus and Muslims were
overlapping and interwoven in several parts of the country as both religious categories were also embedded in the caste system. For example, in the province of Pun-
jab, the majority of Mohammadans also reported their caste status as *Rajput, Jat,
Arain, Gujar, Mochi, Tarkhan and Teli* (Census of India, 1911a). The Census Commissioner of the 1921 census reported to have classified some of the communities
as 'Hindu-Mohammadan' (Census of India, 1921, p. 115). The boundary line between Hindus on the one hand and Sikhs, Buddhists and Jains on the other is even
more indeterminate. Faced with the problem of Jains and Buddhists also wanted to
be recorded as Hindus, but the census schedule did not offer such multiple choice;
the problem was solved through a footnote that was added to the concerned table or
put under the category of 'Others' (Hutton, 1931, pp. 379–380). Census Commissioner E. A. Gait reiterated that the religions of India as we have already seen are
by no means mutually exclusive (Census of India, 1911a). Even the Census Commissioner J. H. Hutton had observed that 'while borderline between tribal religions

and some aspects of Hinduism is not at all easy to draw, it is just as hard to define it between Hinduism and Islam, and even between Hinduism and Christianity in the case of a number of intermediate sects which offer points of identity with both' (Hutton, 1931, p. 380).

However, in spite of enormous difficulties attempts were made to classify religious groups. There was also an attempt to divide religion racially into five such categories, namely, Indo-Aryan, Iranian, Semitic, Primitive and Miscellaneous, and also into sects (Census of India, 1921)

A schematic classification is presented below for illustrative purposes (Census of India, 1921)

I **Indo-Aryan**
  A Hindu
    a Hindu -Brahmanic
    b Hindu -Arya- Vedic Theists
    c Hindu-Brahmo- Eclectic Theists
  B *Sikh*
  C Jain
  D Buddhist
II **Iranian**
  A *Zoroastrian (Parsi)*
III **Semitic**
  A *Musalman*
  B *Christians*
  C *Jews*
IV **Primitive**
  A *Animistic/Tribal*
V **Miscellaneous** (Minor religion and religion not returned)

However, the above scheme of classification was not uniformly followed in every census. Various sects of Hinduism namely Brahmanic, Arya and Brahmo were identified. In the 1881 census, *Kabirpanthis* (a unitary religion without rituals based on love and unity of mankind) and *Satnamis* (true god under the title of Satnam; no idol worship) were kept as separate religious groups (Her Majesty's Stationary Office, 1883), but were later merged with Hindus. Muslims were divided into Shias and Sunnis, and Christians into many sects like Roman Catholic, Anglican, Methodist, Lutherans, and Presbyterian, etc. However, differences in practices and sects were gradually obliterated and dropped from census tables, and, Indian population was classified into ten mutually exclusive religious categories namely Hindus, Muslims, Sikh, Buddhists, Jains, Christians, Zorastrians, Jews, Tribals and Others in colonial India. Census thus shattered the inclusive nature of religious faith and practices and created religious identities from an epistemological gaze hitherto unknown to Indians. Not only has the population been classified by religion, but its size, distribution and growth rate have been explained in terms of fertility and virility, as well as conversions and changes in the inclusion and exclusion of certain groups in the respective religious groups. The first census of 1872 revealed that Hindus constituted 73.5% of the population compared to 21.5% of the Muslims (Her Majesty's Stationary Office, 1881). As such, the new religious communities viewed themselves in terms of number, the idea of majority and minority was ingrained in their mind and also in the political system. The colonial power assured them to grant each of them resources, privileges and a share of power according to their demographic strength in the form of separate electorates and reservations in admission to educational institutions and in government jobs. This was the genesis of demographic communalism in India (Bhagat, 2001). It also brought the two major religious communities namely Hindus and Muslims in sharp competition, clashes broke out and riots took place since the late 19th century, culminating in communal mayhem and partition of India on the religious line in 1947 (Bhagat, 2012). The partition was a bitter experience of numbers by religion creating political polarisation of majority Hindus and minority Muslims which acquired centre stage even in independent India.

This is evident in the continued process of construction of religious identities in independent India. After independence, seven mutually exclusive religious categories namely Hindus, Muslim, Sikh, Christians, Jain and Buddhist found a prominent place in the census questionnaire and the rest of religions and faiths were clubbed and presented in the category Others. The process of mutually exclusive classification was accompanied by internal homogenisation within each religious community by census exercise. For example, when the response on religion was scrutinised in the 2001 census, there were 1,700 religions and faiths reported, but these were suppressed in the Others category in the census tables (Census of India, 2001). Therefore, it may be argued that independent India embarked upon the colonial legacy of religious consolidation through classification, numericisation and homogenisation of religious beliefs and practices in the name of protecting rights in contrast to the colonial policy of divide and rule. On the other hand, as India aspired to eradicate casteism and caste-based social discrimination, it stopped the collection of data on caste since the 1951 census, right after independence, except for the disadvantaged categories like SCs and STs. In most of the periods since independence, caste and religious identities have been competing in the sphere of vote bank politics, and various political parties favoured one over the other. This led to the demand for the inclusion of caste in the recent censuses, in order to know the changing status of castes other than SCs and STs, known as OBCs, declared based on the last complete census conducted in 1931 during the British rule. Until now, the Census of India has not undertaken a full caste census, but it has been a matter of intense political debate in the country. At the time of the 2011 census, the government scuttled the demand for the inclusion of caste in the census but allowed a separate socio-economic caste census to be held outside the aegis of the Census of India. As mentioned earlier, it was conducted by the Ministry of Rural Development and the Ministry of Housing and Urban Poverty Alleviation for the rural and urban areas respectively. The data collected under the socio-economic caste census was very helpful in understanding multi-dimensional poverty (Sahu, 2018). However, the detailed socio-economic data by individual caste status has not been released to the public domain so far. Thus, it does not serve the demand that led to the establishment of the socio-economic caste census, nor does it resolve the political controversy with regard to affirmative action for the OBCs. Some researchers argue that the data deficit is closely related to a democracy deficit, weakening the legitimacy of democratic institutions. In this situation, the conflicting relationship between the state and communities surrounding the issue of numbers has not been resolved by true, reliable and representative data (Agrawal & Kumar, 2020).

The ambivalence of the state, publishing some data while withholding others, is very much evident with respect to religion in independent India. It is worthwhile to mention that demographic data by religion was published, but data on the educational level and occupational characteristics by religion was not published until the 2001 Census. It was only in 2001 that the census began providing data on literacy and educational levels and also the worker and non-worker composition by religious categories. Only since the 2001 Census, we came to know that Jains are the most literate community and the Muslims the least. These figures point to the responsibility of the state in uplifting its downtrodden citizens and reshaping the inter-community relationship. Following the results of the 2001 census on literacy and education by religion, the Government of India constituted a Prime Minister's High-Level Committee known as the Sacher Committee in 2005 to look into the socio-economic backwardness of the Muslim community. The committee used the 2001 census data as a base and compared the conditions of Muslims with those of the SCs on the basis of social justice. The committee also gave several recommendations including the recommendation of reservation of jobs for Muslims in government services (Sacher Committee, 2006).

Although the politics of identity have been controversial in India as in elsewhere, it is noteworthy that SCs and STs have improved their income position to a larger extent than comparable disprivileged groups in the USA or in South Africa (Piketty, 2022, pp. 175–202).

2.2.3   *Recent innovations in statecraft and political controversy*

In the 2011 census, an attempt was also made to prepare the National Population Register (NPR) for ultimately preparing the National Register of Citizenship (NRC) as required by the Citizenship Act 2003. The NPR is a list of all "usual" residents of India, i.e. anyone who has lived in the country for more than six months regardless of their nationality. While the census is a listing and enumeration of household members, NPR is a listing and linking of family members, and NRC will be prepared to identify citizens from the residents listed in the NPR. During the time of the 2011 census, it was intended that the NPR would be used to provide a Unique Identity Number (UID) to the residents of India. However, it did not happen. On the other hand, the preparation for the census originally scheduled for 2021, was preceded by the Citizenship Amendment Act 2019 granting citizenship to those refugees of Hindu, Sikh and Buddhist, Jain, Parsi and Christian religions who came to India prior to December 31, 2014 from the countries of Pakistan, Bangladesh and Afghanistan. In addition to this, the creation of the NRC in Assam was concluded in

2019 in order to identify illegal immigrants, which practically excluded the 1.9 million population of the state of Assam. This led to the unprecedented protest against the preparation of the NPR in the run-up to the 2021 census because of its link to the NRC. The census also proposed to collect data by computer-assisted personal interviews (CAPI). As a consequence of the measures to contain the COVID-19 pandemic, the census process was stopped and has not been rescheduled so far. Although the 2021 census was stopped due to the COVID-19 pandemic, and census operations are crucially linked to the protests related to NPR and NRC, it points out the role of the census as a political instrument of the state. The fact that India as one of the most populous countries in the world has not announced until 2023 when the next census will actually take place re-emphasises this interpretation.

2.3   Practical implications

The tendency of the census to homogenise populations and create mutually exclusive groups obliterating the numerical minorities has serious implications for democracy and human well-being. Do we look at diversity as a strength or a weakness? It is also a deeply epistemological question. An epistemology based on diversity and inclusion would design a census methodology of enumeration and classification of religious and cultural practices different from current practice in India. A mutually inclusive classification would provide, firstly, that one has the possibility to choose more than one option simultaneously. For example, in Japan, answers to the question about religion are not mutually exclusive. One can be a Buddhist and Shinto simultaneously. As a result, Shintos constitute 70.4% and Buddhists 68.8% of the Japanese population, together exceeding 100% (Plecher, 2019). Secondly, there could be allowed as many religious beliefs and faiths as possible. Statistical inclusion of diversity is fundamental to the realisation of the principles of democracy and the protection of statistical minorities.

Further, it is to be realised that identity does not have a singular form. In the opinion of Amartya Sen, each one of us has multiple identities in relation to religious faith, cultural practices and occupational status. In India, however, religious identity (alternatively competing with caste identity) is being politicised to override all other identities. The Hindu majority – invented through the census – is shaping the political imagination of India. It survives and prospers on the idea of a monolithic group of Muslims, who in reality are also divided into various faiths, castes and linguistic groups. We should remember that, "Our religion is not our only identity, nor necessarily the identity to which we attach the greatest importance" (Sen, 2005, p. 56). The state has an enormous role to play in this respect because the census methodology employed shapes identities; but much depends upon the political imagination.

2.4   Conclusions

In the modern India, caste and religion are two competing identities in the electoral politics shaped by colonialism through the instruments of census and demography.

Castes in India historically were localised status groups arranged hierarchically from higher to lower status with little sense of solidarity with similar status groups outside their local habitat and have also been mobile and mutable; but through census enumeration and categorisation castes have been made fixed and immutable. They also acquired a sense of number and demographic weight and also developed solidarity with similar groups outside their own territory. In the sphere of religion, there have also been interwoven practices, whereas census counting made them mutually exclusive. As numbers are known by religion, the demographic idea of majority and minority not only acquired centre stage but also shaped politics and ultimately the partition of India on a religious line in 1947. The census gradually failed to record religious faiths and beliefs and enumerated membership of the various religious communities as self-reported by respondents elicited from a census questionnaire. This chapter elucidated that the census has a deep impact in creating a contrived social reality and unfolding identity politics suitable for statecraft. Although this type of statecraft began during the colonial regime, it was basically not changed after independence but rather re-emerged in various new forms. Apart from ethnic categorisation, enumeration added population size and growth as the basis of the construction of identities in the era of electoral politics influenced by the emerging demographic reality.

Further, there are people with multiple identities but there is no multiple choice given in reporting identities in censuses of India. The prevalent census methodology requires the construction of mutually exclusive homogeneous categories; however, in actual practice definite boundaries are either lacking or overlapping. It is worthwhile to emphasise that Indian identity is regionally embedded but has a pan-Indian value strengthened by Indian Constitution during the last 70 years. Although a change in the census methodology would be required, that would only become possible through a new political imagination of the state.

Note

1 The term "backwardness" is commonly used in public discourse in India. As in many other post-colonial countries, this wide usage can be traced back to a colonial epistemology (Agrawal, Kumar 2020: 37, endnote 57).

References

Agrawal, A., & Kumar, V. (2020). *Numbers in India's periphery: The political economy of government statistics*. Cambridge, England: Cambridge University Press.

Ahmad, I. (1978). *Caste and social stratification among Muslims in India*. Delhi, India: Manohar.

Ahmad, A. (1999). *Social geography*. Jaipur, India: Rawat Publications.

Ahmed, R. (1981). *The Bengal Muslims 1871–1906: A quest for identity*. Delhi, India: Oxford University Press.

Anderson, B. (1991). *Imagined communities: Reflections on the origin and spread of nationalism* (Rev. ed.). London, England: Verso.

Ansari, G. (1959). *Muslims caste in Uttar Pradesh*. Lucknow, India: Ethnographic and Folk Culture Publication.

Appadurai, A. (1993). Number in the colonial imagination. In P. van der Veer & C. Brecken-ridge (Eds.), *Orientalism and the post-colonial predicament* (pp. 314–339). Philadelphia, PA: University of Pennsylvania Press.

Baines, U. A. (1893). *Census of India 1891: A general report.* India Office, London. (Reprinted by Manas Publications, 1985, Delhi).

Bandyopadhyay, S. (1992). Construction of social categories: The role of the colonial census. In K. S. Singh (Ed.), *Ethnicity, caste and people. Proceedings of the Indo-Soviet Seminars held in Calcutta and Leningrad in 1990* (pp. 26–36). Organised by Anthropological Survey of India, New Delhi and Institute of Ethnography, Moscow. Delhi, India: Manohar.

Bandyopadhyay, S. (1997). *Caste, protest and identity in colonial India: The Namasudras of Bengal, 1872–1947.* London, England: Curzon Press.

Bhagat, R. B. (2001). Census and the construction of communalism in India. *Economic and Political Weekly, 36,* 4352–4355.

Bhagat, R. B. (2003). Role of census in racial and ethnic construction: US, British and Indian census. *Economic and Political Weekly,* February 22, 2003, 686–691.

Bhagat, R. B. (2006). Census and caste enumeration: British legacy and contemporary practice. *Genus, 62*(2), 119–134.

Bhagat, R. B. (2007). Caste census: Looking back, looking forward. *Economic and Political Weekly, 42*(21), 1902–1905.

Bhagat, R. B. (2012). Hindu-Muslim riots in India: A demographic perspective. In A. Shaban (Ed.), *Lives of Muslims in India: Politics, exclusion and violence* (pp. 163–186). London, England: Routledge.

Bhagat, R. B. (2013). Census enumeration, religious identity and communal polarization in India. *Asian Ethnicity, 14*(4), 434–448.

Census of India. (n.d.). *Census questions (1872–2011).* Retrieved from https://censusindia.gov.in/census.website/CENSUS_ques

Census of India. (1911). Vol. XV, *United Province of Agra and Oudh,* Report by E. A. H. Blunt. Allahabad, India: Govt. Press. (Reprinted by Usha Publication, New Delhi, 1987).

Census of India. (1911a). Vol. 1, *India, Report,* by E. A. Gait. Calcutta, India: Superintendent Govt. Printing India. (Reprinted by Usha, Delhi, 1987).

Census of India. (1921). *Volume I, India, Part-I, Report.* Calcutta, India: Superintendent Government. Printing.

Census of India. (1961). Vol. I, India, Part V–A (ii) *Special tables for scheduled tribes.* New Delhi, India: Registrar General and Ex-Officio Census Commissioner for India.

Census of India. (1991). *Primary census abstracts: Scheduled castes,* Series 1, India, Part II–B (ii). New Delhi, India: Registrar General and Census Commissioner, India.

Census of India. (2001). *The first report on religion.* New Delhi, India: Registrar General and Census Commissioner.

Cohn, B. (1987). The census, social structure and objectification in South Asia. In B. Cohn (Ed.), *An anthropologist among the historians and other essays* (pp. 224–254). Delhi, India: Oxford University Press.

Das, A. N. (1994). *India invented: A nation in the making.* New Delhi, India: Manohar.

Deshpande, S., & John, M. E. (2010). The politics of not counting caste. *Economic and Political Weekly, 45*(25), 39–42.

Dirks, N. B. (2001). *Castes of mind: Colonialism and the making of modern India.* Princeton, NJ: Princeton University Press.

Gait, E. A. (1913). *Census of India, 1911, Vol. 1, India, Part I-Report.* Calcutta, India: Superintendent Government Printing.

Guha, S. (2003). The politics of identity and enumeration in India C. 1600–1990. *Comparative Study of Sociology and History, 45*(1), 148–167.

Her Majesty's Stationary Office. (1881). *The report on the census of British India taken on 17th February 1881,* Vol 1. London, England: Eyre and Spotisswoode.

Hutton, J. H. (1931). *Census of India 1931* (Vol. 1). New Delhi, India: Reprinted by Gyan Publishing House, 1996.

Kalpagam, U. (2014). *Rule by number: Governmentality in colonial India.* Lanham, MD: Lexington Books.

Kaviraj, S. (1993). The imaginary institution of India. In P. Chatterjee & G. Pandey (Eds.), *Subaltern studies VII* (p. 20). Delhi, India: Oxford University Press.

Kumar, M. (2004). Relationship of caste and crime in colonial India: A discourse analysis. *Economic and Political Weekly,* March 6, 2004, 1078–1087.

Kukutai, T., & Thompson, V. (2015). 'Inside out': The politics of enumerating the nation by ethnicity. In P. Simon, V. Piché and A. Gagnon (Eds.), *Social statistics and ethnic diversity: Cross-national perspectives in classifications and identity politics* (pp. 39–61). Cham, Switzerland: Springer. doi:10.1007/978-3-319-20095-8_3

Piketty, T. (2022). *A brief history of equality.* Cambridge, MA: Harvard University Press.

Plecher, H. (2019). *Religious affiliation in Japan in 2017.* Retrieved from https://www.statista.com/statistics/237609/religions-in-japan/

Risley, H. (1908). *The people of India.* Calcutta, India: Thacker, Spink & Co.

Sacher Committee (Prime Minister's High Level Committee). (2006). *Social, economic and educational status of the Muslim community of India: A report.* New Delhi, India: Cabinet Secretariat, Govt. of India.

Sahu, N. K. (2018). *Modern census to identify the deprived and vulnerable: Technology and institutions in managing socio-economic and caste census (SECC).* New Delhi, India: Academic Foundation.

Said, E. W. (1978). *Orientalism. Western conceptions of the orient.* London, England: Routledge and Kegan Paul. (Rev. ed. Penguin, Harmondsworth, 1995).

Samarendra, P. (2011). Census in colonial India and birth of caste. *Economic and Politically Weekly, 46,* 51–58.

Sen, A. (2005). *The argumentative Indian: Writings on history, culture and identity.* London, England: Penguin Books.

Sharma, R. S. (2005). *India's ancient past.* New Delhi, India: Oxford University Press.

Socio Economic Caste Census. (2011). Retrieved from https://secc.gov.in/welcome

Srinivas, M. N. (1962). Varna and caste. In M. N. Srinivas (Ed.), *Caste in modern India and other essays* (p. 66). Bombay, India: Asia Publishing House.

Srivastava, S. C. (1972). *Indian census in perspective,* Census Centenary Monograph No 1. New Delhi, India: Office of the Registrar General, India.

Tönnies, F. (1957). *Community and society: Gemeinschaft und Gesellschaft,* Translated and edited by Charles P. Loomis. East Lansing, MI: Michigan State University Press.

Upadhya, C. (1997). Social and cultural strategies of class formation in coastal Andhra Pradesh. *Contributions to Indian Sociology, 31*(2), 169–194.

Wyatt, A. K. J. (1998). Dalit Christians and identity politics in India. *Bulletin of Concerned Asian Scholars, 30*(4), 16–23.

3 Education censuses and recognition

The politics of collecting and using data on indigenous students in Latin America

Daniel Capistrano, Christyne C. Silva
and Rachel Pereira Rabelo

3.1 Introduction

Latin America houses about 800 indigenous peoples with a population of around 58.2 million or 10% of the total population (CEPAL, 2020). Since the 1980s, the collective political mobilization of those peoples occupied a significant space in national institutional politics throughout the continent and even in international agendas such as the Sustainable Development Goals (SDGs). One of the achievements of this general mobilisation was a social and political recognition of the ethnic diversity of Latin American societies (Jackson & Warren, 2005; Rice, 2012). This recognition is often expressed in instruments of data collection for official statistics in the region (Angosto-Ferrández & Kradolfer, 2014; Del Popolo, 2008).

The recognition through measurement is far from being a simple inclusion process, it is a multidimensional construct. There is a significant advancement in the understanding of the relationship between official quantifications and the struggles over identity politics, material resources, land, community, or autonomy (Desrosières, 2011; Simon, Piché, Gagnon, 2015). In Latin America, works such as those from Schkolnik and Del Popolo (2013), Bailey, Loveman and Muniz (2013), Angosto-Ferrández and Kradolfer (2014), Loveman (2014) and Telles (2017) document the breadth of this effort in the region, painting a comprehensive map in the complex task of surveying, questioning, describing, and recording the ethnoracial official statistics.

However, as of yet, to the best of our knowledge, there has not been any comparative work in Latin America that either discusses the ethnicity categories in education censuses or explores the political use of official statistics on indigenous students. This occurs despite the increasing importance of education data in Latin America to the funding and design of education policies. Education policies, in turn, are also gaining increasing importance in the region. According to the UNESCO Global Education Monitoring Report (2020), Latin America and the Caribbean recorded the largest increase in spending on education as a percentage of GDP, from 3.9% in 2000 to 5.6% in 2017. In addition, education censuses have gained increasing relevance for other policy areas in Latin America, including migration

DOI: 10.4324/9781003259749-5

and population projections (UNESCO Institute for Statistics, 2016). However, the educational statistics production, their use, and limitations have not received sufficient attention from social scientists in Latin America (Daniel, 2016; Gil, 2019).

Education censuses distinguish themselves from other traditional statistical processes such as population censuses and national household surveys as they allow for the interpretation of very small groups, unlike other demographic household surveys, and they are usually more frequent than population censuses. At the same time, they differ from a simple collection of administrative records from schools as they have unique instruments and standardized methods, painting a cohesive picture of the national population in schools and enabling cross-data consistencies with other national data.

Contributing to filling this gap in the literature, this work addresses the following questions:

— To what extent do the education censuses in Latin America incorporate questions on the ethnicity of primary and secondary school students?
— Have the data generated by those censuses on ethnicity been used in policy documents?

To answer these questions, we conducted a comparative analysis of ethnicity categories used in education censuses of ten countries in Latin America: Argentina, Bolivia, Brazil, Chile, Colombia, Ecuador, Mexico, Paraguay, Peru, and Uruguay. Furthermore, we explored the political use of statistics produced by these censuses in the latest national plans for education from these countries. Although these countries also collect data on the tertiary level of education, the data collection methods are more diverse. Therefore, for this study, we focused on education censuses for primary and secondary levels of education only.

For that, we first discuss the theoretical framework for the construction of official statistics about indigenous students and give an overview of the legal framework guiding the statistical processes in the selected countries. Then, we compare the instruments used in the education censuses with a focus on the questions related to ethnic identity. Finally, we assess to what extent data generated by education censuses on indigenous students are used in the latest national plans of education.

3.2   Theoretical framework

Many countries in Latin America have consolidated their education censuses as a result of educational reforms championing accountability reproduced by the World Bank and the International Monetary Fund (IMF). This is despite the fact that many countries already have established educational statistics of varying degrees, mainly in state agencies, universities, non-governmental organizations, and foreign aid agencies (Akkari & Perez, 1998).

It is from the 1980s Latin America has undergone drastic changes in its educational systems, which can be observed through different theoretical approaches (Beech, 2002; Braslavsky & Cosse, 2006; Guzmán, 2005; Walsh, 2000).

Following the world expansion of mass education (Meyer, Ramirez, & Soysal, 1992), the acceleration of some educational changes had a significant impact on the educational data systems in the region: decentralization and responsibility; increased school autonomy for management and pedagogical practices with a view to interculturality; curriculum drawn up on a list of skills and not for encyclopedic knowledge; creation of evaluation systems; and actions for the professionalization of teachers.

As a consequence of these processes, national educational statistical systems were designed or redesigned not only to meet demands from international organizations but also to increase the management capacity of national states in relation to education policies (Akkari & Lauwerier, 2014). This management capacity was increased in the tension between pressures for education reforms and a "new public management" (Parente, Villar, Parente, & Villar, 2020) on the one hand and the demand of organized civil society for transparency and accountability on the other hand. Across the region, the establishment of statistical systems was mainly driven by the central government. This centralization affected directly education policies that were dislocated from the local context, standardizing local cultures, traditions, and institutions (Rosar & Krawczyk, 2001). As we explore later in this paper, this process influences the scope of identity politics in education.

In relation to identity issues within statistical systems, Rallu, Piché, and Simon (2006) identify four different types of statistical systems: the first case, "counting to dominate", is related to those systems designed to enable and sustain colonial power. The second one, "not counting in the name of national integration", guides the establishment of a system aiming to ignore or erase ethnic diversity in the name of national unique identity. The third case, "counting or not counting in the name of multiculturalism", has to do with an "appreciation" of cultural diversity. The final one, "counting to justify positive action", is based on the recognition of ethnic diversity and a normative intent to provide evidence for the reversal of the perspective put forward in case one. Rallu and colleagues argue that the Latin American historical experience recognizes multiculturalism and the statistical systems either count it in order to reaffirm this cultural diversity (e.g. Brazil) or don't use ethnic categories taking the diversity as given (e.g. Venezuela). This, however, has not been the case historically.

As Loveman (2014) demonstrates, the ethnoracial classification has been the basis of colonial domination in the early establishment of Latin American statistical systems. This approach has changed over the second half of the 20th Century with several countries dropping questions on ethnicity/race/colour from their national censuses. As the author argues, at the turn of the century, the ideologies of national integration were "replaced by a growing recognition that states should acknowledge diversity and address ethnoracial inequality" (Loveman, 2014).

However, the regulatory frameworks that establish the right to education and guide the implementation of policies for indigenous peoples in those countries differ considerably. In a recent study, the UN Economic Commission for Latin America and the Caribbean compared the regulatory frameworks of 17 countries in Latin America in relation to rights for indigenous peoples (CEPAL, 2020). It has been found that 11 out of the 17 countries analyzed have specific laws in their constitutions regarding education for indigenous peoples. 15 countries have addressed indigenous rights for education either in their constitution or in their general law for the educational system. The only two countries that do not have legal provisions for the education of indigenous peoples are El Salvador and Uruguay.

3.3   Methods

To explore the production and use of data on indigenous students, we have analyzed two types of documents. First, instruments of education censuses provide an indication of the type and scope of information that is produced in the region. In addition, we have also assessed education national plans to explore to what extent the data produced by these censuses are being used in these documents which are considered here as education policies themselves (Ball, 1993).

3.3.1   Education censuses

The general recognition of indigenous rights in the national legal framework can also be observed in the development of the measures in the official statistics and their main instruments (Bailey et al., 2013). The format of the education census instruments is somewhat similar to those adopted in population censuses. Therefore, countries vary considerably in the data collection methods and ethnic categories as already observed for population censuses (Del Popolo, 2008; Morning, 2008).

To assess these different methods of ethnic classification in education censuses of the region, we have analyzed the data collection instruments from ten countries that represent almost 82% of the estimated indigenous population of the region (CEPAL, 2020).

In eight of the selected countries, data was collected for each student (individual data), whereas other countries collected information on students aggregated within each school (aggregated data). These countries adopt different criteria to record the ethnicity of students. In most countries, self-identification with an ethnicity, nation, people, or colour/race is the main criterion. However, countries also adopt other criteria such as geographic location and language. Table 3.1 below presents the main criterion used for each country.

For the discussion of the use of this data collection on legal instruments of recognition, we have focused only on the content of the education census instruments. So it is not possible to clearly distinguish between different conventions and methods of data collection, and the level of accuracy or confidence in which the quality of statistics is usually described (Desrosières, 2000). In addition, there are significant differences in the categories used in the education census instruments even when they contain similar methods of "self-identification" or language identification for instance. These differences also have an important impact on the information generated but are more difficult to compare across countries due to the specificity of those categories.

Table 3.1 Identification criteria for indigenous peoples in population and education censuses in Latin America

Country	Population census		Education census		
	Language	*Self-identification*	*Language*	*Self-Identification*	*Individual data*
Argentina	X	X	X[a]	X[a]	
Bolivia	X	X	X	X	X
Brazil	X	X		X	X
Chile	X	X	X	X	X
Colombia	X	X		X	X
Ecuador	X	X		X	X
Mexico	X	X	X		X
Paraguay	X	X	X	X	X
Peru	X		X	X	
Uruguay					X

Source: Elaborated by the authors for Education censuses; World Bank (2015) for population census.
Notes: [a] These variables are combined in the same question.

3.3.2   National education plans

National strategies and policies in education precede the establishment of national education statistical systems. Despite having different trajectories and purposes in different countries of the region, a similar feature is their secondary importance in national policies with the main function of supporting, above all, the national development in particular the economic development (Braslavsky & Cosse, 2006).

However, the national education plans have become an important reference for education policies in Latin America (Casassus, 2001). They achieved this status mainly as evidence of the neoliberal process pushed forward in the 1980s that emphasized educational diagnosis, planning, and strategies for efficiency (Bentancur, 2010). However, several failures of these reforms led to important changes in the nature of the new plans established in the first decades of the 21st Century (Suasnábar, 2017). On the one hand, they kept some features of the education reforms expressed in the standardized tests and governance by numbers. On the other, new efforts of integration of policies and long-term planning changed substantially the role of these plans.

National education plans are the most relevant policy documents to our purpose of investigating the politics of education censuses data. First, they are designed with the intent of indicating policy actions based on a diagnosis of policy issues. In identifying these issues, national plans usually rely heavily on official data as a source of evidence. In addition, unlike documents for specific policies, the national plans have a comprehensive scope, covering all priorities for the educational system. Finally, these plans usually cover a time period of around ten years, which is longer than government terms but not as long as the planned mandate of laws and regulations.

Table 3.2 National education plans analyzed

Country	Document	Period
Argentina	Argentina Enseña y Aprende	2016–2021
Bolivia	Plan Sectorial de Desarrollo Integral de Educación para el Vivir Bien	2016–2020
Brazil	Plano Nacional de Educação	2014–2024
Chile	Plan de Aseguramiento de la Calidad de la Educación	2020–2023
Colombia	Plan Decenal de Educación	2016–2026
Ecuador	Plan Decenal de Educación	2006–2015
Mexico	Programa Sectorial de Educación	2013–2018
Paraguay	Plan Nacional de Educación	2009–2024
Peru	Proyecto Educativo Nacional	2006–2021
Uruguay	Plan Nacional de Educación (Componente ANEP): Aportes para su elaboración	2010–2030

Source: Elaborated by the authors.

The national education plans selected for analysis are detailed in Table 3.2 below. These plans were selected using the following criteria in priority order: (i) Policy reference for the national government, preferably for a period longer than a standard legislative session (four to five years); (ii) established or in implementation during the 2010–2020 decade; (iii) cover transversal issues related to education. All documents selected refer to official plans released by national authorities in the country, except for Uruguay, which has not yet published a national plan. For this case, we have used the contributions for a national plan prepared by the National Administration of Public Education (ANEP).

3.4   Results

In this section we briefly describe the main characteristics of each national plan and how the information on ethnicity from the education census is discussed in each of them. In the next section, we analyze how the absence or presence of this information can be related to multiple facets of the politics of indigenous recognition in the region.

3.4.1   Argentina

Argentina's National Education Law recognizes indigenous peoples with respect to their language and identity, aiming at the promotion of multiculturalism. The national plan, on the other hand, provides more specific guidelines on indigenous students in terms of their interculturality and their language, in addition to pedagogical and socio-educational strategies to ensure school attendance. However, despite the references to indigenous specific issues, there is no mention to statistics on these populations either from the census or from other sources. Furthermore, these guidelines regarding indigenous education do not translate into specific goals or indicators in the national plan.

The census instruments are designed to collect aggregated information for the total of "students of the indigenous population and/or speakers of indigenous/original languages", without distinction. There is a guide for schools to standardize self-identification. At the end of the instrument, the data collection also lists all categories of indigenous peoples that should be considered. However, there is no characterization of the school, or pedagogical materials, or of the teachers who work with those students, either in multicultural or intercultural terms. This absence is a symbolic expression of the persistent situation of dispute and the feasibility of statistics for the national discourse (López, 2006).

3.4.2   *Bolivia*

In Bolivia, as opposed to the other selected countries, indigenous recognition issues are expressed in an education-specific regulatory framework: the Plurinational Education Law. Although this law does not make reference to the education census, it establishes the *"Sectoral Plan for Integral Development of Education for the Vivir Bien"* which sets clear guidelines for intracultural, intercultural and multilingual education of nations, indigenous and Afro-Bolivian peoples. The diagnosis of the current situation of these populations is not defined in terms of the school census. In addition, there is no specific data for those social groups which would allow for an understanding of the challenges faced by each nation/people categorized in the last demographic census.

Nevertheless, the sectoral plan contains goals and indicators (quantitative and qualitative) for the education of these populations aiming at the pluriversity of knowledge and ways of life of Buen Vivir. All of this without explicit mentions of how those goals could be monitored by the established educational information system.

The official education statistics are a hybrid of administrative and demographic data, which provides a relevant set of information from the indigenous family context. This construction of statistics seems to be coherent with the relational cosmovision of the human being and his ancestry: the student's language learned in his childhood; the language spoken by the student (up to three) and the mother's and father's language. However, the instruments do not exactly interact with the quantitative or qualitative conditions related to the school context in which this intra and intercultural education is offered.

Another relevant aspect of the education census instruments is that it provides a number of ethnic groups as categories for identification, despite the fact that it is only possible to mark a single option, in a recognition challenge also expressed in the last demographic census (Schavelzon, 2014). Besides that, the 2010 demographic census offered 41 different categories for respondents, in contrast to the 37 offered in the education census.

3.4.3   *Brazil*

The Brazilian National Law for Education and the National Educational Plan provide guidelines for interculturality and bilingualism for indigenous students, with the support of the communities for the formulation of specific curricular guidelines. It also establishes government support for indigenous education explicitly stating the need for statistical data related to this modality.

However, these regulations are scarce in relation to information on indigenous education and lack specific objectives and indicators for educational support. Only the monitoring framework of the National Education Plan provides a description of indigenous education, and yet without characterizing the specific needs and advances of the hundreds of indigenous peoples living in the national territory.

The education census questionnaires collect information on the colour/race of students and teachers, as well as other information from the indigenous school context: special school's location, special student's location, colour/race of the teachers, specific pedagogical materials, information about teaching of indigenous language and languages spoken in the school. In that case, there are about 330 language categories that can be selected to identify the instruction languages. However, due to the absence of specific objectives for indigenous education in the national plan, it is not possible to know to what extent this comprehensive set of information is used.

3.4.4   Chile

The Chilean education plan emphasizes the educational inequalities that affect students highlighting gender and indigenous background as important factors driving these inequalities. The statistics that illustrate this issue come from the demographic census and no reference is made to the education census. However, at the national level, it is possible to have information on indigenous students. The annual data on education is based on the Students General Information System, which collects data on each individual student. This register contains data on the student's ethnic origin and indigenous language.

3.4.5   Colombia and Ecuador

Both Colombia and Ecuador do not have any references to education or population censuses in their national plans and, therefore, no references to statistics of indigenous peoples. In fact, the references to statistics or data are generally scarce in comparison to other countries such as Argentina and Brazil. However, Bolivia and Colombia do collect data on indigenous students on the education census and there is some coverage on indigenous education on the national plans.

However, for both countries, similar to Bolivia, there is a disconnect between the use of census data and the attention given to indigenous issues in the national plans. In these three countries, the national plans provide a fairly comprehensive assessment of the challenges faced by the education of indigenous peoples and indicate aspirations and policies to deal with these challenges. However, neither the diagnostics nor the policy solutions presented are based on data produced by the education censuses.

3.4.6 México

In 2013, Mexico administered a comprehensive school census that collected information on each student and teacher of all primary and secondary schools in the country. However, given that this census has not been repeated since then, we have analyzed the annual standardized national data collection instruments known as "911 Format" or "911 Questionnaires". Those instruments have a set of different questions depending on the educational level (primary or secondary) and the type of school. The questionnaire designed for generic primary and secondary schools registers the total number of "indigenous students or students who speak an indigenous language" by gender. An additional field also collects the number of indigenous language speakers by grade.

The Mexican sector plan for education includes some statistics in relation to indigenous students. However, none of them are based on the education census. An important argument that provides the basis for the policy plan is that indigenous students tend to have worse performance on national assessments in Spanish and Mathematics. Another relevant aspect of the plan is that it has a particular emphasis on policies for indigenous languages (Strategy 3.4) and indigenous women (Strategy 3) which are the only two possible disaggregations from the education census. This indicates the limitations that the available data pose for policies in the area (Schmelkes, 2013).

3.4.7 Paraguay

In Paraguay, the data collection instruments, both in demographic and education censuses, are aligned with the constitutional recognition of the country's ethnic, cultural and linguistic diversity, formed by 20 ethnic groups and five language families (guaraní, maskoy, mataguayo, zamuco and guaicurú). According to the 2012 census, the indigenous population represents 1.8% of the total population of the country and less than 10% of the population lives in urban areas (ECLAC, 2020).

The relevance of the indigenous peoples for official statistics is found in the history of the population census. The first Indigenous Census, carried out separately from the National Population and Household Census, took place in 1981, still under the authoritarian Stroessner regime. Then, under a democratic regime, the 1992, 2002, and 2012 census were carried out in parallel with the National Population and Household Census, with methodological advances in data collection, such as changing the definition of indigenous identification using the criterion from spoken language in the census 1981 and 1992 to self-declared in the census 2002 and 2012 (Villagra, 2014). The inclusion of self-identification as a census criterion resulted in an increase from 1.2 to 1.7 in the percentage of the indigenous population in the total of the Paraguayan population.

Likewise, the National Educational Plan 2009–2024 highlights policies for indigenous peoples as a key to tackling educational inclusion, bearing in mind that the average number of years of study for the indigenous population aged ten or more is only 2.2 years, which is equivalent to the second grade. To this end, the plan invests in differentiated educational offers that meet specific requirements that allow for achieving educational results similar to those of the rest of the population. At the same time, it recognizes bilingualism and ethnic diversity as major challenges for the educational sector. This concern is clearly represented on the School Census data collection form, which covers which indigenous people the student belongs to, their native/first language and most widely spoken idiom, and whether they have an indigenous identity card.

3.4.8   Peru

In Peru, the 2013 changes in the demographic census on the collection of data from indigenous population had an important role in creating a policy favourable to the social development of indigenous peoples (Villasante, 2018). Until the 2007 Demographic Census, the identification of indigenous peoples in Peru was linked to the use of the native language, and only in 2013 did this identification change to self-declaration by citizens over 12 years of age (based on their customs and ancestry's identity). According to the 2017 census, 26% of the country's population is indigenous and about 65% of that population lives in an urban area (ECLAC, 2020).

The General Education Law of Peru (2003) includes in its chapter on equity a specific article with the recognition and guarantee of the rights of the indigenous population (Art.19) and provides for the elaboration of a National Education Project as a strategic structure for conducting the development of educational policies (article 7). The *Proyecto Educativo Nacional 2006–2021* has six strategic objectives for which they are established, based on a diagnosis, policies, and measures to achieve the expected results. The concern with the effectiveness of policies and the overcoming of exclusions and discrimination is highlighted. The plan includes the promotion of intercultural regional curriculum across the country at different levels and modalities of education, which includes the participation of representatives of indigenous peoples in the preparation of regional curriculum (Objective 5). The education census is characterized by the collection of summarized information according to the students' original language.

3.4.9   Uruguay

According to the 2011 census, 2.4% of the Uruguayan population is indigenous (CEPAL, 2020). Although Uruguayan legislation addresses general concepts of human rights, there is no specific recognition of indigenous peoples and their rights to protect their cultural and linguistic identity.

While the demographic census collects only ethnic-racial ancestry, the collection of education census data shows a complete lack of information on ethnicity. Likewise, there are no specific legal or political provisions for the education of indigenous peoples, as noted in the General Education Law and the Educational Plan (preparatory document elaborated by the National Public Education Administration - ANEP). Even norms aimed at vulnerable populations such as the

Plan Nacional de Educación en Derechos Humanos 2016 (PNEDH), there is an absence of specific policies for indigenous peoples, in contrast to programmes and actions aimed at issues of gender identity, disability, and geographic inequalities or minorities in general.

3.5   Discussion

The identification of indigenous peoples in education censuses differs considerably among countries in Latin America. However, almost all selected countries recognize indigenous peoples and their right to education in their national legal frameworks.

We have observed that identifying indigenous peoples in education censuses serves different purposes. Following Rallu et al.'s (2006) typology, we asserted that the initial motivation for including the identification of indigenous peoples in statistical systems in Latin America is "counting to dominate". This colonialist conception, however, was replaced in the second half of the past century by a pluralist conception that emphasizes counting for inclusion.

The education censuses in Latin America are consolidated in this scenario of multiculturalism and recognition of indigenous peoples. However, our analysis indicates that this recognition does not necessarily translate into specific questions about ethnicity in education census instruments. Despite almost all countries in the region having included indigenous rights in their legal framework (CEPAL, 2020), some countries, such as Uruguay, do not have specific questions for indigenous peoples in education censuses. Even among those countries that include this information, most of the instruments analyzed are restricted to the same questions using broad ethnic categories included in the demographic census, with only a few instruments including questions on language or specific ethnic information.

It is also noted that this absence of questions is not related to the size of the indigenous populations but rather to the political salience of indigenous issues in the country. Among those countries that address ethnicity clearly in their instruments, there is a considerable variation regarding the criteria for the identification of indigenous peoples. At the same time, there are no homogeneous and clear narratives among the countries studied. The normative and political discourses navigate among concepts of "multiculturalism", "inter and intraculturalism", as well as bilingualism and linguistic pluralism.

This apparent lack of conceptual precision is also reflected in the use of these data in major national policy documents. We have analyzed the national plans for the education of ten countries in the region. None of them make significant use of census data on ethnicity. However, there are considerable differences with regard to the way that the evidence is used in those plans to guide policies for indigenous peoples.

Figure 3.1 shows the relative frequency of terms associated with census and indigenous peoples in each of the national plans. A general overview suggests a positive association between census and indigenous issues. However, some countries like Ecuador, Mexico, and Colombia have few or no references to census data but

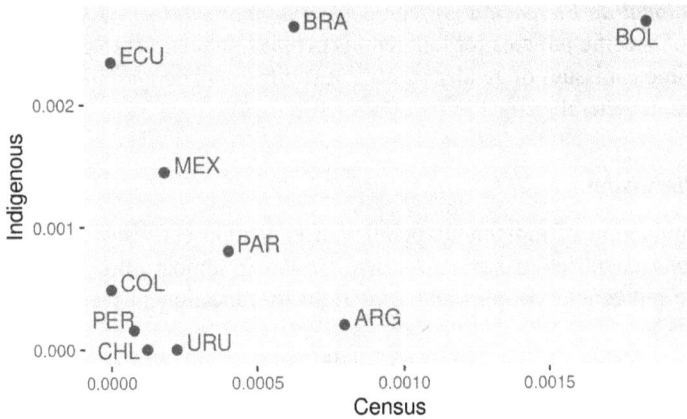

Figure 3.1 Relative frequency of terms associated with "indigenous peoples" and "census" in national education plans of selected countries.

Source: Elaborated by the authors.

have a considerable amount of coverage of indigenous issues. On the other hand, Argentina is a clear case where there is some use of census data but few references to indigenous peoples. Bolivia is an emblematic case where both references to indigenous peoples and census are relatively frequent.

Our qualitative analysis shows that for a group of countries that includes Argentina, Brazil, and Mexico, education for indigenous people is addressed under the general concept of inclusion. This means that the country should focus on participation and completion. For that, census data are used to assess the amount of out-of-school children, to examine their performance in school, and to make sure that the whole population has access to the educational system. Among this group, we observed an extensive use of data, in particular of progress indicators, and constant use of measurable targets.

The second group of countries comprising Bolivia, Colombia, Ecuador, and Paraguay does not focus as much on issues of access and completion. Much more attention is given to the content of teaching and learning, i.e. an intercultural curriculum, pedagogy, and education practice. Therefore, as traditional instruments of education census do not have information on these aspects, the use of data on ethnicity is much more limited. In addition, there seems to be an intentional effort diverging from the traditional use of statistics and proposing a different perspective on what matters in education. The third group, Chile, Peru, and Uruguay, does not cover considerably any of these two dimensions.

Our analysis suggests that the first group shares a conception of production and use of statistics that mixes the "counting to dominate" and "counting to justify positive action" approaches. On one hand, the narrow and fixed focus on inclusion denotes a preoccupation with control and participation in society under the dominant terms. Despite that, this conception is also concerned with practical and feasible ways of guaranteeing the right to education for all recognizing the unequal access to the educational system.

It is rather difficult to classify the second group's conception under the typology proposed by Rallu et al. (2006). There is a concern in relation to identifying indigenous peoples in the census instruments but, at the same time, there is no clear use of these data in national plans. This absence may be seen negatively as it ignores the evidence posing challenges to the plan, monitoring, and effectiveness of education policies. However, there is also an implicit scepticism in relation to the dominant categories of ethnic classifications, and to the general rationale of evidence-based policy focused on a narrow view of inclusion. In this broader perspective, inclusion does not mean attending schools with the same colonial curricula, pedagogy, and practices. Instead, it establishes a plan for the revision of what is traditional schooling, enabling the actual inclusion of indigenous peoples. This sceptical view on the current education statistics may also foster the conditions in which indigenous people can reflect and act on the production of statistics about themselves aiming towards data sovereignty (Walter, Kukutai, Carroll, & Rodriguez-Lonebear, 2020).

3.6   Summary and conclusion

To summarize, our analysis indicates that ethnicity data have been collected through education census in most of the selected countries. This phenomenon results from historical struggles of a diverse group of political actors and social movements. Rallu et al.'s (2006) typology on statistical systems is useful to identify different periods and trajectories of consolidation and adaptation of those data in education census in the region. At least in the normative pieces analysed, the justification for production of those data has been moving in the past decades from a "counting to dominate" perspective towards a "counting to justify positive action". This movement itself has also promoted the further expansion and consolidation of data collection on ethnicity across the region.

However, we have found that the widespread existence of data on ethnicity does not produce a widespread use of those data in national education policy plans. Considerable risk of the absence of ethnicity data in policy plans is the weakening or loss of political relevance of these data. As Loveman (2021, p. 147) ponders, the boom in the production of ethnoracial statistics in Latin America is an important achievement but is "simultaneously a politically contentious and tenuous accomplishment that could well be short-lived".

At the same time, we have found that Rallu et al.'s (2006) framework has limitations in explaining the omission of ethnicity data in policy plans. Either to justify positive action or to highlight multiculturalism, one would expect that there would be a connection between the production of these ethnicity data and the national education policy plans. We argue that the lack of political use of data on ethnicity suggests that education policy plans are either uninterested in bringing evidence as part of policy-making for inclusion and interculturality or they are sceptical about the relevance or reliability of these data. The latter points out to important questions in relation to the focus of education policy on measurement and pedagogy as well as the feasibility of representing cultures and identities through those statistical instruments. Whichever the case is, whether for lack of interest or for critical scepticism, the reflection on the reasons for the lack of use of ethnicity data is necessary for the improvement of policy and planning in education.

References

Akkari, A., & Lauwerier, T. (2014). The education policies of international organizations: Specific differences and convergences. *Prospects, 45*(1), 141–157. doi:10.1007/s11125-014-9332-z

Akkari, A., & Perez, S. (1998) Educational research in Latin America: Review and perspectives. *Education Policy Analysis Archives, 6*(7), 1–10.

Angosto-Ferrández, L. F., & Kradolfer, S. (2014). The politics of identity in Latin American censuses: Introduction to a special issue. *Journal of Iberian and Latin American Research, 20*(3), 321–327. doi:10.1080/13260219.2014.995871

Bailey, S. R., Loveman, M., & Muniz, J. O. (2013). Measures of "race" and the analysis of racial inequality in Brazil. *Social Science Research, 42*(1), 106–119. doi:10.1016/j.ssresearch.2012.06.006

Ball, S. J. (1993). What is policy? Texts, trajectories and toolboxes. *Discourse: Studies in the Cultural Politics of Education, 13*(2), 10–17. doi:10.1080/0159630930130203

Beech, J. (2002). Latin American education: Perceptions of linearities and the construction of discursive space. *Comparative Education, 38*(4), 415–427. doi:10.1080/0305006022000030739

Bentancur, N. (2010). *Los Planes Nacionales de Educación: ?Una nueva generación de políticas educativas en América Latina?* V Congreso Latinoamericano de Ciencia Política. Retrieved from http://cdsa.aacademica.org/000-036/26.pdf

Braslavsky, C., & Cosse, G. (2006). Las Actuales Reformas Educativas en América Latina: Cuatro Actores, Tres lógicas y Ocho Tensiones. *REICE. Revista Iberoamericana sobre Calidad, Eficacia y Cambio en Educación, 4*(2), Article 2. Retrieved from https://revistas.uam.es/reice/article/view/10077

Casassus, J. (2001). A reforma educacional na América Latina no contexto de globalização. *Cadernos de Pesquisa, 114*(2), 7–28. doi:10.1590/S0100-15742001000300001

CEPAL. (2020). *Los pueblos indígenas de América Latina – Abya Yala y la Agenda 2030 para el Desarrollo Sostenible: Tensiones y desafíos desde una perspectiva territorial.* CEPAL. Retrieved from https://www.cepal.org/es/publicaciones/45664-pueblos-indigenas-america-latina-abya-yala-la-agenda-2030-desarrollo-sostenible

Daniel, C. (2016). La sociología de las estadísticas. Aportes y enfoques recientes. *Contenido. Cultura y Ciencias Sociales, 7*, 3–24. Retrieved from https://ri.conicet.gov.ar/handle/11336/45805

Del Popolo, F. (2008). *Los pueblos indígenas y afrodescendientes en las fuentes de datos: Experiencias en América Latina.* CEPAL. Retrieved from https://www.cepal.org/es/publicaciones/3616-pueblos-indigenas-afrodescendientes-fuentes-datos-experiencias-america-latina

Desrosières, A. (2000). Measurement and its uses: Harmonization and quality in social statistics. *International Statistical Review, 68*(2), 173–187. doi:10.1111/j.1751-5823.2000.tb00320.x

Desrosières, A. (2011). *The politics of large numbers: A history of statistical reasoning.* Cambridge, MA: Harvard University Press.

Gil, N. L. (2019) Estatísticas e Educação: considerações sobre a necessidade de um olhar atento. *Pensar a Educação Em Revista, 5*(2), 1–29. Retrieved from https://pensaraeducacaoemrevista.com.br/2019/12/17/estatisticas-e-educacao/

Guzmán, C. (2005). Reformas educativas en América Latina: Un análisis crítico. *Revista Iberoamericana de Educación, 36*(8), 1–12. doi:10.35362/rie3682779

Jackson, J. E., & Warren, K. B. (2005). Indigenous movements in Latin America, 1992–2004: Controversies, ironies, new directions. *Annual Review of Anthropology, 34*(1), 549–573. doi:10.1146/annurev.anthro.34.081804.120529

López, L. C. (2006). De transnacionalizacion y censos. Los 'afrodescendientes' en Argentina. *AIBR: Revista de Antropología Iberoamericana, 1*(2), 265–286.

Loveman, M. (2014). *National colors: Racial classification and the state in Latin America.* Oxford, England: Oxford University Press. doi:10.1093/acprof:oso/9780199337354.001.0001

Loveman, M. (2021). A política de um cenário de dados transformado: estatísticas etnorraciais no Brasil em uma perspectiva comparativa regional. *Sociologias, 23*(56), 110–153. doi:10.1590/15174522-109784

Meyer, J. W., Ramirez, F. O., & Soysal, Y. N. (1992). World expansion of mass education, 1870–1980. *Sociology of Education, 65*(2), 128–149.

Morning, A. (2008). Ethnic classification in global perspective: A cross-national survey of the 2000 census round. *Population Research and Policy Review, 27*(2), 239–272. doi:10.1007/s11113-007-9062-5

Parente, J. M., Villar, L. B. E., Parente, J. M., & Villar, L. B. E. (2020). Educational systems in the context of the transition from new public management to the post-new public management: Comparative study between Brazil and Spain. *Educar Em Revista, 36*, 1–23. doi:10.1590/0104-4060.67115

Rallu, J.-L., Piché, V., & Simon, P. (2006). Demography and ethnicity: An ambiguous relationship. In G. Caselli, J. Vallin, & G. J. Wunsch (Eds.), *Demography: Analysis and synthesis.* Boston, MA: Elsevier. Retrieved from http://public.ebookcentral.proquest.com/choice/publicfullrecord.aspx?p=269539

Rosar, M. F. F., & Krawczyk, N. R. (2001). Diferenças da homogeneidade: elementos para o estudo da política educacional em alguns países da América Latina. *Educação & Sociedade, 22*(75), 33–43. doi:10.1590/S0101-73302001000200004

Schavelzon, S. (2014). Mutaciones de la identificación indígena durante el debate del censo 2012 en Bolivia: mestizaje abandonado, indigeneidad estatal y proliferación minoritaria. *Journal of Iberian and Latin American Research, 20*(3), 328–354. doi:10.1080/13260219.2014.995872

Schkolnik, S., & Del Popolo, F. (2013). *Pueblos indígenas y afrodescendientes en los censos de población y vivienda de América Latina: Avances y desafíos en el derecho a la información.* Retrieved from https://repositorio.cepal.org//handle/11362/35946

Schmelkes, S. (2013). Educación y pueblos indígenas: problemas de medición. *Revista internacional de estadística y geografía, 1*, 5–13.

Simon, P., Piché, V., & Gagnon, A. A. (Eds.). (2015). *Social statistics and ethnic diversity: Cross-national perspectives in classifications and identity politics.* Cham, Switzerland: Springer International Publishing. doi:10.1007/978-3-319-20095-8

Suasnábar, C. (2017). The cycles of educational reform in Latin America: 1960, 1990 and 2000. *Revista Española de Educación Comparada, 30*, 112–135. doi:10.5944/reec.30.2017.19872

Telles, E. E. (2017). Multiple measures of ethnoracial classification in Latin America. *Ethnic and Racial Studies, 40*(13), 2340–2346. doi:10.1080/01419870.2017.1344275

UNESCO Institute for Statistics. (2016). *The effect of varying population estimates on the calculation of enrolment rates and out-of-school rates.* Information paper; 36. UNESCO Institute for Statistics. doi:10.15220/978-92-9189-208-2-en

Villagra, S. P. C. (2014). Los censos indígenas en Paraguay: Entre el auto-reconocimiento y la discriminación. *Journal of Iberian and Latin American Research, 20*(3), 423–435. doi:10.1080/13260219.2014.995878

Villasante, M. (2018). Censo étnico peruano: ¿cómo conciliar la diversidad étnica y la ciudadanía nacional? *Revista Ideele, 273,* 1–10. Retrieved from https://revistaideele.com/ideele/content/censo-%C3%A9tnico-peruano-%C2%BFc%C3%B3mo-conciliar-la-diversidad-%C3%A9tnica-y-la-ciudadan%C3%ADa-nacional

Walsh, C. (2000) Políticas y significados conflictivos. *Nueva Sociedad, 165,* 121–133. Retrieved from https://nuso.org/articulo/politicas-y-significados-conflictivos/

Walter, M., Kukutai, T., Carroll, S. R., & Rodriguez-Lonebear, D. (Eds.). (2020). *Indigenous data sovereignty and policy.* Abingdon, VA: Routledge. doi:10.4324/9780429273957

World Bank. (2015). *Indigenous Latin America in the twenty-first century: The first decade.* Washington, DC: World Bank. Retrieved from https://openknowledge.worldbank.org/handle/10986/23751

II

The politics of institutional autonomy

4 Population census – large-scale project of public statistics in transition

Walter J. Radermacher

4.1 Introduction

The historical, present and future development of population censuses can be characterised by many aspects from a wide spectrum of statistical methodology and practice. In this way, however, one can only very inadequately approach two questions that are crucial for understanding and an overall picture: How can the establishment and the trajectories of national censuses be described and explained? How can the relative institutional autonomy of census taking be adequately described and explained? The following contribution aims to approach these questions on two levels, first the level of the statistical system and second that of the census.

More than 20 years ago, the report "Statistics - A Matter of Trust" (HM Treasury, 1998) was published in the United Kingdom which laid the foundations for overcoming the spreading crisis of confidence through a solidly structured statistical system. This report is not globally unique. Instead, it is one of a series of international, European and national measures and agreements which, since the fall of the Berlin Wall in 1989, have strengthened official statistics as the backbone of policy in democratic societies - two prominent examples being the UN Fundamental Statistical Principles (UNSD, 2018) and the EU Statistics Code of Practice (Eurostat, 2018).

If we want to solve our current and future difficulties, we need to address precisely those points that have emerged as key factors in determining the quality of statistics by asking the three questions

- What (statistical products, quality profile)?
- How (methods)? and
- Who (institutions)?

Statistics can and should facilitate the resolution of conflicts. The goal must be to stop quarrelling about the facts and instead quarrel about the conclusions drawn from them.

In the past, a crisis of the COVID dimension would have led relatively quickly to a situation where the need for information would have been directed to official statistics as the preferred provider; this has changed recently for many reasons. On the one hand, there is the danger that the much-cited data revolution and learning

DOI: 10.4324/9781003259749-7

algorithms (so-called AI) are presented as an alternative to official statistics (which are perceived as too slow, too inflexible and too expensive), instead of emphasising possible commonalities and cross-fertilisation possibilities. On the other hand, after decades of austerity policies, official statistics are in a similarly defensive situation to that of public health systems in many respects and in many countries: There is a lack of financial reserves, personnel and know-how for the new and innovative work now so urgently needed.

In January 2017, half a year after the Brexit referendum and at the beginning of Donald Trump's presidency, William Davies published his widely noted and discussed article *"How statistics lost their power – and why we should fear what comes next"* in the Guardian. (Davies, 2017) He expressed his concerns that nothing less than the end of a statistical era has come, with serious consequences for the public discourse, trust in experts and politics, and opportunities for populist politicians to use for their own purposes. According to Davies' scenario one might ask the question of whether the statistics logic is being replaced by a data logic. This question is particularly urgent for official statistics. To be able to answer it adequately, it is necessary to reflect on what the specific mandate and unique selling proposition of official statistics is and where its place can be in such a data logic. In addition to the already introduced quality dimensions of Who, What and How, it is necessary to be clearer about the guiding principle that quality is "fitness for purpose". A fourth quality characteristic referring to the intended purposes and potential users is therefore introduced:

- For Whom/What (public discourse and political decisions)?

The following questions arise: First what positioning will emerge for the system of public statistics from the current rapid developments regarding new data sources, new statistical methods and new societal issues? Second, what will be the role and function of the Population Census in this context, and within the system of public statistics? For both topics, there are aspects that are located more at the national or at the international level; these are also closely interrelated. The following article will address the first questions before examining the specific role of the census. Both are not purely scientific topics at the beginning of a research study. Rather, they refer to the three driving forces relevant to evolutions or revolutions (Mac-Feely & Nastav, 2019) in statistics (science, statistics, society) and aim to derive current and future developments in the informational environment of statistics with the aim of identifying and counteracting possible risks.

4.2   System of public statistics

4.2.1   *History of official statistics in fast motion*

The term *statistics* has the same linguistic roots as *state*. In every form of governance, rulers needed statistical information, information that allowed them to target policy. The current form of statistics has experienced its first great blossoming with the birth of the nation-state and has since developed along with this in all its variants, ups and downs. Since the Enlightenment period, statistics has been closely married to the state and its various forms, each with characteristic imprints on the programme and role of statistics. Over the past 2 centuries, there has been a constant interplay of impulses from statistical-methodological innovations, new data and data processing as well as socio-political framework conditions and influencing factors. This evolution has been largely continuous but has also been marked by major upheavals and abrupt changes.

An interesting feature in the history of statistics has to do with statistics and statisticians being children of their own time. What is important or unimportant, which questions are pursued scientifically or empirically, and which mental attitude or conviction plays a role in the work of a statistician, is generally not decided objectively, but bears the stamp of the historical episode.

Desrosières describes the history of statistics[1] in relation to that of economic theories and that of economic policies in a very condensed overview of crucial phases and forms of governance as well as their impact on statistics (see Figure 4.1). The history of statistical methods is linked *"with the history of issues placed on the agenda for official decisions which themselves subsume: (1) ways of conceptualizing society and the economy, (2) modes of public action, and (3) different forms of statistics and of their treatment."* (Desrosières, 2011) From this overview of historical stages of development one can also deduce how the current official statistics programme has developed over the different forms of governance. According to this overview, we are currently in the phase of neo-liberal governance, which seeks

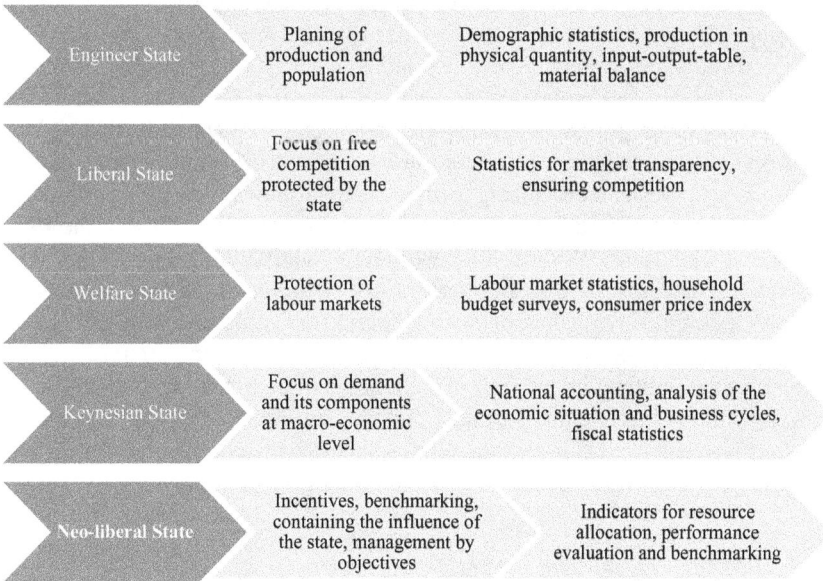

Engineer State	Planing of production and population	Demographic statistics, production in physical quantity, input-output-table, material balance
Liberal State	Focus on free competition protected by the state	Statistics for market transparency, ensuring competition
Welfare State	Protection of labour markets	Labour market statistics, household budget surveys, consumer price index
Keynesian State	Focus on demand and its components at macro-economic level	National accounting, analysis of the economic situation and business cycles, fiscal statistics
Neo-liberal State	Incentives, benchmarking, containing the influence of the state, management by objectives	Indicators for resource allocation, performance evaluation and benchmarking

Figure 4.1 The state, the market, and statistics.

Adapted from source: Desrosières (2011, p. 45).

to achieve self-regulation of individual actors using measures and indicators rather than instructing them with regulations. This approach promises no less than modernity, democratisation and transparency. Furthermore, flattening of hierarchies and auditability[2] are to be ensured as well as a general quality improvement. Improving efficiency is the overall goal of this approach. For a long time now, evidence-based decision-making has become a good governance standard, not only in the private but also in the public sector. Statistical indicators are of central importance for (evidence-based) decision-making and management as well as for communication. (Desrosières, 2011; Prutsch, 2020; Radermacher, 2021; Umbach, 2020).

For a few years, however, there has been growing resistance to the widening of data-based governance (as part of a liberal, Western-type form of modernity), which unfortunately is driven by the advocates of populism. Growing scepticism about all forms of experts does not stop at journalists or at statisticians. Coupled with a lack of statistical literacy and the impression of being at the mercy of the representatives of a perceived technocratic regime, resistance is formed, which not only reduces the side effects of the medication "statistics", but – comparable to the resistance to general vaccination – argues from a gut feeling, that completely negates the existence of facts and thus throws the benefits and progress of enlightenment overboard (Davies, 2017; Pullinger, 2017). So, when talking about the "Data Revolution" (Nastav & MacFeely, 2020), the second revolution, that of the neo-liberal governance, and the third, the populist resistance to expert opinion, should all be mentioned in the same breath.[3] All three developments are related to each other and from each other in a certain way. All three together must be perceived as a social framework for official statistics in the present and in particular in the future.

Abrupt changes that fundamentally alter the course of events are rare in the history of statistics. Among them is undoubtedly the end of the Cold War and the collapse of the Soviet Empire. Statisticians in the Eastern European countries were by no means lacking in sophisticated methodologies for data collection and data processing. Statisticians in the Eastern European countries did not necessarily have to learn new survey and processing methods in the following transition phase. This was therefore not the main challenge for them in the transition phase of the 1990s. Rather, the framework conditions for working in statistical institutions according to Western standards were as different, as they only could have been. Statistics occupies a completely distinct place in a liberal state structure, with different objectives, rules, competencies and responsibilities. While in (communist) planned economies the political importance of statistics is high, but their autonomy is low, in Western liberal states this is exactly the opposite. In this phase of transition, it was thus proven once again that the relationship between the state and its citizens is mirrored in official statistics.

4.2.2   Statistics – the role of the market and the state

One can now get the impression that a change is taking place comparable to the end of the 1980s. The neo-liberal model, which has been the standard of governmentality since the end of the Cold War, has developed cracks and is increasingly being questioned in the face of growing crises. With the COVID pandemic and the mechanisms of its political management, the relationship between market and state is critically examined in terms of its effectiveness and crisis resilience. While for decades there has been nothing as a generally accepted policy direction other than that the state should be shrunk if possible and its services handed over to the market, this doctrine is increasingly being questioned and possible reviews are being discussed.

It does not yet seem as if a new political doctrine has yet emerged that enjoys broad acceptance. In this respect, it is important for official statistics to follow the developments in this transition phase attentively and with an awareness of the importance of their own task to be able to actively advocate for their own position.

With the possibilities of omnipresent, Big Data and the seemingly endless options for their use (individually and without governmental regulatory authority), the landscape of data and information is changing fundamentally. The statistical logic of the past 2 centuries is being replaced by data logic.

> With the authority of statistics waning, and nothing stepping into the public sphere to replace it, people can live in whatever imagined community they feel most aligned to and willing to believe in. Where statistics can be used to correct faulty claims about the economy or society or population, in an age of data analytics there are few mechanisms to prevent people from giving way to their instinctive reactions or emotional prejudices.
>
> (Davies, 2017)

These anxieties about statistics in light of the growth of data science were resonated in the report of the American Statistical Association (He, Madigan, Yu, & Wellner, 2019) "Statistics at a Crossroads – Who is up for the challenge?", which was the result of a workshop[4] by the American Statistical Association:

> With the fast establishment of various Data Science entities across campuses, industry, and government, there is a limited time window of opportunity for a successful transformation that we must not miss. We must effect this change now by reimagining our educational programs, rethinking faculty hiring and promotion, and accelerating the cultural change that is required.
>
> (He et al., 2019)

After four years of increasingly intense populist politics and after one year in the pandemic state of emergency, it is time to return to the discussion, which was launched by William Davies' article. In the pandemic, it became abundantly clear that there was still a need for government services and functions that seemed to have disappeared from our radar screen (Radermacher, 2020a). This includes – besides public healthcare – the provision of statistical information, indicators, dashboards, etc. with solid quality, comprehensibility and trustworthiness. What became equally clear, however, is the high importance of sufficient education among the general population to enable them to understand, classify and interpret such information for their own needs.

4.2.3   *Which crises?*

It is required, as in the 1990s after the fall of the Berlin Wall, to ask the fundamental question again, namely, do we (still and again) deserve official statistics as the backbone of democratic decision-making, and if so, what should their tasks be, how should they be financed and anchored in the political system? To do this successfully, it is necessary to approach the field of interactions between the world of data, facts, information, indicators, etc. on the one hand, and the world of opinion-forming, decision-making, media, etc. on the other, with a somewhat more precise and differentiated view (Radermacher, 2019). It is important to understand how statistics and society are co-created and co-produced, including in particular "governance by the numbers" and "informational governance".[5]

Given the urgency and stress of the current situation, it seems natural to focus the statistical issues of Corona on the decisions that are about to be taken. At this point, however, a broader perspective and longer horizon are advocated[6]. Why does a more strategic and long-term view seem so essential right now, when we hardly have time for the most dramatic questions?

There are two important arguments for this

First, the most pressing risks identified by the World Economic Forum over the last years have not simply disappeared because we are now fighting a pandemic: The Global Risks Report 2021 (World Economic Forum, 2021) analyses

> Risks from societal fractures—manifested through persistent and emerging risks to human health, rising unemployment, widening digital divides, youth disillusionment and geopolitical fragmentation. ... Environmental degradation—still an existential threat to humanity—risks intersecting with societal fractures to bring about severe consequences.
>
> (World Economic Forum, 2021)

Second, it must be borne in mind that more substantial changes, extensions or accelerations of official statistics cannot be realised overnight. Official statistics are a kind of ocean liner whose course can only be changed with considerable advance planning and preparation. It is therefore time to look far ahead now in order to be able to adapt to the information needs of the future, two to five years from now.

4.2.4   *Which statistics?*

Let us approach the question of the future of official statistics in such a way that we unfold the trinity of statistical quality and start with the statistical programme, i.e. the

What: The traditional domains of economic and social statistics will be consulted, but in a way which will call into question their division into specialised domains. This is illustrated by the example of agricultural statistics. Whereas in the past the aim was to quantify farmers' production performance as quickly and accurately as possible, the aim will be to be able to cast statistical light on sustainable food production from the cradle to the grave, including agricultural production in connection with its inputs, outputs as well as its impact on biodiversity, water protection, etc., and international trade in agricultural goods. Let us approach the COVID exit with a second example, by asking ourselves what effects the financial support measures have not only on public finances, unemployment or inflation, but also on the conformity of industrial production and consumption with the goals of climate protection. In addition, however, it will also have to deal with areas which do not yet belong to the programme of official statistics or which will be given higher priority in it. Current examples of this are migration (development especially since 2015), health (current pandemic) and biodiversity (highly classified as a risk), for which new statistics or statistics with improved quality (speed, level of detail, representativeness, etc.) are clearly needed.

How: The need for the reform of statistical production processes is very clearly summarised in the Bucharest Memorandum of European Statistics of 2018 (European Statistical System Committee, 2018), which among other things states, *"the variety of new data sources, computational paradigms and tools will require amendments to the statistical business architecture, processes, production models, IT infrastructures, methodological and 1quality frameworks, and the corresponding governance structures."* As a response to the existence of important new data sources and methods of data science and artificial intelligence, the processes of statistics are fundamentally changing

* Smart Open Statistics' are seen as a goal and guideline for the entire business architecture,
* Primacy of the use of existing data (multiple sources – mixed modes),
* Integrated data management,
* Flexible offers for the use of data (multipurpose standard offers, customised services, access to microdata for research).

In addition, it will also be a matter of official statistics increasingly acting not only as a producer of information but also as a service provider. Such services should generally have to do with available competencies, strengths and experience: One could generally help manage the quality of statistics, even if they are produced outside the factory walls of official statistics. For example, statisticians could be helpful in the design of measurement concepts as well as in the use of imputation and estimation methods or in the application of national accounting to other subject matters. They could assist in improving communication and providing information. They could take over the role of a standard setter and certifier of statistical standards.

Who: Institutions of official statistics are part of the public administration of their country (of the supranational or international level, respectively). They enjoy a status of professional independence that is more or less guaranteed by the respective state governance. Furthermore, official statistics are committed to adhering to principles of good public administration in terms of e.g., citizen participation, accountability, etc. What is important in the near future, therefore, is to emphasise these strengths of a public institution through which trust can be strengthened and

maintained. At the same time, networking with the scientific community and partnerships with other producers of statistics is to be further intensified, open data
access is to be created for everyone and, at the same time, the confidentiality of individual data is to be safeguarded. Finally, there is an urgent need to invest more in
the general education of understanding, skills and abilities to handle facts, graphics
and maps. Only if the population (and of course the political scene in particular) is
sufficiently aware of the differences in the quality of statistical information, if one
is capable of distinguishing between fake and fact, then a basis of trust can grow
and flourish.

For Whom/What: The mission of public statistics is essentially to provide facts
that are fit for the specific purpose of public discourse and policy decisions. In the
following section, we will look at the different aspects that need to be considered
in order to fulfil this task.

4.2.5   *Official statistics: public good, infrastructure, product, language, authority*

In this paper, we are less concerned with statistics as a scientific discipline, but
with its application for the purposes of public discourse and policymaking. Evi-
dently, there are two sides to this matter, that of the statistical experts who generate
facts about a country's population, society, economy and environment, and that of
those who use them; producers and consumers, to put it simply. Statistical information, in the sense of convention theory, also known as "Économie des conventions" (Diaz-Bone & Didier, 2016), are artefacts that are designed and produced
(Radermacher, 2019). For such informational products, the same rules apply as for
other products. Their design must meet the needs of the customers, they must be
manufactured with good quality and a professional distribution is just as important
as an educated clientele.

4.2.5.1   *The factory*

The public infrastructure that provides society, politics and the economy with elementary facts is official statistics. Where would we be without GDP, inflation
rates, mortality tables, population figures, etc.? International comparisons, based
on which momentous political decisions are made, are based on international
methodological standards and classifications. Official statistics must necessarily
work differently than individual, fast and flexible data uses can (Radermacher,
2020b). It works with a long lead time, industrial production lines, international
standards (this is production capital) and democratically decided programmes. In
this way, internationally, infra-nationally and temporally comparable indicators
of high quality are produced with efficient use of resources (value for taxpayers'
money) and a minimum of response burden. This infrastructure must be regularly
maintained and adapted to new requirements. Let's compare it with transport and
mobility: a data strategy aims at individual mobility, its rules, promotion, etc. In
parallel, however, it is necessary to make public rail transport fit for the future with

data and statistics. This requires investments in the infrastructure because new areas are to be opened up and modern high-speed trains are to be run on them. Individual data use alone can be inefficient and ineffective. In the 1960s, we thought that promoting and regulating individual transport was the only option. Today we know that this policy has led us into congested cities and roads because we did not at the same time push the expansion of the rail infrastructure with sufficient political weight.

If public statistics infrastructure is not modernised, geared to new technology (Shinkansen-like fast statistics) and new terrains are not opened up (COVID, …), there will be parallel infrastructures both in the public and private sectors, which will develop their own standards. To continue the picture: we will have an outdated, unattractive public railway with multiple rail widths (partly public, partly private) and incompatible industry standards; a setback for trust, transparency and education.

To prevent such a situation from occurring, the integration of the various producers of a country under one umbrella, into a well-coordinated statistical system is crucial. Roles need to be attributed; responsibilities have to be defined in order for citizens to be able to rely on government statistics to meet the highest quality standards.

4.2.5.2 *Dissemination and communication*

Is there a profile comparable to retailers when it comes to informational products? Those who specialise in analysing the data can of course take on this task. One can also imagine leaving this role to others, because they are more familiar and experienced with it, such as journalists (data journalists, science journalists). Further, we would need as "retailers" and multipliers the professionals in education, i.e., teachers, etc.

Informational product marketing must be based on the results of empirical market research so that one can target for example statistical training in order to be effective and efficient for the respective group of recipients. Additionally, appropriate means of promotion should be applied. Branding and labelling are undoubtedly part of a broader marketing strategy to achieve a positive attitude and culture towards data and statistics.

Seal of quality: To simplify consumer information, a certificate should be introduced that provides trustworthy information about the quality profile of the informational product. The effectiveness of such a quality label depends in turn on whether trust can be placed in it and whether citizens actually trust it. By way of example, let's be more aware of the point: In the area of ecological product labels, there is a suspicion by consumers that these only serve greenwashing and are intended to generate demand in the corresponding markets. Certification and quality labelling require a neutral and trustworthy institution behind the certificate and label that checks compliance with quality standards and awards a seal on this basis. The choice of an institution (or its establishment) that has appropriate authority and credibility, can and will shape informational governance in a country (or internationally).

Once again, the question arises at this point about the alternatives, namely market or state. History has taught us that there is no unconditional guarantee on either side that such certification would actually be carried out in a credible, neutral and reliable manner. In this respect, especially in the current situation of political upheaval and in places, where the primacy of the rule of law does not seem to be fully respected by politics, there may be a need for greater commitment on the part of civil society in the monitoring of statistical quality and, above all, (professional) independence (Diaz-Bone & Horvath, 2021).

4.2.6   Users and literacy

The fact that initiatives to improve data literacy are gaining momentum, supported not only by business but also by politics and science, is extremely urgent and very welcome. Data literacy serves to promote maturity in a modern digitalised world and is important for all people - not only for specialists. This education, like other education, is about several dimensions of competence: knowledge, skills and values (Schüller, 2020).

For the citizen, the entrepreneur, the teacher, the student, etc., who wants to understand and apply the indicators of the public statistical sources, mathematical and technical skills are of secondary importance. If you want to buy furniture for your flat, it wouldn't make much sense to learn carpentry beforehand. Rather, it is important to understand enough about the product and its properties to be able to judge its quality in the light of personal application goals and questions. This indeed already requires a lot of knowledge and experience in dealing with quantitative information. Such competencies do not necessarily belong to the field of mathematics but demand practice in interpreting indicators in their context, an assessment of the reliability of sources and processes, experience with graphical representations of statistics (including the flaws that may appear in them) and practice in assessing uncertainties, etc.

There has been ample opportunity in the pandemic crisis of recent months to observe the lack of successful communication between data experts, policy makers and citizens. In order to communicate the facts of statistics to the general public in a comprehensible way, a language, a communication is needed, which in turn requires (more) effort on both sides, so that the existing gap between producers and users can be bridged. For someone to understand what facts from the factory of statistics can tell us, courses in data analytics or in-depth knowledge of probabilities do not provide much help. Rather, some understanding is required of descriptive statistics. It is of great advantage to have gained some experience with these empirical methods. In this sense, it could also be helpful to involve citizens more in the production of statistics. Such co-production could – besides possible provision of data sources – actually contribute to bridging the gap between producers and users of statistical indicators (König, Pickar, Stankiewicz, & Hondrila, 2021).

4.2.7 *Public statistics – governance and policy*

If there is data for policy, there is also a policy for data (and statistics). In principle, policymakers must be expected to act quickly now and create the conditions for the statistical infrastructure to develop as described above in the coming months and years. It is essentially a matter of giving the status of generating evidence relevant to decision-making in the public sector the status that this has long since been given in the private sector. The creation of trust, efficiency and effectiveness in the public information sector will only succeed with official statistics as a powerful actor.

Public statistics require a political framework, if only because they embody a public infrastructure, maintained by public institutions that carry out a public mandate financed with tax money. This is hardly a new insight. As a consequence, most countries (and international institutions) have a corresponding statistical governance, consisting of a body of laws, rules, principles, codes of conduct, programmes, in place (Howard, 2021). What is new, however, is that more comprehensive informational governance (Soma, MacDonald, Termeer, & Opdam, 2016) is in the process of emerging, which deals with the political framework for data, i.e. rules, rights, obligations, institutions etc. with the focus of the debate being on the difficult issues of so-called Big Data.

This should in no way mean that a review and corresponding amendments or modernisations of the governance of public statistics are already included and dealt with; quite the contrary. In some respects, it may not even be a matter of changing existing governance, but of fully implementing it in the first place (Sæbø & Andersen, 2021). The European Statistics Code of Practice explains what is particularly important in this respect as follows:

> Institutional and organisational factors have a significant influence on the effectiveness and credibility of a statistical authority developing, producing and disseminating European Statistics. The relevant Principles are professional independence, coordination and cooperation, mandate for data collection, adequacy of resources, quality commitment, statistical confidentiality, impartiality and objectivity.
>
> (Eurostat, 2018)

For official statistics to be able to cope successfully with such a rapid and radical development of its role, the appropriate conditions must be created in terms of governance, finances, personnel, etc. Beyond the canon of the already existing criteria of the Code of Practice, further demands result from the introductions of this article, e.g. with regard to the introduction of quality labelling and certification of statistical information as well as with regard to initiatives for the improvement of statistical literacy.

Adherence to these principles and quality standards is essentially outside the sphere of influence of the statistical institutions themselves. If there is a lack of political attention and will to address this issue, public statistics will sooner or later fall behind and will no longer be able to meet the demands. The tragedy of the commons applies especially to public infrastructure. If bridges, roads, canalisations (and public statistics) are not maintained for a certain period of time, it is hardly noticeable. Later on, however, the resulting damage and repair costs are even higher.

4.2.8   Summary of important steps and developments for public statistics

- Smart Open Statistics are seen as a goal and guideline for business architecture,
  - Primacy of the use of existing data (multiple sources - mixed modes),
  - Integrated data management,
  - Flexible offers for the use of data (multipurpose standard offers, customised services, access to microdata for research),
- Public' statistics need to be organised and coordinated by the head of the National Statistical Institute as Chief Statistician,
- A forward-looking programme planning and product development (long-term) is necessary combined with more flexible evaluations and iterations (short-term),
- Standards for the information quality, their assurance and certification have to be established for the entire system of public statistics,
- Close cooperation between science and statistics production, access to data for research; promotion of networks/partnerships are important,
- International cooperation, data exchange under strict conditions, use of international data sources will play an ever-bigger role,
- Improvement of data literacy shall become part of the mandate of public statistics.

4.3   Census

4.3.1   Population census – (the most) important component of the statistical system

In order to be able to explain and describe the establishment and trajectories of the census, one must understand it in its dual nature: as a singular major statistical event as well as part of a statistical system. The framework conditions, the political culture and circumstances, etc. must not be ignored in the analysis. In Anglo-Saxon countries, for example, where administrative registration of persons is not carried out, it is still mandatory to conduct a complete registration of the resident population from time to time; the same (still) applies in many countries of the developing world. In contrast, in the countries that have reliable registers (such as the Scandinavian countries), a change has long since taken place in which a classic census in the sense of a knock-on-the-door survey has been replaced by the evaluation of administrative sources. Between these two poles, there are as many different frame conditions as there are solutions and methodological approaches (Kukutai, Thompson, & McMillan, 2015; Ventresca, 1995). New data sources, such as geocoded information of buildings, further increase the diversity. Modern business architecture in statistics means the use of different sources and available technology ("smart open statistics").

This diversity does not make it easy at the international level to develop and agree on statistical standards for the census. This is reflected in the annotations to the current United Nations guidelines:

> The Handbook provides guidelines, mainly, for population and housing censuses based on traditional field enumeration. ... Information on the use of registers and administrative sources for censuses can be found in the Principles and Recommendations for Population and Housing Censuses.
>
> (United Nations Department of Economic
> and Social Affairs, 2017, 2021)

For statistics in the European Union, the census represents a very special challenge in that a convention is needed which, on the one hand, guarantees sufficient comparability of the results, but on the other hand, leaves the national statistics sufficient freedom in the organisation and implementation. It was not until the last census round in 2011 that a compromise and a legal regulation could be agreed upon in this regard. This compromise defines a set of variables and regulates the delivery of results to the European level with sufficient granularity and delay. This approach of harmonising outputs was supported by technical and methodological innovations such as the Census Hub (provision of Hypercubes) and the possibilities offered by geocoding (1km^2 grid). In the 2020 census round, this approach is being consistently pursued, for example by providing so-called 'mesodata' from the national level that *"open new angles of analysis, for example by tabulating data for sufficiently small areas and/or small subgroups of the population."* (Eurostat, 2019, p. 14) Apart from the diversity of organisational approaches, in the field of demographic statistics, there is also the fact that these are variables which are partly codified legally/administratively at the national level, and which are politically sensitive. In this light, it should also be seen that there has only been a legal convention for the area of demographic statistics in Europe since 2013, which contains a uniform definition of the *"usually resident population"* (European Union, 2013, p. Art. 2(c))

In the statistical community, the question of the possibility of changing the method of the census is fraught with considerable divergence of opinion. According to the introduction to the special focus on population censuses in the Statistical Journal of the International Association for Official Statistics (IAOS) (Durr, 2020), some authors question even the theoretical soundness of the register-based census.[7] The criticism is ammunitioned with nothing less than questioning the scientific basis of the census alternatives: *"The efforts to replace the census are not based on scientific or philosophical deficiencies of the former, but because of technological developments and political, financial and social considerations"* (MacDonald, 2020). Such an approach is characterised by a one-sided natural science understanding of what the scientific method is and should be for statistics: *"Any empirical science observes physical objects (people, animals, celestial objects, rocks, etc.) and their characteristics at a clearly delimited location and specific time"* (MacDonald, 2020, p. 19). Such naïve realism (Lupton, 2013, p. 49), which is neither adequate nor helpful for a description of what (official) statistics is, may however relatively often characterise the views of professional statisticians, possibly more often than the view that statistics as informational products are both "made" and objective (Porter, 1995; Saetnan, Lomell, & Hammer, 2011). Such a narrow understanding of science is even more astonishing, since it would fit – if at all – at best to survey-based statistics and would exclude large parts of official statistics (including national accounts and foreign trade statistics), which are based on the evaluation of existing data sources, from being able to refer to a scientific basis (Radermacher & Körner, 2006). Any effort to further develop statistics by using new data sources and integrating data science methods based on them would also be discredited generally in this way.

What plays a further role in the positioning of the census within the national statistical system is its prominent position and significance. In the past, the censuses, which were carried out at longer intervals, were subject to very special requirements, but also to expectations, be it with regard to the quality of the statistics (precision, detail), be it with regard to the risks of these large-scale projects or also with regard to media attention or financial budgets. In this periodically established microcosm within the statistical world, professional careers could make or break. This has also distilled an understanding of what the specific strengths are in terms of the quality of information provided by the census. Such a domain-specific definition of quality can also be observed in other areas of statistics, especially in the contrast between survey statistics and national accounts. Yet, why is this a noteworthy aspect in this context? Because in a transition from a traditional census to (whatever) alternative methods, fundamental beliefs are shaken. These beliefs have much to do with the fact that statisticians want to have a foundation that they build (and control) themselves without depending on other sources (and their properties), an inventory that can be used to correct the errors of incremental updates from time to time; "ground truth" so to say. As understandable and comprehensible as this view may be, it nevertheless harbours the danger of naïve realism, in that the correspondence of the census results with (unknown) reality are idealised because they are supposedly based on the simple (and presumably only scientific) method of enumeration. In contrast, the procedures of merging data from several sources are perceived as having something artificial, model-like about them, which makes them appear less credible, less observation-based, and less scientific, not least in the eyes of the public. The fact that even a traditional census is based on conventions in many respects and can by no means provide a one-to-one portrayal of the reality of an entire nation at a moment in time is not brought to the fore in contrast. For these reasons, a transition to alternative census methods is difficult over and above the technical and methodological challenges that need to be solved.

In order to assess whether there is an alternative to the traditional census, it is decisive how narrow or broad one draws the horizon of consideration. A narrowly defined framework (a) would focus the analysis on whether the individual statistical variables collected in the census can be generated with sufficiently good quality from other sources, without, however, putting the essential design parameters of the census (time, granulation in terms of content and geography, response rates, budget, etc.) up for disposal. In a broader framework (b), on the other hand, one would allow the entire design of the census to change, e.g. to mix representative surveys with an evaluation of other sources, to associate components at other points in time, and to consider and weigh the costs and benefits (i.e. quality) of these alternatives holistically. In a very broad approach (c), one would consider the entire system of demographic and social statistics over a longer time horizon and ask how this entire system can be optimised under the given framework conditions (data availability, information needs, the burden on respondents, costs, staff capacities, ...) (Espinasse, Le Palud, Prévot, ël Solard, & Vanotti, 2021; Roux, 2020).

In view of the options contained in multiple data sources, the current information needs and the technical possibilities, an optimal overall solution for the statistical areas should be sought, which in the past the census has served for several years. It is necessary to ask what information this statistical instrument is supposed to satisfy and how these goals can be achieved most effectively and efficiently given the given circumstances and framework conditions (Kyi, Knauth, & Radermacher, 2012). There are clearly defined specific needs (such as providing the basis for elections) but also requirements that arise from the use of census results in the official statistics system (the basis for random samples and imputations). Finally, it is necessary to consider what quality of statistical information over the course of several years is adequate for the information needs of modern societies (high dynamics due to migration, etc.). In a traditional alternation between censuses taking place at intervals of several years and sample collections and annual updates in between, very uneven qualities are obtained over time and major corrections have to be made after each census (with corresponding retrospective adjustment of time series). This problem has a greater impact today than in times of more static population structures. In this respect, there is a case for integrating the classical census methodologically (as far as possible) and aiming for continuous statistical reporting on population, households, etc. Where this optimum lies in each case, how best quality can be achieved, etc., are questions that must be answered at the national level. In European countries, the move is towards integrating censuses into a continuous system of demographic and social statistics. In other countries, however, the solution may look different.

4.3.2 *Population census in Germany*

The statistical recording of the population and the associated determination of sociological-demographic characteristics is an equally important and sensitive area of statistics for politics. It is therefore not surprising that the census and the characteristics contained in its questionnaire were of great interest to National Socialist politics. Without being able to go into these connections here (for more details see Wietog, 2001), however, this period before the end of the Second World War also shapes the history of the census afterwards to a certain extent.

Two noteworthy aspects should be pointed out in this context

- The census planned for 1983 in the Federal Republic of Germany was stopped by a ruling of the Constitutional Court (more on this below). The background to this was objections raised by civil society concerning the protection of the privacy of individual data. The consequences of this ruling were extremely far-reaching, both politically and legally, as it laid the foundation for the development of data protection law as we know it today. In Germany, a corresponding right was even incorporated as a legal provision into the constitution.
- Censuses were conducted in the former federal territory in 1950, 1961, 1970 and 1987, and in the former German Democratic Republic (GDR) in 1950, 1964, 1971 and 1981. The 2011 census was the first all-German census since reunification. This time series is very astonishing, as it refers to a time gap of more than 20 years in which there have been no new census results and even though there would have been sufficient reasons to conduct a census after the German Reunification.

From both historical facts, it can be seen that the census in Germany is conducted under great political and public attention and that each element is determined by the legislator and reviewed by the courts.

Against this history, the census conducted in 2011 was politically very important on the one hand, but difficult on the other, as a compromise had to be reached between information needs on the one hand and reservations about burdening respondents, secrecy concerns and objections to high project costs. The compromise[8] was to forego a full census and replace it with an evaluation of administrative registers combined with the conducting of a (rather large) random sample (see below). The future development of censuses in Germany is predetermined and will move step by step towards greater use of register data and integration of the census (see Chapter 10 of this book).

4.3.3   *Census in court*

In a lawsuit, the Federal Constitutional Court of Germany deals with the complaint of some local authorities and municipalities, which cast doubt on the results of the census 2011. On the deadline of 9 May 2011, a nationwide population, building and housing census took place. This was for the first time a combined census, based on registers and a sample survey of the population. In contrast to the previous censuses, only about 10% of the population was surveyed. To reduce the number of required surveys, data already collected in registers has been used; these were supplemented by primary statistical surveys. Among other things, the census established the official population figures for all communities in Germany. In addition to a double counting test in municipalities with at least 10,000 inhabitants, the procedure provided for a random sample survey in order to correct inaccuracies in the register of migrants, as well as a so-called "survey to clarify discrepancies" in smaller municipalities. Among other things, the results led to the population of

Berlin being determined to be around 180,000 lower than the population update, and the Hamburg population to be around 82,800 fewer. In particular, the applicants allege that the rules in the budget sample infringe on certainty requirements. In addition, the statistical basis of the sampling procedure was not determined accurately enough. A sufficiently precise population investigation was not assured. The application of various methods depending on the size of the municipality was incompatible with the requirements of inter-municipal and federal equal treatment, as it led to a disadvantage for larger communities (BVerfG, 2017).

This was not the first trial before the Federal Constitutional Court. Previously, in 1983, the census had been stopped by the court. At that time, it was essentially about the demarcation of the statistical sphere and the fundamental right to informational self-determination, which has laid one of the foundations for modern data protection (BVerfG, 1983).

The lawsuit census 2011 is well suited to tracking the questions of truth and reality in statistics. On the one hand, there is a lot of nostalgia, which transfigures the old-fashioned door-to-door census as an error-free and "correct" methodology, because supposed counting is simple, robust and not burdened with statistical-technical machinery. There is a widespread, almost naïve idea that one could simply count the population of a whole country as a herd of sheep. Since the days of Emperor Augustus,[9] however, a traditional census has been a mammoth undertaking, solving a variety of technical and methodological problems, using estimation techniques and making inaccuracies and mistakes unavoidable. Even if it does not appear so to non-statisticians, the questions of how accurately and undistorted a statistic reflects the reality are essentially the same, regardless of whether a traditional census is used or a multiple source approach, which is based on other existing information and sampling techniques, thus being able to reduce costs in this way. What remains are subjective question marks, impressions and reservations, which seem to be more critical regarding modern statistical methods compared to the supposedly simple methods of counting (Bubrowski, 2017; Davies, 2017).

As the German court case and the comments in public show, it is essential that such misconceptions, prejudices and reservations towards more sophisticated methodological procedures are taken seriously and that they are already anticipated and taken into account during the preparatory work (as far as this is possible, of course) (BVerfG, 2018).

The *lessons learned* are not primarily related to the production of these statistics, but especially to the design and also to communication. When designing methodically and technically demanding statistics with far-reaching political consequences, as in this case of the census, it is crucial to leave absolutely no doubt that the design (of course, in the context of the exogenously given conditions) meets the current scientific and technical standards. Transparency, (explicit and detectable) sound scientific advice, and active public relations work are crucial in a modern statistical process to familiarise a critical public with the procedures and results. In any case, when there is criticism of supposedly incorrect results, we should at first focus on the concrete and answerable questions about the adequacy of the design of statistics and its error-free implementation (see in particular the considerations and guidelines concerning the census 2011 court case (BVerfG, 2018).

4.4　Conclusion

Hardly ever has the lack of adequate statistics for making essential political decisions and gaining the support of the general population for their consequences been as visible and painful as it is now, during the pandemic crisis. For official statistics, these times are challenging in several ways. At the same time, new products, services, programmes and methods must be developed without jeopardising the proven quality concept, although the budgetary and capacity framework conditions are often not very suitable for such an endeavour. Consequently, the decisive objective must be to regain support for official statistics. For this to succeed, new conventions are needed in which the production model of public statistics is adapted to the changed framework conditions by using all available data sources and technical-methodological procedures on the one hand and to modified information needs of modern societies on the other. These demands apply first and foremost to the major projects and areas within the statistical system, of which the censuses are one. For the census in particular, the objective is that it should be policy-relevant, without being politically driven.

Notes

1　"Statistics is not only, as a branch of mathematics, a tool of proof, but is also a tool of governance, ordering and coordinating many social activities and serving as a guide for public action. As a general rule, the two aspects are handled by people of different specializations, whose backgrounds and interests are far apart. Thus, mathematicians develop formalisms based on probability theory and on inferential statistics, while the political scientist and sociologist are interested in the applications of statistics for public action, and there are some who speak of 'Governing by indicators'. The two areas of interest are rarely dealt with jointly." (Desrosières, 2011, p. 41)
2　Michael Power sees even "audit as intrinsic to modern society," "a constitutive principle of social organizations" and an "institutional norm" (Power, 1994).
3　Ulrich Beck has emphasised that the struggle among rationality claims follows from a division of the world between experts (in rationality) and non-experts (deviating from rationality) in the non-reflexive modernity (Beck, 1998, p. 57).
4　https://www.amstat.org/ASA/News/Statistics-at-a-Crossroads-Recommendations-Are-Released.aspx
5　https://officialstatistics.com/news-blog/why-should-there-still-be-need-elaborate-official-statistics-future
6　See also the discussion platform in https://officialstatistics.com/news-blog/crises-politics-and-statistics
7　See 'The Third Discussion on the SJIAOS Discussion Platform: The Definition, Methodology and Relevance of Census Taking'. 1 Jan. 2020: 13–14. https://content.iospress.com/articles/statistical-journal-of-the-iaos/sji209003
8　https://www.zensus2011.de/EN/2011Census/Methodology/Methodology_node.html
9　"In those days Caesar Augustus issued a decree that a census should be taken of the entire Roman world. (This was the first census that took place while Quirinius was governor of Syria.) And everyone went to their own town to register" (Luke 2 NIV, 2).

References

Beck, U. (1998). *Risk society towards a new modernity*. London, England: Sage.

Bubrowski, H. (2017, October 24). Das geschätzte Volk. *Frankfurter Allgemeine*. Retrieved from http://www.faz.net/aktuell/politik/inland/bundesverfassungsgericht-untersucht-zensus-2011-15261904.html

BVerfG. (1983). *Urteil des Ersten Senats vom 15. Dezember 1983 (Volkszählungsurteil)* (1 BvR 209/83, Rn. 1–215). Karlsruhe. Retrieved from Bundesverfassungsgericht website: https://www.bundesverfassungsgericht.de/SharedDocs/Downloads/DE/1983/12/rs19831215_1bvr020983.pdf

BVerfG. (2017). Mündliche Verhandlung in Sachen "Zensus 2011" am Dienstag, 24. Oktober 2017, 10.00 Uhr (Pressemitteilung No. 73/2017). Karlsruhe. Retrieved from Bundesverfassungsgericht website: https://www.bundesverfassungsgericht.de/SharedDocs/Pressemitteilungen/DE/2017/bvg17-073.html

BVerfG. (2018). Urteil des Zweiten Senats vom 19. September 2018 (2 BvF 1/15, 2 BvF 2/15 – Rn. 1–357). Karlsruhe. Retrieved from Bundesverfassungsgericht website: http://www.bverfg.de/e/fs20180919_2bvf000115.html

Davies, W. (2017, January 19). How statistics lost their power – and why we should fear what comes next. *The Guardian*. https://www.theguardian.com/politics/2017/jan/19/crisis-of-statistics-big-data-democracy

Desrosières, A. (2011). Words and numbers - For a sociology of the statistical argument. In A. R. Saetnan, H. M. Lomell, & S. Hammer (Eds.), *The mutual construction of statistics and the society*. New York, NY: Routledge.

Diaz-Bone, R., & Didier, E. (Eds.). (2016). *Conventions and quantification – Transdisciplinary perspectives on statistics and classifications* (Historical Social Research, 41, [2, Special Issue]). Köln: GESIS.

Diaz-Bone, R., & Horvath, K. (2021). Official statistics, big data and civil society. Introducing the approach of "economics of convention" for understanding the rise of new data worlds and their implications. *Statistical Journal of the IAOS, 37*(1), 219–228. doi:10.3233/SJI-200733

Durr, J.-M. (2020). Guest editorial. *Statistical Journal of the IAOS, 36*, 5–10. doi:10.3233/SJI-200619

Espinasse, L., Le Palud, V., Prévot, J., ël Solard, G., & Vanotti, L. (2021). Accuracy of French census population estimates. *Statistical Journal of the IAOS, 37*(4), 1105–1124. doi:10.3233/SJI-210849

European Statistical System Committee. (2018). *Bucharest memorandum on official statistics in a datafied society (trusted smart statistics). 104th DGINS Conference, Bucharest, 10th and 11th October 2018. As adopted by the European Statistical System Committee (ESSC) meeting on the 12th October 2018*. Retrieved from https://ec.europa.eu/eurostat/documents/13019146/13239158/The+Bucharest+Memorandum+on+Trusted+Smart+Statistics+FINAL.pdf

European Union. (2013). Regulation (EU) No 1260/2013 of the European Parliament and of the Council of 20 November 2013 on European demographic statistics, 1260/2013 C.F.R. (2013). Retrieved from https://eur-lex.europa.eu/LexUriServ/LexUriServ.do?uri=OJ:L:2013:330:0039:0043:EN:PDFEurostat. (2018). *European statistics code of practice – For the national statistical authorities and Eurostat* (p. 20). Luxembourg: Publications Office of the European Union.

Eurostat. (2019). *EU legislation on the 2021 population and housing censuses*. Luxembourg: Publications Office of the European Union.

He, X., Madigan, D., Yu, B., & Wellner, J. (2019). *Statistics at a crossroads – Who is for the challenge?* Retrieved from https://www.amstat.org/ASA/News/Statistics-at-a-Crossroads-Recommendations-Are-Released.aspx

HM Treasury. (1998). *Statistics: A matter of trust.* Retrieved from https://www.gov.uk/government/publications/statistics-a-matter-of-trust

Howard, C. (2021). Government Statistical Agencies and the Politics of Credibility. Cambridge: Cambridge University Press.

König, A., Pickar, K., Stankiewicz, J., & Hondrila, K. (2021). Can citizen science complement official data sources that serve as evidence-base for policies and practice to improve water quality? *Statistical Journal of the IAOS, 37*(1), 189–204. doi:10.3233/SJI-200737

Kukutai, T., Thompson, V., & McMillan, R. (2015). Whither the census? Continuity and change in census methodologies worldwide, 1985–2014. *Journal of Population Research, 32*(1), 3–22. doi:10.1007/s12546-014-9139-z

Kyi, G., Knauth, B., & Radermacher, W. (2012, May 22–23). *A census is a census is a census?* UNECE-Eurostat Expert Group Meeting on Censuses Using Registers, Geneva, Switzerland. Retrieved from https://www.unece.org/fileadmin/DAM/stats/documents/ece/ces/ge.41/2012/use_of_register/WP_15-IP_Eurostat_01.pdf

Lupton, D. (2013). *Risk* (2nd ed). London, England: Routledge.

MacDonald, A. L. (2020). Of science and statistics: The scientific basis of the census. *Statistical Journal of the IAOS, 36*(1), 17–34. doi:10.3233/SJI-190596

MacFeely, S., & Nastav, B. (2019). "You say you want a [data] revolution": A proposal to use unofficial statistics for the SDG Global Indicator Framework. *Statistical Journal of the IAOS, 35*(3), 309–327. doi:10.3233/SJI-180486

Nastav, B., & MacFeely, S. (2020). You say you want a [data] revolution: Reflections one year on. *Statistical Journal of the IAOS, 36*(4), 1299–1306. doi:10.3233/SJI-200722

Porter, T. M. (1995). *Trust in numbers: The pursuit of objectivity in science and public life.* Princeton, NJ: Princeton University Press.

Power, M. (1994). *The audit society.* Oxford, England: Oxford University Press.

Prutsch, M. J. (2020). Science, numbers and politics in a "post-truth" world. *Statistical Journal of the IAOS, 36*(4), 1035–1041.

Pullinger, J. (2017, January 24). Statistics are even more important in a 'post-truth' world. *The Guardian.* Retrieved from https://www.theguardian.com/politics/2017/jan/24/statistics-are-even-more-important-in-a-post-truth-world

Radermacher, W. J. (2019). Governing-by-the-numbers - Reflections on the future of official statistics in a digital and globalised society. *Statistical Journal of the IAOS, 35*(4), 519–537. doi:10.3233/SJI-190562

Radermacher, W. J. (2020a). How statistics can help — Going beyond COVID-19. *Data & Policy.* Retrieved from https://medium.com/data-policy/how-statistics-can-help-going-beyond-covid-19-22bb2ce92440

Radermacher, W. J. (2020b). *Official statistics 4.0 - Verified facts for people in the 21st century.* Cham, Switzerland: Springer.

Radermacher, W. J. (2021). Guidelines on indicator methodology: A mission impossible? *Statistical Journal of the IAOS, 37*(1), 205–217. doi:10.3233/SJI-200724

Radermacher, W. J., & Körner, T. (2006). Fehlende und fehlerhafte Daten in der amtlichen Statistik. Neue Herausforderungen und Lösungsansätze. *AStA Advances in Statistical Analysis, 90*(4), 553–576.

Roux, V. (2020). The French rolling census: A census that allows a progressive modernization. *Statistical Journal of the IAOS, 36*, 125–134. doi:10.3233/SJI-190572

Saetnan, A. R., Lomell, H. M., & Hammer, S. (2011). *The mutual construction of statistics and society*. London, England: Routledge.

Schüller, K. (2020). *Future skills: A framework for data literacy* (Working Paper No. 53). Retrieved from https://hochschulforumdigitalisierung.de/sites/default/files/dateien/HFD_AP_Nr_53_Data_Literacy_Framework.pdf

Soma, K., MacDonald, B. H., Termeer, C. J., & Opdam, P. (2016). Introduction article: Informational governance and environmental sustainability. *Current Opinion in Environmental Sustainability, 18*, 131–139.

Sæbø, H. V., & Andersen, M. (2021). Coordination and quality assurance through a programme for official statistics: The Norwegian case. *Statistical Journal of the IAOS, 37*(1), 361–369. doi:10.3233/SJI-200738

Umbach, G. (2020). Of numbers, narratives and challenges: Data as evidence in 21st century policy-making. *Statistical Journal of the IAOS, 36*(4), 1043–1055. doi:10.3233/SJI-200735

United Nations Department of Economic and Social Affairs. (2017). *Principles and recommendations for population and housing censuses* (ST/ESA/STAT/SER.M/67/Rev.3). New York, NY.

United Nations Department of Economic and Social Affairs. (2021). *Handbook on the management of population and housing censuses*. New York, NY: United Nations.

UNSD. (2018). *Fundamental principles of national official statistics*. UNSD. Retrieved from https://unstats.un.org/unsd/dnss/gp/fundprinciples.aspx

Ventresca, M. J. (1995). *When states count: Institutional and political dynamics in modern census establishment, 1800–1993*. Stanford, CA: Stanford University. Retrieved from https://books.google.de/books?id=nF1FAQAAIAAJ

Wietog, J. (2001). *Volkszählungen unter dem Nationalsozialismus*. Berlin, Germany: Duncker & Humblot.

World Economic Forum. (2021). *The global risks report*. Retrieved from https://www.weforum.org/reports/the-global-risks-report-2021

5 Population censuses in crisis

United States, Brazil, and Ecuador in comparative perspective

Byron Villacís[1]

5.1 Introduction

Population censuses are technologies of governance that help to coordinate social activities and guide public actions (Desrosières, 2012, p. 41). They usually are interpreted as objective devices that produce comprehensive statistics under the veil of scientificity (Porter, 1996, pp. 33–35, 41–43). However, their process of production evolves in interaction within and between diverse political contingencies framed in procedures, usually closed to external interpretations (Prévost, 2019; Rottenburg, Merry, Park, & Mugler, 2015, p. 8). While there is substantial literature studying these contingencies in a historical sense for specific countries (Anderson, 1988; Gill, 2007; Loveman, 2009; Prewitt, 2010a; West & Fein, 1990) there are fewer analyses trying to understand the contingencies in a comparative way, and particularly, studying the tensions between social agents that influence directly and indirectly the operation (Baffour, King, & Valente, 2013; Emigh, Riley, & Ahmed, 2016). The 2020 census round represented an invaluable opportunity to fill this void: The United States (US), Brazil and Ecuador faced the arrival of the COVID-19 pandemic and, at the same time, their agencies in charge of the census suffered political interventions from their respective central governments. This chapter deploys a description and analysis of how internal and external political and social forces affected and reacted to the supposed balanced statistical operation, and how the interventions impacted dissimilar infrastructures.

The investigation is based on archival research and in-depth interviews. In the first stage, methodological archives, public announcements, and media reports were collected, coded, and analyzed. This procedure was complemented by 15 in-depth interviews with key census administrators, national and international experts, and local academics observing the operations in the selected countries. Together, it allowed the reconstruction of historical contexts, statistical institutional frameworks, and details about the processes of crises and intervention. I paid particular attention to the identification of key actors, mechanisms of intervention, and outcomes affecting the census.

Three points are salient across the analysis. First, all three cases experienced political interventions tied to changes in political power at the national level. This was true in mature and stable statistical systems, such as the case of the US, but

DOI: 10.4324/9781003259749-8

also in the cases of Brazil and Ecuador, countries with dissimilar weaknesses in their statistical systems in terms of autonomy, professionalization, and stability. Second, the capacity to react against interventions depends on two factors. On the one hand, groups interested in the census' outcome need previously constituted social spaces to make visible the demands and to organize reactions. These spaces can be workers' unions, associations of ethnic minorities, observatories of public statistics, or groups of study within academic circles. Brazil and the US have a long tradition of civil society organizations that entered into action during the 2020 census crisis; however, in the case of Ecuador, there were fewer spaces able to embrace concerns. On the other hand, these demands take the form of political struggles only if there is a juridical system able to materialize complaints. Through a functional and prompt legal reaction, the US census avoided interventions from the federal government in the form of attempts to include of a question with political intentionality. In the cases of Brazil and Ecuador, the absence of a juridical power capable of legitimize a space for struggle provoked that, either noisy (Brazil) or discrete (Ecuador) complaints, never took the form of concrete legal reactions. Third, the COVID-19 pandemic played a justificatory transversal role: in the case of the US, it served as an excuse to the government for political intervention, in the case of Brazil functioned as a justification to find a temporal agreement between the conflicting parts; and in the case of Ecuador, it was used as a pretext to delay the operative and to justify internal organizational mistakes present before the arrival of the pandemic.

Together, the chapter vindicates the need to understand the contexts where censuses are produced before attempts to analyze its statistical outcomes. The naïve exploitation of their data or – even worse – the merge between several censuses to then compare results without this consideration, leads to implicit assumptions that are far from realistic. Censuses are artifacts of the societal and institutional context in which they are collected, thus, their statistical exploitation and technical investigation require a broader panorama of reasoning and action. Additionally, the chapter argues in favor of the need for comparative analysis not only at a statistical level but also at an operative one. Only then is possible to make visible social forces interacting with the census from inside and outside its organization. This perspective is essential to understand the impacts of the quantitative outcomes of the census, usually couched in a language of objectivity (Emigh, Riley, & Ahmed, 2020, p. 290).

The rest of the chapter is divided into five sections. In the next two, I describe the conceptual framework and methodology. In the subsequent section, I explain the results from the analyzed cases: US, Brazil, and Ecuador. Then, I proceed with a comparison, to finalize with a discussion.

5.2 Conceptual framework

From a disciplinary perspective, the study of population censuses traditionally has taken four paths. Demographers and statisticians usually pay attention to the outcomes of the operation itself, either producing fertility, mortality, or migration commensurations (Condran, 1984; Eriksson, Niemesh, & Thomasson, 2018; Jaadla, Reid, Garrett, Schürer, & Day, 2020; Retherford, Cho, & Kim, 1984), creating geographic profiles and historical evolutions (Hirschman, Alba, & Farley, 2000; Ruggles & Magnuson, 2020) or using its results as inputs for frequentist or Bayesian modeling (Vanella, Deschermeier, & Wilke, 2020; Voutilainen, Helske, & Högmander, 2020). Most societies know the census thanks to this knowledge production, which usually is inserted afterwards into systems of governance (Murray, 1992). A smaller but significant number of demographers diagnose quality characteristics and operational technicalities such as coverage, costs, inaccuracies, respondent errors, and age heaping, among others (United Nations, 2008). These analyses usually take the form of evaluations where the intention is to identify operative gaps (Brown 1998; Hogan, Cantwell, Devine, Mule, & Velkoff, 2013).

In the realm of economics, the census has been the target of analysis searching for associations and causalities between the evolution of population and its components with economic growth (Howitt, 1999), business cycles (Losch, 1937; Simon, 2019), development (Coale & Hoover, 2015), labor (Durand, 2015), or aging (Maestas, Mullen, & Powell, 2016). When economists look at the census as an administrative operation, usually they apply a cost-benefit analysis (Roseth, Reyes, & Yee Amézaga, 2019; Spencer, May, Kenyon, & Seeskin, 2017). A third and popular approach comes from political scientists who usually study representation and apportionment (Kaiser, 1968; Kastellec, Lax, Malecki, & Phillips, 2015), quality of vote tallies (Challú, Seira, & Simpser, 2020), the independence of census offices (Prévost, 2019; Prewitt, 2003, 2010a), and privacy policies (Singer, Van Hoewyk, & Neugebaue, 2003).

Sociologists usually pay attention to the role they play in relation to social forces (Alonso & Starr, 1989; Saetnan, Lomell, & Hammer, 2012), the construction of ethnic categories (Loveman, 2014; Mora, 2014), methodological designs and implications (Sullivan, 2020), the way they serve to discipline societies (Schweber, 2006), institutional changes (Ruggles & Magnuson, 2020), its relation with state formation (Anderson, 1983; Curtis, 2001; Loveman, 2005), the constitution of conventions and consent (Boltanski & Thévenot, 2006; Desrosières, 1998, 2011; Porter, 2011; Rodríguez-Muñiz, 2017), and historicizations (Anderson, 1988). In the vision of sociologists and political scientists, censuses are not only interpreted solely as technical artifacts. They are understood as political gears that play a role in the political strategies of social actors (Bourdieu, 2014, p. 142). These contingencies imply that the analysis of a census is incomplete if we do not problematize the broader political economy in which they are undertaken: the incidents inside and outside the state during the execution, the social conditions surrounding it, the actors involved, the implications of how governments interpret its functions (Desrosières, 2012, p. 53; Kukutai, Thompson, & McMillan, 2015) and the conditions of institutionalization (Loveman, 2014).

Taking these elements as a reference, this chapter embraces the arguments from the sociology of statistics (Desrosières, 2012; Porter, 1996; Starr, 1987), deploying two types of conceptual and analytical tools. First, it complements analyses that usually retain a strong and implicit state-centered bias because they see the census as a state-centered process implying a top-down exercise of state formation and control (Emigh, Riley, & Ahmed, 2015, p. 486). Instead, it embraces the notion that the census is a complex progression that finishes producing statistical outcomes through interactions between and within social forces that shape the census and, at the same time, shape the society (Baffour et al., 2013; Emigh et al., 2015; Saetnan et al., 2012, p. 13). However, complementing these critical positions, it problematizes the role of censuses not only as the tension generated by the construction of particular categories, for example, race (Mora, 2014). Instead, pays attention to the tensions inside and outside the organization in charge of the census to understand how social forces influence administrative and organizational processes. Second, and to achieve the objective mentioned in the previous point, it uses the framework to understand how the demands of census users are affected by internal and external operative procedures that usually are neither visible nor problematized. These contingencies happen in the middle of a progression of activities where governments attempt to gain partisan advantage by shaping the statistical production against the judgment of the statistical agency (Prewitt, 2010, p. 228). Therefore, the aim is to identify and interpret the interferences through two dimensions: the mapping of social agents inside and outside the state, and through a systematic comparison across countries, the identification of mechanisms of intervention and the effects of the operation.

5.3   Research design

The chapter is supported by document analysis and in-depth interviews. In the first case, I collected methodological and public documentation in relation to the 2020 census round from US, Brazil, and Ecuador. This includes institutional reports from the respective offices in charge of the census (Census Bureau [CB] for the US, the *Instituto Brasileiro de Geografia e Estatística* [IBGE] for Brazil, and the *Instituto Nacional de Estadística y Censos* [INEC] for Ecuador). I also collected national and international reports from organisms with the function of supporting census programs, such as the United Nations,[2] the World Bank (WB), and the Inter-American Development Bank (IADB). These reports are complemented by analysis and investigations from academic circles under the epistemological communities described in the previous section.

Public documentation helped to understand formal and official positions regarding the census; however, analysis of this kind is limited at the moment of identifying political actors, contingencies, positions, and points of conflict inside and outside the state. This gap was covered through two sources: the collection of media reports, analysis, public opinion, and specialized news reports.[3] After this initial collection key actors and institutions that played a role in methodological discussions were identified. This included pundits, representatives of organizations, and academics. I carried out 15 in-depth interviews[4] divided into two stages. In the first one, an exploratory interview was developed to identify (1) points of conflict between the official execution of the census and external positions usually materialized in attempts of intervention from the central government, (2) strengths and weaknesses of the operation, and (3) agents that have interest in the execution of the process. This initial step mapped the actors with interests and their possible agreements or conflicts. The second stage of interviews was focused on conclusive arguments about the position of each identified actor, the role of the institution in charge, mechanisms, and interpretations regarding their respective crisis, and reactions from the actors in the field. Together, the interviews and media reports helped to understand the external narratives that are less controlled by the state, the relation between and within actors in society, and the way those relations influenced the census operation.

US, Brazil, and Ecuador were included as case studies because they share three essential characteristics: (1) their censuses were affected, although at different levels, by the common crisis of the pandemic COVID-19, either by delaying or cancelling their operations, (2) their censuses were affected by significant attempts of political intervention from their respective governments, and (3) their governments share ideological conservative positions: Donald Trump from the Republican Party in the US, Jair Bolsonaro from the extreme right conservative political field in Brazil, and Lenin Moreno from conservative alliances in the right in Ecuador.[5] This last commonality helps to signal the limitation of the chapter: the interventions and crises in censuses happened in diverse political settings and under a variety of governmental contexts (Prewitt, 2010b). The findings deployed here need to be read to understand that, while the comparison among right-wing governments helps to weigh crises under similar political administrations, the mechanisms can differ under dissimilar settings. This diversity in the typology of selected cases provides the ground for a reflexive, non-essentialist engagement with global and historical diversity, which at the same time, is capable of uncovering patterns in the production of statistical systems that emerge when contrasting variously positioned systems of relational political practices within the global power structures (Fourcade, Lande, & Schofer, 2016, p. 16).

5.4   A polemic question testing the juridical system: the case of the US

The history of federal censuses in the US goes back to 1790. The first census was taken under the supervision of Thomas Jefferson. The questionnaire included six questions: name, relation with the head of the family, age, condition of freedom or slavery, sex, and race (US Census Bureau, 2019). Until 1870, District Marshals were responsible for data collection, then, the need for standardization and the incremental complexity of operations opened a period of institutionalization of the statistical system. The CB was created in 1902, as part of the Department of Interior, and one year later, it shifted to become part of the Department of Commerce and Labor (Census Bureau, 2005). The last legal reform happened in 2012 when President Obama enacted the Presidential Appointment Efficiency and Streamlining Act. This reform, among other things, set the term of the CB Director at five years, with a maximum of two periods. Additionally, the position requires demonstrated ability in managing large organizations, and experience in the collection, analysis, and use of statistical data (GOVTRACK, 2012). The CB Director must be appointed by the President and confirmed by the Senate.

The planning of the 2020 census started in 2016 under the direction of John H. Thompson, a tenured functionary since 1975. He was supposed to be in charge of the entire operation; however, he resigned in May 2017 (Sweetland, 2017). Analysts and specialists argued that the main reason for the abrupt exit was the announcement of budgetary cuts (Bromwich, 2017; Marshall, 2017). This was the result of political pressures from the Republican party who usually claim excessive budgets and invasion of privacy in the census (Weyl, 2012). Ron S. Jarmin, another CB career functionary, replaced Thompson for one year and a half. In January 2019, by the nomination of President Trump, Steven Dillingham arrived at the CB direction. Due to his extensive public career at the federal level, Dillingham received support from academic organizations, experts in the field, and the Republican and Democratic parties (Mervis, 2018). His arrival calmed the waters respecting the upcoming census; however, concerns would soon come from another front.

By legal requirement, the CB must submit the list of proposed questions to Congress. The first official release was in March 2018, including seven questions.[6] Despite this initial formality, the Secretary of Commerce Wilbur Ross[7] promoted an extra one: "Is this person a citizen of the United States?" He argued that the inclusion of this question was a requirement to put into effect the Voting Rights Act, as solicitude of the Department of Justice (NPR, 2018a). However, Democrat representatives, who at the time had recently regained control of the House, demonstrated that it was a White House initiative and thus, a politically motivated operation (Wire, 2019). Additionally, President Trump intensively promoted the inclusion of the question through public declarations arguing the need for improved district excluding non-citizens (Baumgaertner, 2018). The reaction from state and civil society organizations was immediate and methodical, arguing that the proposal was attempting to discourage the participation of immigrants in the census, which will result in undercounting and reducing political spaces for Democrats and racial minorities (Lo Wang, 2018a).

The reaction from social forces was diverse, firm, and rapid. Complaints and concerns appeared from at least 131 groups, including ethnic associations (Muslim Advocates, Asian American Associations, and Latino Groups), Unions and Conferences (Conference on Civil and Human Rights and the American Civil Liberties Union), and academic organizations (ASA, 2018; Gamboa, 2018; LCCHR, 2018). Protests were based on technical reports from the very CB (Census Bureau, 2019), and even Ron S. Jarmin, the CB Director who lasted only a year and a half in office, expressed public concerns (Lo Wang, 2018c). Soon, the conflict arrived in the juridical field: at least three federal lawsuits emerged from the states of California, New York, and Maryland (de Vogue & Hartfield, 2019). In a matter of weeks, what started with a political intention to intervene in the census, promptly installed a legal battle at a federal level.

The conflict faced two points of interest: at the core, the discussion was the inclusion of the question; at the border, the timing of the juridical resolution. The latter was converted into a government tactic to delay the juridical result and tacitly permit the inclusion of the question in the census. On the other hand, the plaintiffs were looking for urgent determinations; not only to avoid the inclusion of the question, but also to leave the census calendar unaffected. A big portion of the trial was spent retracing the origin of the citizenship question: Secretary Ross argued that it was the result of a deep and conscious process considering all legal procedures. Democrats and plaintiffs argued that the inclusion was political and influenced by actors like Steve Bannon, an ultra-conservative strategist (Cornwell, 2018). The procedure allowed to have access to personal communications between the parties; which revealed, five key facts: (1) One of the main tasks delegated from President Trump to Ross as Secretary of Commerce was to include the question in the census (Lo Wang, 2018b), (2) former CB Director Ron S. Jarmin and CB technicians formally disagreed with the inclusion of the question. Six months after Jarmin's disagreement, he was replaced by Steven Dillingham (Elliott, 2017; Lo Wang, 2017; NPR, 2018a, 2018b), (3) there were communications between Steve Bannon and Secretary Ross with the sole intention of the inclusion of the question (Lo Wang, 2017), (4) Thomas B. Hofeller, a strategist known for gerrymandering electoral districts, wrote a report justifying the inclusion because it would be advantageous for Republicans and non-white Hispanics (Wines, 2019), and (5) new appointed CB Director Dillingham preferred to have an unvoiced position during the conflict (Lo Wang, 2019b). Even a year later, he avoided taking a position, generating energetic criticisms and concerns in the Congress (Mervis, 2020).[8] This documentation confirmed the politically motivated inclusion of the question and revealed key actors involved in the intervention.

Just a few months before the census, in January 2019, Jesse M. Furman, District Judge, Court for the Southern District of New York, blocked the inclusion of the question claiming that Secretary Ross's decision violated the Administrative Procedure Act (Lo Wang, 2019a). However, Furman also rejected the plaintiffs' claim that adding the question violated the Enumeration Clause of the US Constitution (Wines & Benner, 2019). The Supreme Court agreed and confirmed Furman's position, and the Trump administration decided, on July 2019, to withdraw the question (Lo Wang, 2019c). The census continued its march without the polemic question, nevertheless, analysts and activists consider that the damage was already done, particularly affecting the quality and representativeness of the census (Williams, 2019).[9] On the other hand, the Trump administration continued its efforts to obtain citizenship data through administrative records outside the census (Mangan & Breuninger, 2019). Later, the COVID-19 pandemic arrived, affecting the schedule of operations, including the decision to finish operations one month earlier than planned. According to some analysts, this meant that immigrants and minority communities were not sufficiently included in the census (Wines, 2020). Finally, a big portion of the tensions was liberated after the departure of Trump from power: in the first day of office, President Joe Biden decided to stop the intention of counting undocumented immigrants and immediately the CB announced a more relaxed calendar of production of indicators (Lo Wang, 2021). The case of the US exemplifies (1) how an intervention from the central government involved mostly political actors outside the arena of a professionalized CB, and (2) how the reaction from social actors achieved political and judicial attention at a national level. The following case will show what happens when, in a similar intervention, the role of the judicial system is absent.

5.5 Austerity as the face of intervention: the Brazilian Case

Brazil has executed population censuses since 1872. Nevertheless, the formal institutionalization of the agency in charge arrived in 1936 with the creation of IBGE. The institute is recognized nationally and internationally for its significant level of autonomy and isolation from political conjuncture.[10] Until 2020, Brazil executed 12 population censuses, one of the longest experiences in the region (Dargent et al., 2018; Villacis & Thome, 2020). Over time, the institution gained prestige due to methodological improvements, technological advances, and partnerships with international statistical offices (IBGE, 2020).

The president of IBGE is appointed by the President of the Brazilian government. Historically, this procedure has not been an impediment to the technical autonomy of the organization. For example, since 2003, three of four presidents belonged to tenured positions within the organization (Dargent et al., 2018). Despite this context of relative autonomy, since 2016 the country faced a severe political instability period that affected IBGE and the census. The first woman holding the presidency of the country, Dilma Rousseff, was impeached, opening the door for Michel Temer's interim government. Subsequently, Jair Bolsonaro, a former military officer, and a recognized conservative extremist, arrived at power. The political crisis implied a radical transformation in the vision of the government and its organizations: from a progressive left to an extreme right.

This virulence did not leave the tranquility of the statistical system unaltered. During the 14 years Worker's Party government (2002–2016) the Presidents of the IBGE were two tenured functionaries: Eduardo Pereira Nunes (2003–2011) and Wasmália Vivar (2011–2016). The arrival of Temer to the Presidency of Brazil in 2016 meant the arrival to IBGE of Paulo Rabello de Castro, an orthodox economist from the "Chicago School" recognized by its neoclassical views (Brender, 2010). After 11 months in charge, he was appointed president of the strategic public bank *Banco Nacional de Desenvolvimento Econômico e Social* (BNDES). His replacement, Roberto Luis Olinto Ramos, would only last one year and a half. President Bolsonaro appointed, in February 2019, Susana Cordeiro Guerra as the head of IBGE.

Cordeiro Guerra is an American-Brazilian political scientist with degrees from Harvard and MIT. Before accepting her appointment, she was living in the US for more than 20 years, lately working at the WB as an economist. She came at the recommendation of the Minister of Economy Paul Guedes, a key political figure in Bolsonaro's government. Guedes is another economist from the Chicago School and is responsible for the austerity and privatization agenda implemented since 2019 (Chicago Maroon, 2018; ISTOE, 2019). In the first three months of her presidency, Cordeiro Guerra announced that the priority of the administration would be the "increment of productivity and austerity". Immediately, she ordered a reduction of 25% of the 2020 census budget and a significant reduction of the size of the questionnaire. The excluded questions were part to the 2010 official questionnaire and all of them were necessary for the application of key social programs, such as housing deficit, the population residing abroad, type of education (public vs. private), income from other members of the household (keeping only income from the head), number of hours working per week, and number of babies who died during pregnancy (Rossi, 2019). These changes provoked a reconfiguration of the planning of the census and a strong reaction from several social organizations.

One of the first entities reacting was the Brazilian Association of Population Studies (ABEP), expressing a "great concern regarding the decision of the federal government cutting the budget", and warning that "this decision does not imply only a reduction in the number of questions, but a compromise in the quality of the entire operation" (ABEP, 2019). Cordeiro Guerra answered that the changes pursue a "better census, not a larger one", and that decisions were driven by productivity principles (Globoplay, 2019). Nevertheless, concerns grew systematically. For example, labor unions inside and outside of IBGE organized panels to defend the original design of the census, the association of collective health (ABRASCO) announced "incommensurable" damages if pursuing the methodological modifications, scholars from Universities such as UNICAMP and *Universidade Federal do Rio Grande do Norte* (UFRN) published lengthy complaints, and observatories such as the *Observatorio das Metropoles* published formal protests (ABEP, 2019; Observatório das Metrópoles, 2019). Additionally, external census experts argued that the reduction in the number of questions pointed not only to austerity motivations but also to ideological ones.[11] In the words of one of the interviewees: "it is a tacit and silent privatization. You weaken the public system, and by necessity, consumers of statistics must find the information somewhere else".[12]

Unlike the US case, the Brazilian intervention provoked also struggle inside the statistical office: IBGE workers, consultants, and experts took divided positions, key technicians resigned, and some denounced that the organization was under attack and that changes represented a raucous ideological shock (Cavenaghi, 2019; ISTOÉ, 2019; O Globo, 2019). The discussion arrived in the political field through public hearings at the Chamber of Deputies, the Federal Attorney for Citizens' Rights, and the Human Rights Commission of the Federal Senate (Globo, 2019; Senado Noticias, 2019). To complete the picture of actors, and as a second significant difference with the US case, the authorities of IBGE announced cooperation from international organizations such as the IADB. This decision was interpreted as an attempt to gain legitimacy and support for their positions (Parana Cooperativo, 2019).

Despite these mobilizations, the conflict never arrived in the judicial field, and the president of IBGE declared closed the opportunities to revise her decisions regarding the questionnaire (Agencia IBGE, 2019). However, in August 2019, the same IBGE president, as a reaction to the announcements of new budget cuts to the census, sent a communication to the Minister of Economy warning that the census was at risk due to a lack of resources (Metro 1, 2019). Curiously enough, the arrival of COVID-19 meant an agreement between the parties: the census had to be suspended either due to priorities of health expenditures in the government, or to have more time to organize a reprogramming of the operation without affecting the methodology (Globo, 2020). Due to increasing disagreements with the government, President Cordeiro Guerra resigned from her position and the interim presidency is currently trying to continue the planning of the census under even more threats of budget cuts from the central government.

The case of Brazil exhibits three differences and one similarity with the US. First, it shows a type of intervention from the central government that involved intermediaries that affected the process from *within* the IBGE and shows how the reaction from social actors achieved political attention but failed in the process of judicialization. Additionally, external actors, in the figure of multilateral financial institutions, were used as a mechanism of legitimization. However, in the same way as in the US, the political intervention managed to maintain professional profiles in the heads of the statistical office which, as we will see later, it is not always the case. The consequences of the Brazilian census still need to be evaluated, but they are far from not affecting the process originally designed by IBGE.

5.6   An IMF agreement weakening an already fragile system: the case of Ecuador

The Ecuadorian Directorate of Statistics and Censuses (Dirección General de Estadística y Censos) was created in 1944, partially as a consequence of international attempts to institutionalize public statistics in the Latin American Region. In 1950, as a part of the continental initiative labeled as "Census of the Americas", the country executed the first national census (Villacís, 2021). The National Institute of Statistics was created in 1970 under a military regime, and six years later, the Statistical Law created the National Institute of Statistics and Censuses (INEC) (Villacis & Thome, 2020). Since then, the country has executed seven population censuses, frequently affected by political and institutional instabilities (Villacis, Thiel, Capistrano, & da Silva, 2022). In the last two decades, there have been three significant institutional reforms. First, in 2007, the institute changed its organizational dependency from the Minister of Finance to the National Secretary of Planning and Development (SENPLADES). This change meant an organizational strengthening due to improvements in the budget and expansion of autonomy. Then, in 2016 the law "Código Ingenios" comprised the requirement that the head of INEC must have a PhD in the field and relevant experience.[13] Finally, in 2019, a Presidential decree changed the organizational dependency from SENPLADES to the Presidency of the Republic, provoking a loss of autonomy and damages to the public credibility of INEC (El Comercio, 2019; INEC, 2015a; Presidencia de la República del Ecuador, 2007). This tumultuous history frames the context of the Ecuadorian statistical system: an unstable space usually affected by frequent political interventions.

INEC started to organize the 2020 census in 2017; however, plans were dramatically affected because of accumulated consequences of organizational instability, technical mistakes, and political interventions. In the period 2012–2021, the entity had ten Directors, that is, an average nine months per administration. None of the heads of INEC in the last 20 years had tenure.[14] This instability was combined with polemic methodological changes that affected the public image of the organization.[15] For example, in 2014, the National Household Survey of Employment (ENEMDU) suffered major methodological transformations in the categories of underemployment (Villacis, 2014). This was interpreted as a consequence of public complaints from the then President of the Republic, Rafael Correa, who criticized that underemployment was "too high" and that the category should be revised. A few months later, INEC took a submissive attitude and decided to change the categories without public consultation or technical discussion. The change resulted in a substantial reduction in the proportion of underemployment and the subsequent creation of a new category: "Inadequate employed"[16]. However, President Correa, publicly criticized one more time the new categories, arguing that they make no sense, and explicitly ordering that INEC should revise them (SECOM, 2016, p. 27, 13). Again, INEC changed the methodologies, this time creating the label "Non-Complete Employed" and eliminating the "Inadequate employed" (Villacis, 2016a, 2016b). A few months later, INEC changed another key indicator of economic performance: the level of informality; which, again, the change resulted in a substantial reduction in its proportion (INEC, 2015b). These changes did not go unnoticed by the public opinion that interpreted INEC as an organization that politically intervened and passively disposed to change methodologies according to political moods (Carrión Sanchez, 2018).

Unfortunately, conditions worsened after 2016. Several policies concerning access to public data started to be ignored, users complained due to difficulties in accessing to information, questions in official surveys were changed without participative processes, and the calendar of publications started to be informally administrated[17]. In 2018, the methodology of ENEMDU was changed one more time. In this case, INEC modified the sample in a way that implied the loss of statistical representation at the provincial level; breaking a statistical series that used to be available for more than 15 years. INEC justified the change due to the reduction of the budget for the survey and redesign of the questionnaire due to efficiency evaluations[18] (Villacis, 2019). Then in 2019 temporal heads of the institute changed one more time the methodology affecting the expansion factors,[19] to then change one more time the methodology to come back to the version of 2019. In sum, the most sensitive survey for the public opinion suffered five methodological changes in the last six years. This virulence solidified the public concerns regarding explicit political intervention, loss of autonomy, and dismantling of the statistical system (Diario La Hora, 2019).

Concurrently to all these events, President Moreno decided to implement in the country an aggressive agenda of austerity, reduction of the size of the state, and privatization of public services (Salgado, 2019). The government depended mainly on an agreement with the International Monetary Fund (IMF), which in practical terms, implied the elimination of the public planification system (SENPLADES), and a subsequent dismantling of public statistics (Diario La Hora, 2021; El Universo, 2019b). Public workers and functionaries got fired, and for those who stayed, their salaries were reduced; all the new positions were eliminated, and public investments were suspended (El Universo, 2019a). Under these circumstances, the 2020 census was supposed to be planned. By no surprise, its budget, was affected,[20] a significant portion of workers of the project was laid off, the questionnaire received criticism due to the informality in the process of construction of questions and the elimination of essential variables,[21] and the stage of cartographic update (a crucial component of the operative) presented major delays.[22] It is in this context – that is, an already weak and erratic system – that the pandemic of COVID-19 arrived, delaying the execution of the census (El Comercio, 2020). In December 2020, the then Director announced that the census was suspended and that INEC was looking for external financial support (CEPAL, 2020), especially from the IADB.[23] Same as the Brazilian case, the help from international financial institutions was interpreted as a mechanism to desperately obtain some external legitimacy. Meanwhile, media reports and opinion experts confirmed that the census will not be executed even in 2021, generalizing the idea that the institution is in a structural crisis. In the words of a national newspaper front-page report about INEC: "the statistical system and the census is stuck in a swamp" (Diario La Hora, 2021).

Unlike the cases of the US and Brazil, these interventions, beyond some isolated opinion pieces, did not generate reactions from civil society. Academics and analysts were absent or only voiced their opinions in private.[24] The case of Ecuador exemplifies a structurally intervened statistical system combined with weak (or even absent) reactions from civil society. In addition to this, the Ecuadorian case shows a further difference: the heads of the institute have a traditionally less professional background and less legitimation of credentials when compared with tenure statisticians, such as in the case of Brazil or, endorsement of externally validated professionals, such the case of the US.

5.7   A comparative perspective

The clearly differentiated space of production of censuses in these countries fabricates statistical outcomes that later are naively interpreted as equal objects of analysis. In frequent cases, they are merged as if they were produced under homogenous administrative processes. In this section, I deploy a systematic comparison of the described cases to highlight dimensions that should be considered to avoid superficial understandings of these operations. I argue that there are three points of interest at the moment of comparing the crises which, at the time, help to make visible differentiated spaces of census production: the identification of actors, the mechanisms of the intervention, and the way in which societies react. Nevertheless – and as emphasized in the chapter – the point is to make a comparison according to each context, to interpret differences and convergences only then. This context is problematized by the background of each statistical system, by the nature of the political intervention, and the arrival of the pandemic of COVID-19.

The construction of a dispersed and stable statistical system in the US is a consequence of an institutionalized network of public organizations with defined rules and administrative autonomy (Sullivan, 2020). However, this background did not liberate the country from political interventions (Emigh et al., 2016; Mora, 2014; Porter, 1996, pp. 41–43). In the case reported, the federal government attempted influence using legal and organizational weapons: (1) assigning the task of intervention to the Secretary of Commerce, (2) deploying a political order inside the state apparatus, and (3) fighting in the judicial field. Although Trump lost the legal battle, it is not enough to conclude that his defeat was complete. First, the change of

Ron S. Jarmin for Steven Dillingham in the CB Direction effectively silenced the voice that should have played a technical role. Second, the operation of the census is already affected: there are significant concerns about the impact that the intervention will have on the participation of minorities. Nonetheless, in this case, the capacity of mobilization of activists and civil organizations avoided major damages and conquered legal reactions.

If there is one significant difference between the US and Brazil, is that in the latter, the capacity to capitalize the complaints in the legal field is absent. Despite the mobilization of NGOs, academia, unions, and even workers inside IBGE, the protests did not solidify in a place of resolution where an independent third party or a judge could decide. The place of struggle was predominantly the political field where the government had a clear advantage. This happened in a country where its statistical office is recognized as autonomous and exemplary for the Latin American Region. Even with this traditional prestige, the institution could not block the intervention. For some analysts, this occurred, at least in part, as a consequence of the political polarization of justice: the crisis of the census happened at a moment when the change of presidents of the country was linked to political interventions into the judicial system (Nunes & Melo, 2019). Another difference between the US and Brazil is that in the latter the reaction and struggle was present outside *and* inside the statistical office, showing that society has established mechanisms to defend the autonomy organically inserted in the statistical system. The case of Ecuador is more concerning: the country does not have a legitimized statistical system. Its history with population censuses is recent, usually affected by international influences and organizational instabilities with ephemeral periods of strengthening. What is more significant, however, is the absence of reaction from social groups, either through organized collectives, such as in the case of Brazil, or through legal channels, such as the US. High levels of instability and weak governments depending on the agendas of international financial agencies seem to complicate the panorama. One commonality between Ecuador and Brazil is precisely the attempted role of legitimation that plays these international organizations; however, the external intervention seems to generate an effect, intentional or not, of silencing internal actors. Finally, there is one key difference between the Ecuadorian case with the US and Brazil. In the former the level of professionalization is significantly reduced: despite attempts to improve the profile of the heads of INEC, the system still struggles to have high-level professionals that avoid intervention and procure an autonomous agenda.

Table 5.1 deploys a comparison among the cases through three categories of analysis. In the first four rows, it compares the country and organizational contexts. The size of the population is included to understand the institutional and logistical complexity. The age of the office in charge and the number of censuses help to understand the experience and trajectory of the entities. The appointment of the authorities helps to explain how easy a direct intervention from the central government is. This first section confirms the matureness of the US system and the frailer variations of Brazil and Ecuador. The second set of seven rows explains the nature of the crisis in each case: the specificity of intervention, using Prewitt's (2010)

Table 5.1 Crisis of the census in US, Brazil, and Ecuador – Contexts, actors, and outcomes

		US Census Bureau	IBGE Brazil	INEC Ecuador
Context of the Organism in charge of the Census	Country population (MM, 2019)	328	209	17
	Statistical office antiquity (# Years)	118	84	44
	Number of censuses Executed	23	12	7
	Appointment of authorities	Nominated by the president and confirmed by the Congress	Designated by the President of the Republic	Designated by the President of the Republic
Context of Conflict	Type of intervention	Political and judiciary: Central government introducing a question and rejection in the judicial field	Organizational: new government inserts a neoliberal agenda that affects the statistical system	Institutional: Within a fragile background government dismantled the statistical system due to the implementation of the neoliberal agenda
	Agents intervening in the census	Minister of Commerce (active), Department of Justice (reactive)	President of IBGE (active), President of the Republic	President of the Republic
	Agents defending the census	NGOs, State Governors, Ethnic Associations, among others	Unions, Academics, NGOs	NA
	Designated adjudicator	Supreme Court of Justice	NA	NA
	Type of battle	Legal	Bureaucratic	No public conflict
	Intervention linked to change of political ideology of government	Yes (Shift from Democrats to Conservative Republicans)	Yes (Shift from progressive left to extreme right)	Yes (Shift from progressive left to conservative right)

(*Continued*)

Table 5.1 (Continued)

	US Census Bureau	IBGE Brazil	INEC Ecuador
Impact and role of COVID-19	Used as an excuse from the government to delay legal resolutions attempting to affect the implementation of the calendar.	Used as an argument from both conflicting parties to delay the resolution.	Used as an excuse by INEC to delay the operation by at least one year and justify internal mistakes.
General Outcome	Controversial question not included, but affection in quality due to fear of segments of the population.	Questionnaire reduced and operative under risk due to lack of funds.	Census under risk due to lack of funds and loss of public trust in the institution.

Source: Author.

definition, the agents intervening, the existence (or not) of a ruler solving the conflict, the field where the conflict took relevance, the association between the intervention with the change of political ideology in the government, and the effect that COVID-19 caused in the operation. The last row describes the consequences of the crisis and general specifications for each country. Despite the three cases departing from a differentiated context and went through diverse types of interventions, they converged in the timing associated with changes in the ideological vision of the government. The shift from Democrats to Republicans in 2017 in the US and the shift from progressive to conservative right governments in Brazil (2019) and Ecuador (2017) marked a turning point that materialized the crisis.

5.8 Discussion

Population censuses, when compared with other public statistics, have the appearance of technicality, and reduced political bias. This chapter has documented the ample possibilities of political intervention even within mature and stable statistical systems. The identification of actors, mechanisms, effects, and convergences vindicates the need to understand the contexts where censuses are produced *before* attempts to analyze statistical outcomes. In the same way, it showed that political interventions could take the form of direct methodological modifications, but also, the form of defunding and dismantling statistical systems and public trust. In this latter scenario, when no census is produced or when questions are eliminated, it deploys two effects: the uncertainty is socialized, and the possibilities of statistical production get reduced to those who have the economic means to produce them in the private sector. In other words, an intervention in the statistical system in the form of dismantling public capacities is a latent but effective form of privatization. From these elements emerge two points of discussion.

First, the performance of public statistics requires a body of expertise institutionalized in the responsible agencies. However, there are at least, two additional components to procure their appropriate performance. On the one hand, a functional legal infrastructure is needed to facilitate the activation of protections against interventions. This infrastructure needs to be autonomous and to have sufficient capacity to resolve conflicts within time restrictions of statistical operations. This implies that knowledge about the role of public statistics should be internalized in the bureaucratic apparatus. If the effectiveness of juridical resolutions is absent – or even if they are unhurried – political interventionists will take advantage of the system.

On the other hand, if social spaces inside and outside the state are not able to voice claims from civil society, the interventions will be easily executed. This is evident thanks to the comparative perspective of the investigation and coherent with the notion that, by far, one the most pregnant dimensions of variation when comparing statistical systems has to do with the sheer volume of political engagement, whether it is of a contentious or a more institutionalized type (Fourcade et al., 2016, p. 16). In the same way, the comparison exhibits that the distribution of political and economic capital between professionals, activists, journalists, judges, bureaucrats, and elites are essential to understand how societies react when population census are politically intervened and what institutions are essential to resist the intervention. The most dramatic case is Ecuador, where the legal system was not tested because there were no attempts to formalize complaints against the dismantling of the statistical system. The dispossession of knowledge regarding the most elemental categories that define social groups is represented by the passive acceptance of severe modifications in questions, categories, methodologies, and bureaucratic procedures; a neat demonstration of symbolic violence deployment (Bourdieu, 2014, p. 66, 125, 145). In this sense, the construction of a strong statistical system, and state formation, implies the strengthening of mechanisms of social mobilization, and even the recognition of the existence of spaces for collective action. This also points to the notion that interventions can be explicit and noisy, but also silent and latent.

Second, when analyzing institutional conditions of statistical systems is necessary to pay attention to the organizations performing a formal role in the state but also in the social structures in society. What is the role of academia? What do civil organizations have to say? What is the role of workers and unions? What is the position of those who are not in power? Population censuses belong to the family of public statistics embedded in societal complexities; this implies that they are constructed – or destroyed – within social contexts. Consequently, the global scripts of census that attempt to define (or associate) them with the idea of objectivity are inevitably the result of political and social struggles, whether apparent or unobtrusive, which social sciences has the role of documenting and interpreting.

Notes

1 I am very grateful for the comments from Mara Loveman, Laura Enriquez, and participants of the Genealogies of Data Junior Scholars Workshop (May 2021) supported by the UC San Diego Institute for Practical Ethics, the USF Center for Applied Data Ethics, and the UC San Diego International Institute.
2 This includes the subsidiary offices of UN: United Nations Population Fund (UNFPA) and the Economic Commission for Latin American and the Caribbean (ECLAC).
3 Media reports were systematized through the three major newspaper of each country and complemented by segmented searches through Factiva Dow Jones ® under the license of UC Berkeley.
4 In the case of the US the interviews include: one former CB census manager, one director in charge of communications and public relations, two academics specialized in census methodology and one journalist covering in detail the 2020 census round. In the case of Brazil, the interviews included two directors from the IBGE, one external census analyst (and former IBGE manager), one journalist covering the census with weekly reports and one academician with extensive investigations on the Brazilian census. The case of Ecuador included one former director in charge of the census, one academician studying the census and one head of an NGO investigating the 2020 census round. Finally, I interviewed two directors of regional offices in charge of supporting and founding Latin American census operations through offices of United Nations. All interviews were recorded in audio, transcribed, coded, and analyzed.
5 While there is vast literature regarding the ideological position of Trump and Bolsonaro, there are fewer contributions analyzing the case of Moreno. Becker and Riofrancos (2018) and Chiasson-LeBel (2019) offer a detailed description on how he implemented a radical neoliberal agenda in Ecuador through alliances with right-wing elites.

6 Number of people living or staying at home, type of ownership, sex, age, race, ethnicity, and relation among members of the household.

7 The Secretary of Commerce supervises the CB.

8 Interviews with academic specialized in census methodology (June 2019) and journalist covering in detail the 2020 census round (August 2019).

9 Interview with academic specialized in census methodology (January 2020) and Interview with former CB census manager (March 2020).

10 IBGE headquarters are in Rio de Janeiro, not in the political capital Brasilia.

11 Interview with external census analyst and former IBGE manager (November 2019).

12 Interview with journalist (October 2019) and interview with external census analyst and former IBGE manager (November 2019).

13 This theoretical advance in the autonomy of the institute only worked in paper because, since its conception, the law has not been put in practice: on the one hand the government ignores the requirement, and on the other hand there are no complaints from civil society actors or politicians.

14 At the end of 2012 the statistician David Vera was temporarily commissioned for three months, then President Correa designated the economist Jose Rosero, who stayed in office four years; then Jorge Garcia was temporarily commissioned for 22 days, then President Correa designated David Vera, who lasted six months; then Jorge Garcia was temporarily commissioned for eight months, then President Moreno designated Reinaldo Cervantes who lasted six months, then Rodrigo Castillo was temporarily commissioned for one year, and then President Moreno officially designated Diego Andrade, who abruptly quit on January 2021, delegating the direction to Victor Bucheli who was temporarily commissioned until the arrival of the new government.

15 Interviews with regional technical advisor from international organization (January 2020) and former head of the census operative (September 2019).

16 Before this change the official survey used to publish employment indicators dividing the total of the Economic Active Population in three big groups: unemployed, underemployed, and employed. After the change it appeared a new category "Inadequate employed", splitting the previous underemployed in two big subgroups (underemployment and inadequate employment). The decision caused that the underemployment now was about 50% of its previous value; however –and this is essential – the name of the category "underemployed" did not change with the methodological variation, causing obvious confusion among users.

17 The three interviewees from Ecuador (director in charge of the census in September 2019, academician studying the census in October 2019, and one head of an NGO investigating the 2020 census in December 2019) pointed to these complaints at some point of the interviews.

18 Interviews with an academician (October 2019), a head of an NGO investigating the 2020 census (December 2019).

19 These factors are sets of numbers used to expand, or in simple words multiple, the individual responses up to an estimate for the entire population.

20 According to interviews to the internal director and the head of the NGO investigating censuses in Ecuador, the budget for the first stage was cut in 35% and 42% for the second stage.

21 This included the attempt of eliminate the category of "mestizos" from the alternatives in the question of ethnic self determination. The elimination would have immense impact in the ethnic depiction of the country; this is because according to previous census the category of mestizos has around 70% of the population (Villacis, 2020).

22 Interview with former director in charge of census (September 2019).

23 Interview with regional officer in charge of supporting and founding Latin American census operations (January 2020).

24 This was a consensus in the interviews with regional technical advisor from international organization (January 2020), with the former head of the census operative (September 2019), with academician studying the census (October 2019), and the head of an NGO investigating (December 2019).

References

ABEP. (2019). *Censo 2020*. Associação Brasileira de Estudos Populacionais. Retrieved from http://www.abep.org.br/site/index.php/noticias-censo-2020?start=115

Agencia IBGE. (2019, July 1). Com questionário definido, conheça as perguntas que serão feitas no Censo 2020. *Agencia IBGE*. Retrieved from https://agenciadenoticias.ibge.gov.br/agencia-noticias/2012-agencia-de-noticias/noticias/24914-com-questionario-definido-conheca-as-perguntas-que-serao-feitas-no-censo-2020

Alonso, W., & Starr, P. (1989). *The politics of numbers*. New York, NY: Russell Sage Foundation.

Anderson, B. (1983). *Imagined communities: Reflections on the origin and spread of nationalism*. London, England: Verso.

Anderson, M. J. (1988). *The American census: A social history* (3rd pr). New Haven, CT: Yale University Press.

ASA. (2018, March 28). ASA fights against adding citizenship question to census. *American Sociological Association*. Retrieved from https://www.asanet.org/census-citizenship-question

Baffour, B., King, T., & Valente, P. (2013). The modern census: Evolution, examples and evaluation. *International Statistical Review/Revue Internationale de Statistique, 81*(3), 407–425.

Baumgaertner, E. (2018, March 26). Despite concerns, census will ask respondents if they are U.S. citizens. *The New York Times*. Retrieved from https://www.nytimes.com/2018/03/26/us/politics/census-citizenship-question-trump.html

Becker, M., & Riofrancos, T. N. (2018). A souring friendship, a left divided: In Ecuador, ideological differences, corruption allegations, and a wider debate over term limits and political control underscore a feud between President Lenín Moreno and his predecessor Rafael Correa. *NACLA Report on the Americas, 50*(2), 124–127. doi:10.1080/10714839.2018.1479452

Boltanski, L., & Thévenot, L. (2006). *On justification: Economies of worth*. Princeton, NJ: Princeton University Press.

Bourdieu, P. (2014). *On the state: Lectures at the Collège of France 1989–1992*. Cambridge, England: Polity Press.

Brender, V. (2010). Economic transformations in Chile: The formation of the Chicago Boys. *The American Economist, 55*(1), 111–122. doi:10.1177/056943451005500112

Bromwich, J. E. (2017, May 10). Census director to resign amid worries over 2020 head count. *The New York Times*. Retrieved from https://www.nytimes.com/2017/05/10/us/politics/john-h-thompson-quits.html

Brown, L. D. E. (1998). Statistical controversies in census 2000. *Jurimetrics, 39*, 347.

Carrión Sanchez, D. (2018). Los Números tambien mienten: Subempleo y Estadística Laboral en el Ecuador. *Revista Economía, 70*(112), 121–136.

Cavenaghi, S. (2019). Por um Censo Demográfico de qualidade em 2020. *Scribd*. Retrieved from https://es.scribd.com/document/408169248/Por-um-Censo-Demografico-de-qualidade-em-2020

Census Bureau. (2005, January 3). *Agency history—History—U.S. Census Bureau*. Retrieved from https://www.census.gov/history/www/census_then_now/

Census Bureau. (2019, January 24). *2020 census barriers, attitudes and motivators study*. The United States Census Bureau. Retrieved from https://www.census.gov/programs-surveys/decennial-census/2020-census/research-testing/communications-research/2020_cbams.html

CEPAL. (2020). *Informe de la XIX Reunión del Comité Ejecutivo de la Conferencia de Estadística de las Américas de la CEPAL* (Reporte Interno LC/CE,19/4). Retrieved from https://repositorio.cepal.org/bitstream/handle/11362/46530/S2000952_es.pdf

Challú, C., Seira, E., & Simpser, A. (2020). The quality of vote tallies: Causes and consequences. *American Political Science Review, 114*(4), 1071–1085. doi:10.1017/S0003055420000398

Chiasson-LeBel, T. (2019). Neoliberalism in Ecuador after Correa. *European Review of Latin American and Caribbean Studies/Revista Europea de Estudios Latinoamericanos y Del Caribe, 108*, 153–174.

Chicago Maroon. (2018). UChicago Ph.D. Paulo Guedes may become next finance minister of Brazil. *Chicago Maroon*. Retrieved from https://www.chicagomaroon.com/article/2018/10/16/uchicago-alum-paul-guedes-may-become-next-finance/

Coale, A. J., & Hoover, E. M. (2015). *Population growth and economic development.* Princeton, NJ: Princeton University Press.

Condran, G. A. (1984). An evaluation of estimates of underenumeration in the census and the age pattern of mortality, Philadelphia, 1880. *Demography, 21*(1), 53. doi:10.2307/2061027

Cornwell, S. (2018, November 14). House democrat to probe census citizenship question. *Reuters*. Retrieved from https://www.reuters.com/article/us-usa-congress-census/house-democrat-to-probe-census-citizenship-question-idUSKCN1NJ1FW

Curtis, B. (2001). *The politics of population: State formation, statistics, and the census of Canada, 1840–1875.* Toronto, Canada: University of Toronto Press.

Dargent, E., Lotta, G. S., Mejía, J. A., Moncada, G., Inter-American Development Bank, & Innovation in Citizen Services Division. (2018). *¿A quién le importa saber? La economía política de la capacidad estadística en América Latina.* Washington, DC: Banco Interamericano de Desarrollo.

Desrosières, A. (1998). *The politics of large numbers: A history of statistical reasoning.* Cambridge, MA: Harvard University Press.

Desrosières, A. (2011). The economics of convention and statistics: The paradox of origins. *Historical Social Research/Historische Sozialforschung, 36*(4 (138)), 64–81.

Desrosières, A. (2012). Words and numbers: For a sociology of the statistical argument. In A. R. Saetnan, H. M. Lomell, & S. Hammer (Eds.), *The mutual construction of statistics and society* (pp. 41–63). New York, NY: Routledge.

de Vogue, A., & Hartfield, E. (2019, April 5). Third federal judge blocks census citizenship question. *CNN*. Retrieved from https://www.cnn.com/2019/04/05/politics/census-citizenship-question-ruling/index.html

Diario La Hora. (2019, May 14). INEC pasa a ser una instancia adscrita a la presidencia. *La Hora*. Retrieved from https://lahora.com.ec/loja/noticia/1102243369/inec-pasa-a-ser-una-instancia-adscrita-a-la-presidencia

Diario La Hora. (2021, January 29). Ecuador vuela a ciegas sin estadísticas fiables y oportunas de empleo y pobreza. *La Hora*. Retrieved from https://lahora.com.ec/noticia/1102339327/ecuador-vuela-a-ciegas-sin-estadisticas-fiables-y-oportunas-de-empleo-y-pobreza

Durand, J. D. (2015). *The labor force in economic development: A comparison of international census data, 1946–1966.* Princeton, NJ: Princeton University Press.

El Comercio. (2019, May 13). Lenín Moreno dispone que el INEC pase a la Presidencia y crea "Planifica Ecuador" para sustituir a la Senplades. *El Comercio*. Retrieved from http://www.elcomercio.com/actualidad/inec-presidencia-planifica-ecuador-senplades.html

El Comercio. (2020, March 25). Procesos para el censo de población se reprogramarán. *El Comercio*. Retrieved from http://www.elcomercio.com/actualidad/censo-poblacion-ecuador-reprogramacion-coronavirus.html

El Universo. (2019a, May 1). Trabajadores marcharon en contra del FMI, la política laboral y gobierno de Lenín Moreno. *El Universo*. Retrieved from https://www.eluniverso.com/noticias/2019/05/01/nota/7311824/trabajadores-marcharon-contra-fmi-politica-laboral-gobierno-lenin

El Universo. (2019b, May 13). Gobierno suprime Senplades y crea Secretaría Técnica Planifica Ecuador. *El Universo*. Retrieved from https://www.eluniverso.com/noticias/2019/05/13/nota/7329230/gobierno-suprime-senplades-crea-secretaria-tecnica-planifica

Elliott, J. (2017, December 29). Trump Justice Department pushes for citizenship question on census, Alarming experts. *ProPublica*. Retrieved from https://www.propublica.org/article/trump-justice-department-pushes-for-citizenship-question-on-census-alarming-experts

Emigh, R. J., Riley, D., & Ahmed, P. (2015). The racialization of legal categories in the first U.S. census. *Social Science History, 39*(4), 485–519.

Emigh, R. J., Riley, D., & Ahmed, P. (2016). *Changes in censuses from imperialist to welfare states how societies and states count.* New York, NY: Palgrave Macmillan.

Emigh, R. J., Riley, D., & Ahmed, P. (2020). The sociology of official information gathering: Enumeration, influence, reactivity, and power of states and societies. In T. Janoski, C. De Leon, J. Misra, & I. W. Martin (Eds.), *The new handbook of political sociology* (pp. 290–320). Cambridge, England: Cambridge University Press.

Eriksson, K., Niemesh, G. T., & Thomasson, M. (2018). Revising infant mortality rates for the early twentieth century United States. *Demography, 55*(6), 2001–2024. doi:10.1007/s13524-018-0723-2

Fourcade, M., Lande, B., & Schofer, E. (2016). Political space and the space of polities: Doing politics across nations. *Poetics, 55*, 1–18. doi:10.1016/j.poetic.2015.12.002

Gamboa, S. (2018, October 4). Trump has created a census conundrum for Latino groups with planned citizenship question. *NBC News*. Retrieved from https://www.nbcnews.com/news/latino/trump-has-created-census-conundrum-latino-groups-planned-citizenship-question-n916461

Gill, M. S. (2007). Politics of population census data in India. *Economic and Political Weekly, 42*(3), 241–249.

Globo. (2019, July 1). MPF pede esclarecimentos do IBGE sobre cortes no Censo 2020. *g1.globo*. Retrieved from https://g1.globo.com/economia/noticia/2019/07/10/mpf-pede-esclarecimentos-do-ibge-sobre-cortes-no-censo-2020.ghtml

Globo. (2020, March 17). Censo é adiado para 2021 por avanço do coronavírus; concurso fica suspenso. *g1.globo*. Retrieved from https://g1.globo.com/economia/noticia/2020/03/17/censo-e-adiado-para-2021.ghtml

Globoplay. (2019). Presidente do IBGE diz que corte no orçamento do censo não vai prejudicar a pesquisa. *Globoplay*. Retrieved from https://globoplay.globo.com/v/7586812/

GOVTRACK. (2012, October 10). Presidential appointment efficiency and streamlining Act of 2011 (2012—S. 679). *GovTrack*. Retrieved from https://www.govtrack.us/congress/bills/112/s679

Hirschman, C., Alba, R., & Farley, R. (2000). The meaning and measurement of race in the U.S. census: Glimpses into the future. *Demography, 37*(3), 381. doi:10.2307/2648049

Hogan, H., Cantwell, P. J., Devine, J., Mule, V. T., & Velkoff, V. (2013). Quality and the 2010 Census. *Population Research and Policy Review, 32*(5), 637–662.

Howitt, P. (1999). Steady endogenous growth with population and R. & D. Inputs growing. *Journal of Political Economy, 107*(4), 715–730.

IBGE. (2020). Population census. *IBGE*. Retrieved from https://www.ibge.gov.br/en/statistics/social/population/22836-2020-census-censo4.html

INEC. (2015a). *Una mirada histórica a la estadística del Ecuador.* Instituto Nacional de Estadística y Censos.

INEC. (2015b). *Actualización metológica de Empleo Informal.* Instituto Nacional de Estadistica y Censos. Retrieved from https://www.ecuadorencifras.gob.ec/documentos/web-inec/EMPLEO/2015/Junio-2015/Metogologia_Informalidad/notatecnica.pdf

ISTOÉ. (2019). Paulo Guedes tratará de transição na presidência do IBGE em reunião no Rio. *ISTOÉ Independente.* Retrieved from https://istoe.com.br/paulo-guedes-tratara-de-transicao-na-presidencia-do-ibge-em-reuniao-no-rio/

ISTOÉ. (2019, May 1). IBGE sob ataque. *ISTOÉ Independente.* Retrieved from https://istoe.com.br/ibge-sob-ataque/

Jaadla, H., Reid, A., Garrett, E., Schürer, K., & Day, J. (2020). Revisiting the fertility transition in England and Wales: The role of social class and migration. *Demography, 57*(4), 1543–1569. doi:10.1007/s13524-020-00895-3

Kaiser, H. F. (1968). A measure of the population quality of legislative apportionment. *American Political Science Review, 62*(1), 208–215. doi:10.1017/S0003055400115734

Kastellec, J. P., Lax, J. R., Malecki, M., & Phillips, J. H. (2015). Polarizing the electoral connection: Partisan representation in supreme court confirmation politics. *The Journal of Politics, 77*(3), 787–804. doi:10.1086/681261

Kukutai, T., Thompson, V., & McMillan, R. (2015). Whither the census? Continuity and change in census methodologies worldwide, 1985–2014. *Journal of Population Research, 32*(1), 3–22.

LCCHR. (2018, July 25). *131 Groups fight to remove 2020 census citizenship question.* The Leadership Conference on Civil and Human Rights. Retrieved from https://civilrights.org/2018/07/25/131-groups-fight-remove-2020-census-citizenship-question/

Losch, A. (1937). Population cycles as a cause of business cycles. *The Quarterly Journal of Economics, 51*(4), 649. doi:10.2307/1881683

Loveman, M. (2005). The modern state and the primitive accumulation of symbolic power. *American Journal of Sociology, 110*(6), 1651–1683. doi:10.1086/428688

Loveman, M. (2009). The race to progress: Census taking and nation making in Brazil (1870–1920). *Hispanic American Historical Review, 89*(3), 435–470. doi:10.1215/00182168-2009-002

Loveman, M. (2014). *National colors: Racial classification and the state in Latin America.* Oxford, England: Oxford University Press.

Lo Wang, H. (2017, December 22). *Page 6659 of administrative record for census citizenship question lawsuits.* Retrieved from https://www.documentcloud.org/documents/5022599-0006659.html

Lo Wang, H. (2018a, January 10). Adding citizenship question risks "bad count" for 2020 census, experts warn. *NPR.* Retrieved from https://www.npr.org/2018/01/10/575145554/adding-citizenship-question-risks-bad-count-for-2020-census-experts-warn

Lo Wang, H. (2018b, June 21). Census overseers seeded DOJ's request to add citizenship question, memo shows. *NPR.* Retrieved from https://www.npr.org/2018/06/21/622409505/before-doj-request-commerce-secretary-considered-adding-census-citizenship-quest

Lo Wang, H. (2018c, July 11). Citizenship question controversy complicating census 2020 work, bureau director says. *NPR.* Retrieved from https://www.npr.org/2018/07/11/627350553/citizenship-question-controversy-complicating-census-2020-work-bureau-director-s

Lo Wang, H. (2019a, January 15). Judge orders Trump administration to remove 2020 census citizenship question. *NPR.* Retrieved from https://www.npr.org/2019/01/15/671283852/judge-orders-trump-administration-to-remove-2020-census-citizenship-question

Lo Wang, H. (2019b, April 1). Census bureau must be "totally objective" on citizenship question, director says. *NPR*. Retrieved from https://www.npr.org/2019/04/01/707628958/census-bureau-must-be-totally-objective-on-citizenship-question-director-says

Lo Wang, H. (2019c, July 11). Trump backs off census citizenship question fight. *NPR*. Retrieved from https://www.npr.org/2019/07/11/739858115/trump-expected-to-renew-push-for-census-citizenship-question-with-executive-acti

Lo Wang, H. (2021, January 20). Biden ends Trump census policy, ensuring all persons living in U.S. are counted. *NPR*. Retrieved from https://www.npr.org/sections/inauguration-day-live-updates/2021/01/20/958376223/biden-to-end-trump-census-policy-ensuring-all-persons-living-in-u-s-are-countedMaestas, N., Mullen, K. J., & Powell, D. (2016). *The effect of population aging on economic growth, the labor force and productivity*. National Bureau of Economic Research.

Mangan, D., & Breuninger, K. (2019, July 3). Trump says he is "absolutely moving forward" with census citizenship question, contradicting his own administration. *CNBC*. Retrieved from https://www.cnbc.com/2019/07/03/trump-says-absolutely-moving-forward-with-census-citizenship-question.html

Marshall, A. (2017, May 11). Bad news for everyone! The 2020 census is already in trouble. *Wired*. Retrieved from https://www.wired.com/2017/05/bad-news-everyone-2020-census-already-trouble/

Mervis, J. (2018, July 20). Census bureau nominee becomes lightning rod for debate over 2020 census. *Science*. Retrieved from https://www.sciencemag.org/news/2018/07/census-bureau-nominee-becomes-lightning-rod-debate-over-2020-census

Mervis, J. (2020, July 30). Census director dodges legislators' questions about Trump memo on undocumented residents. *Science*. Retrieved from https://www.sciencemag.org/news/2020/07/census-director-dodges-legislators-questions-about-trump-memo-undocumented-residents

Metro 1. (2019, August 1). IBGE alerta Guedes que Censo 2020 está ameaçado por falta de recursos. *Metro 1*. Retrieved from https://www.metro1.com.br/noticias/economia/79070, ibge-alerta-guedes-que-censo-2020-esta-ameacado-por-falta-de-recursos

Mora, G. C. (2014). *Making hispanics: How activists, bureaucrats, and media constructed a new American*. Chicago, IL: The University of Chicago Press.

Murray, M. P. (1992). Census adjustment and the distribution of federal spending. *Demography, 29*(3), 319. doi:10.2307/2061820

NPR. (2018a, January 19). Commerce department's administrative record for census citizenship question lawsuits. *NPR*. Retrieved from https://apps.npr.org/documents/document.html?id=4500011-1-18-Cv-02921-Administrative-Record#document/p1289/a428453

NPR. (2018b, February 6). Page 3460 of administrative record for census citizenship question lawsuits. *NPR*. Retrieved from https://apps.npr.org/documents/document.html

Nunes, F., & Melo, C. R. (2019). Impeachment, political crisis and democracy in Brazil. *Revista de Ciencia Política, 37*(2), 281–304.

Observatório das Metrópoles. (2019). Manifesto Observatório das Metrópoles pela realização integral do Censo Demográfico de 2020.*—Observatório das Metrópoles*. Retrieved from https://www.observatoriodasmetropoles.net.br/manifesto-observatorio-das-metropoles-pela-realizacao-integral-do-censo-demografico-de-2020/

O Globo. (2019, May 1). Demissão de diretor de Pesquisa não foi motivada pelo censo, diz presidente do IBGE. *O Globo*. Retrieved from https://oglobo.globo.com/economia/demissao-de-diretor-de-pesquisa-nao-foi-motivada-pelo-censo-diz-presidente-do-ibge-23646382Parana Cooperativo. (2019, June 7). *Entrevista: IBGE mira melhora de dados após acordo tecnológico com BID*. Retrieved from http://paranacooperativo.coop.br/ppc/

index.php/sistema-ocepar/comunicacao/2011-12-07-11-06-29/ultimas-noticias/122748-entrevista-ibge-mira-melhora-de-dados-apos-acordo-tecnologico-com-bid

Porter, T. M. (1996). *Trust in numbers: The pursuit of objectivity in science and public life.* Princeton, NJ: Princeton University Press.

Porter, T. M. (2011). *The rise of statistical thinking: 1820–1900.* Princeton, NJ: Princeton University Press.

Presidencia de la República del Ecuador. (2007). *Decreto Adscripcion INEC a Presidencia de Republica.* Retrieved from https://www.ecuadorencifras.gob.ec/wp-content/descargas/%20Informacion-Legal/Normas-de-Creacion/Decreto-Ejecutivo-de-Creacion/Decreto+Ejecutivo+No.+490-Adscripcion+del+INEC+a+SENPLADES.pdf

Prévost, J.-G. (2019). Politics and policies of statistical independence. In M. J. Prutsch (Ed.), *Science, numbers and politics* (pp. 153–180). Cham, Switzerland: Springer. doi:10.1007/978-3-030-11208-0_8

Prewitt, K. (2003). *Politics and science in census taking* (Reprinted in The American People: Census 2000 ed. R. Farley and J. Haaga). New York, NY: Russell Sage Foundation.

Prewitt, K. (2010a). The U.S. decennial census: Politics and political science. *Annual Review of Political Science, 13*(1), 237–254. doi:10.1146/annurev.polisci.031108.095600

Prewitt, K. (2010b). What is political interference in federal statistics? *The ANNALS of the American Academy of Political and Social Science, 631*(1), 225–238. doi:10.1177/0002716210373737

Retherford, R. D., Cho, L.-J., & Kim, N. (1984). Census-derived estimates of fertility by duration since first marriage in the Republic of Korea. *Demography, 21*(4), 537. doi:10.2307/2060914

Rodríguez-Muñiz, M. (2017). Cultivating consent: Nonstate leaders and the orchestration of state legibility. *American Journal of Sociology, 123*(2), 385–425. doi:10.1086/693045

Roseth, B., Reyes, A., & Yee Amézaga, K. (2019). *The value of official statistics* (Technical Note IDB-TN-1682). Interamerican Development Bank. Retrieved from https://publications.iadb.org/publications/english/document/The_Value_of_Official_Statistics_Lessons_from_Intergovernmental_Transfers_en.pdf

Rossi, A. (2019, August 4). O que revelavam sobre os brasileiros as perguntas que serão cortadas do Censo 2020 do IBGE. *BBC News Brasil.* Retrieved from https://www.bbc.com/portuguese/brasil-48931662

Rottenburg, R., Merry, S. E., Park, S., & Mugler, J. (Eds.). (2015). *The world of indicators: The making of governmental knowledge through quantification.* Cambridge, England: Cambridge University Press.

Ruggles, S., & Magnuson, D. L. (2020). Census technology, politics, and institutional change, 1790–2020. *Journal of American History, 107*(1), 19–51. doi:10.1093/jahist/jaaa007

Saetnan, A. R., Lomell, H. M., & Hammer, S. (Eds.). (2012). *The mutual construction of statistics and society.* New York, NY: Routledge.

Salgado, W. (2019, October 14). Ecuador: Society's reaction to IMF Austerity package. NACLA. Retrieved from https://nacla.org/news/2019/10/14/ecuador-societys-reaction-imf-austerity-package-indigenous

Schweber, L. (2006). *Disciplining statistics: Demography and vital statistics in France and England, 1830/1885.* Durham, NC: Duke University Press.

SECOM. (2016, February 5). Discurso Presidente Correa Lanzamiento Pobreza Multidimensional. *Youtube.* Retrieved from https://www.youtube.com/watch?v=frSEhyoNG-g

Senado Noticias. (2019, August 16). CDH vai discutir cortes no Censo 2020. *Senado Federal.* Retrieved from https://www12.senado.leg.br/noticias/materias/2019/08/16/cdh-vai-discutir-cortes-no-censo-2020

Simon, J. L. (2019). *The economics of population growth.* Princeton, NJ: Princeton University Press.

Singer, E., Van Hoewyk, J., & Neugebauer, R. J. (2003). Attitudes and behavior. *Public Opinion Quarterly, 67*(3), 368–384. doi:10.1086/377465

Spencer, B. D., May, J., Kenyon, S., & Seeskin, Z. (2017). Cost-benefit analysis for a quinquennial census: The 2016 population census of South Africa. *Journal of Official Statistics, 33*(1), 249–274. doi:10.1515/jos-2017-0013

Starr, P. (1987). The sociology of official statistics. In W. Alonso & P. Starr (Eds.), *Politics of numbers* (pp. 7–58). New York, NY: Russell Sage Foundation.

Sullivan, T. A. (2020). *Census 2020 understanding the issues.* Cham, Switzerland: Springer . Retrieved from https://link.springer.com/10.1007/978-3-030-40578-6

Sweetland, H. (2017, May 12). The head of the census resigned. It could be as serious as James Comey. *Time.* Retrieved from https://time.com/4774288/census-bureau-john-thompson-resigned/

United Nations. (Ed.) (2008). *Principles and recommendations for population and housing censuses (Rev. 2).* Department of Economic and Social Affairs, Statistics Division. United Nations.

US Census Bureau. (2019). *Timeline census history.* Retrieved from https://www.census.gov/history/img/timeline_census_history.bmp

Vanella, P., Deschermeier, P., & Wilke, C. B. (2020). An overview of population projections—Methodological concepts, international data availability, and use cases. *Forecasting, 2*(3), 346–363.

Villacis, B. (2014, October 28). Lo adecuado de lo ilegal: Errores del INEC en torno al nuevo subempleo. *El Universo.* Retrieved from https://www.eluniverso.com/opinion/2014/10/28/nota/4158746/adecuado-ilegal-errores-inec-torno-nuevo-subempleo

Villacis, B. (2016a, July 24). *Nuevos cambios de metodología de empleo.* Retrieved from https://byronvillacis.files.wordpress.com/2016/07/cambios-de-inec-julio-2016.pdf

Villacis, B. (2016b, July 24). *Sobre los nuevos cambios del INEC en las categorías de empleo.* Retrieved from https://byronvillacis.org/2016/07/24/sobre-los-nuevos-cambios-del-inec-en-las-categorias-de-empleo/

Villacis, B. (2019, February 3). Advertencia sobre indicadores de empleo y pobreza del Ecuador. *El Universo.* Retrieved from https://www.eluniverso.com/opinion/2019/02/03/nota/7168895/advertencia-sobre-indicadores-empleo-pobreza-ecuador

Villacis, B. (2020, January 17). Los mestizos y el censo de población del 2020. *El Universo.* Retrieved from https://www.eluniverso.com/opinion/2020/01/17/nota/7693956/mestizos-censo-poblacion-2020

Villacís, B. (2021). Experticia estadística en la administración pública ecuatoriana: Mecanismos de emergencia y legitimación. *Íconos, 71,* 81–102. doi:10.17141/iconos.71.2021.4841

Villacis, B., Thiel, A., Capistrano, D., & da Silva, C. C. (2022). Statistical innovation in the global south: Mechanisms of translation in censuses of Brazil, Ecuador, Ghana and Sierra Leone. *Comparative Sociology, 21*(4), 419–446. doi:10.1163/15691330-bja10060

Villacis, B., & Thome, D. (2020). Gender politics in Latin American censuses: The case of Brazil and Ecuador. In M. T. Segal, K. Kelly, & V. Demos (Eds.), *Gender and practice: Knowledge, policy, organizations* (Advances in Gender Research 28, pp. 119–140). Bingley, England: Emerald Publishing.

Voutilainen, M., Helske, J., & Högmander, H. (2020). A Bayesian reconstruction of a historical population in Finland, 1647–1850. *Demography, 57*(3), 1171–1192. doi:10.1007/s13524-020-00889-1

West, K. K., & Fein, D. J. (1990). Census undercount: An historical and contemporary sociological issue. *Sociological Inquiry, 60*(2), 127–141. doi:10.1111/j.1475-682X.1990.tb00134.x

Weyl, B. (2012, June 16). No more questions: Census irks republicans. *CQ*. Retrieved from http://public.cq.com/docs/weeklyreport/weeklyreport-000004107499.html

Williams, T. (2019, June 27). What you need to know about the census citizenship question. *The New York Times*. Retrieved from https://www.nytimes.com/2019/06/27/us/citizenship-question-census.html

Wines, M. (2019, May 30). Deceased G.O.P. strategist's hard drives reveal new details on the census citizenship question. *The New York Times*. Retrieved from https://www.nytimes.com/2019/05/30/us/census-citizenship-question-hofeller.html

Wines, M. (2020, April 13). Knocked off track by Coronavirus, census announces delay in 2020 count. *The New York Times*. Retrieved from https://www.nytimes.com/2020/04/13/us/census-coronavirus-delay.html

Wines, M., & Benner, K. (2019, July 9). Judge rejects Justice Dept. request to change lawyers on census case. *The New York Times*. Retrieved from https://www.nytimes.com/2019/07/09/us/census-citizenship-question.html

Wire, S. (2019, March 14). Democrats confront Commerce Secretary Ross, saying he lied about census question. *Los Angeles Times*. Retrieved from https://www.latimes.com/politics/la-na-pol-census-hearing-citizenship-question-wilbur-ross-20190314-story.html

6 The Latin American Observatory of Population Censuses

Increasing statistical literacy through an academia-civil society network

Gabriel Mendes Borges, Nicolás Sacco and Byron Villacís[1]

6.1 Introduction

Governments around the globe count their population. They have been known since almost five thousand years ago, from the dawn of organized societies (Grajalez, Magnello, Woods, & Champkin, 2013), and nowadays, countries conduct them around every ten years. Likewise, censuses have historically been fundamental for the state, science, and civil society. Today's cornerstone of statistical systems provides essential information for developing small area knowledge, sample frameworks for specialized surveys, and securing considerable technological and methodological investment in official statistics.

Despite the innovations in different information systems, such as alternative methods of demographic data, the relative ease of access, and specialized studies, modern censuses still represent a unique and powerful tool in Latin America to quantify and investigate demographic, social, and economic facts toward providing invaluable knowledge about population structure and dynamics. Users exploit census results, assuming they are valid and reliable due to homogenous and technical procedures. This traditional approach to census data has two limitations: first, it usually disregards the extra-statistical elements of the census, that is, its preparation, (political) contingencies, and additional technical factors that affect the outcomes; and second, even if users go beyond a purely instrumental approach to census results and pay attention to the socio-political context of their production, they generally analyze production processes in a framework of methodological nationalism, that is, neglecting the possibility of studying them across countries or regions. Usually, we understand population censuses as tools utilized by state agencies where governments extract information that later is employed to design public policies. As we know from valuable contributions such as Desrosières (2012) and Emigh, Riley, and Ahmed (2016), the census bases its definitions on social realities, and the outcomes are co-produced by societies.

How are censuses interpreted in spaces that are not academia or government? This chapter describes the experience of a project developed to minimize these limitations with a transdisciplinary tactic. The *Latin American Observatory of*

DOI: 10.4324/9781003259749-9

Population Censuses (OLAC) was formed in 2015 to analyze Latin American censuses' technical and non-technical matters. We describe its experience contributing insights that divulge what is frequently missing when we observe censuses in an isolated and exclusively statistical way by displaying the point of view of a third party that documents and analyzes the heterogeneity of population censuses regarding the social actors involved. At the same time, the initiative shows that there are open spaces in the Global South to discuss ideas about the census in conversational ways.

The rest of this chapter divides into three sections. First, we synthesize the conceptual *corpus* from where the Observatory[2] develops. Second, we describe the project's historical context and to which part of the literature it contributes and explains the mission accomplishments after seven years of work. Finally, we depict the crisis we faced during the 2020 census round, closing these thoughts with theoretical and practical questions for the future.

6.2   Theory of modern census

Paul Starr (1987), in his seminal work on the Sociology of Public Statistics, pointed out five types of interrogations that illuminate the objectification of products of statistical systems: the origins and development of the systems, their social organization, the cognitive organization, the uses and effects, and their current system change. The observation of state action implies the possibility of registering what states want to make visible in the socio-technical design of their efforts. The relative transparency of state actions is amplified when the object observed seems to have the appearance of a technical and objective device (Saetnan, Lomell, & Hammer, 2012). When we face statistical devices such as population censuses, we tend to assume that a significant proportion of its production is guided by balanced, technical, and scientific criteria (Prewitt, 2010) but quantifying governance technologies is also affected by political struggles (Davis, Fisher, Kingsbury, & Merry, 2015; Prévost, 2019). In the case of population censuses, the object produced by National Statistical Offices (NSOs) is the methodological documentation that formally claims to embrace the statistical production process. The task of the external observer is to realize that these documents are the very selective result of a diverse set of forces that interact when a census is conducted. Hence, where to start the critical observation of population censuses?

A relevant line of research highlighted the effects of governments on official statistics, particularly social and economic statistics (Desrosières, 2002). This work showed that official numbers shape government decisions. In specific cases, the literature displays the relationship between public policies and official statistics during modern censuses.[3] It has documented, under long and medium-term perspectives, the growing position that official statistics were taking place in the political arena and the design and implementation of public policies (Ho, 2019) – for example, Otero (2006) in the case of Argentina. Although the literature is divided on how to conceptualize the census as a tool of democratic governance on the one hand and as an object of micropolitics on the other (Aragão & Linsi, 2020; Prévost,

2019, 2020; Prewitt, 2010), historically, official statistics are recognized as a pillar of democracy (Sullivan, 2020), underpinned by maximum transparency. Moreover, this recognition is essential since the construction of statistics is expensive for the public and companies.

In addition to providing legal frameworks for conducting censuses and financing, national states, many of them quite diverse, provide logistical support to carry them out. However, the participation of federal and local authorities in census operations coexists with many strains. On some occasions, the independence of the census can be compromised, and the population perceives the census as an administrative operation rather than a statistical one, diluting its primary objective. This situation may imply a lack of public confidence in using data and the guaranteed confidentiality of the information provided (Emigh et al. 2016). Despite having works of this nature for the current historical cycle, the panorama of the contemporary situation in the region shows the difficult task of building transparent and accessible data in Latin America, still in the 21st century.

This theoretical framework facilitated *OLAC* to deploy a systematic observation of population censuses in Latin America, considering the specificities of each statistical system and, especially, the social conditions behind them. The observation of censuses became a diverse set of text pieces published on an open access blog post that reported, analyzed, and interpreted the production of population censuses considering the particularities of each country. Additionally, the challenge was to embrace official actors and agents out of the state that likewise shapes the production of censuses, for example, unions, academicians, experts, and students. In this sense, to open the discussion with a broad audience, the objective of developing content related to the dissemination of science and following the idea of "public demography" by borrowing proposals from the field of sociology, in particular, those that aim to advance in a "public" discipline (Buroway, 2005).

6.3   Modern censuses in Latin America

While Latin America is changing rapidly, its demographic, social, and economic transformations require quality in public statistics. As economies, societies, and environments change, official statistics evolve and develop and have often grown in line with those changes (Schweinfest, 2020). However, at the same time, the reaction to the disruptive innovations arising from globalization, the structural adjustment of the economies, the solid technological advances, cultural changes, and the rate of change of official statistics seems not to accelerate in the NSOs of the region. The reasons for that can be diverse, but the common characteristic is the budgetary restrictions of the state organisms, where motivations for change are usually scarce, given the uncertainty of the working conditions of institutes workers and the constant political pressure to which they are subjected.

Due to restricted resources in Latin American nation-states and a lack of interest from political power holders, censuses and registration statistics were irregular before the first half of the 20th century. During the second half of the 20th century, a new period was reached, characterized by its technological development and methodological standardization thanks to international organizations that promoted and even financed census operations. When global financial markets became more integrated, nation-states developed potential recipients of private and multilateral international credit. As potential debtors, states had to prove their creditworthiness, including the size of the economy, development, and poverty, among other indicators. For this, national accounts and censuses were needed. At the same time, there was growing attention to the development concept that came mainly from the United Nations (UN). To coordinate the creation of the required statistics, transnational statistics commissions such as the Economic Commission for Latin America and the Caribbean (ECLAC) and the Center of Demography for Latin America and the Caribbean (CELADE in Spanish) were formed. Together, economic and humanitarian forces created a homogenizing environment for states (Alonso & Starr, 1987; Prevost & Beaud, 2015; Schweber, 2006). Also endorsed by organizations that supported population control (Connelly, 2008), these two movements remained especially strong during the social and economic shifts that accompanied the demographic transition in Latin America and other regions within the Global South.

In the 2020 census round, plural social pressures on those responsible for carrying out census operations are perhaps more significant than in the previous four decades (Schweinfest, 2020). Fiscal austerity, increasingly advanced user expectations, technological innovation in data gathering, COVID-19, and the pressure to produce statistics using administrative records are just the tip of the iceberg in a long list of challenges facing the NSOs in Latin America. The discussion on methodological innovations has been going on for quite a while in the literature, primarily focusing on the cases of some countries in Europe and other high-income countries (Abbott, Tinsley, Milner, Taylor, & Archer, 2020; Coleman, 2013; Dygaszewicz, 2020; Scholz & Kreyenfeld, 2016; Valle, Jiménez, & Julián, 2020). In particular, the COVID-19 pandemic has brought attention to the difficulties of the traditional approach of census taking, implying gathering information on all people living on a specific territory at a given time by visiting them face-to-face. Consequently, many countries in Latin America have implemented population censuses with a combination of methods or plans to do so (for more information, please see the Appendix to the Introduction of this book).

In contrast with other regions like North America or Europe, there is a field to explore as an adequate input for formulating public policies or as an instrument for socioeconomic development. Even though its scope is challenging to measure, censuses are underused in the region. While there has been substantial attention to the construction of ethnoracial categories in Latin American censuses (e.g., Loveman, 2014), little attention has been paid to the systematic evaluation of census quality, in-depth data analysis, data dissemination in different formats, the institutional and political context under which they carried out methodological decisions, and the institutional correlation of the NSOs pre- and post-census. Many modern censuses (from the 1960s to the present day) tend to have metadata with insufficient documentation and inadequately systematized files, which makes such data unavailable in practice for sophisticated statistical analysis.

As if this was not enough, in recent census rounds in countries like Argentina, Brazil, and Ecuador, for example, alarms have been raised about the role of political power in the operations of the NSOs concerning issues such as confidentiality, reading, and use of individual data and the contexts and uses of census operations. Beyond the political dispute behind it, mistrust of the possibility of exposing personal data to internal and external abuse by various data producers has reoriented contemporary public perceptions regarding producers of official statistics from census information.

The actors involved in census taking in Latin America must be understood as an unequally structured field of hierarchies and networks (Bourdieu, 2014). Therefore, the Observatory has aimed to identify actors that play a role in the region and their background in terms of institutionalization, history, and political power. In Latin American countries, the role of expertise in censuses have been traditionally concentrated in the contributions of the ECLAC and its office CELADE and the Inter-American Development Bank (IDB). One of OLAC's missions' is to be critical with this concentration of knowledge, recognizing contributions from academia and non-affiliated experts. That is why we consider censuses an endeavor that includes social actors with distinctive distances to centers of power, including internal agents with unique interests across the field of experts.

6.4   The Latin American Observatory of Population Censuses (OLAC)

6.4.1   OLAC mission

OLAC organizes an independent, specialized, and participatory space to discuss the processes inherent to population censuses in the region (in particular, those of the 2020 round) from a civil society perspective. The operational objectives are:

- Promote population data in Latin America and expand the user base and understanding of population dynamics.
- Observe, propose, and update theories, methodologies, and tools that make it possible to understand and use population data in the region, including promoting transparency and creating a critical contribution to the production, analysis, and dissemination of census data, with an emphasis on the design of the 2020 round.
- Compare and compile methodologies and population information to contribute to a better understanding of the demographic dynamics of the region.
- Compile and promote studies on population issues in Latin America based on censuses.
- Create a collaboration network on population issues for academics in the area.
- Promote access to information and open data.

Under the premise that transparent and high-quality statistics benefit everyone, OLAC had, in principle, a project horizon of five years (2015–2020), but continues until the 2020 census round finishes and has addressed groups of users and producers from different Latin American countries (Argentina, Brazil, Colombia, Ecuador, and Uruguay). With this interest in common, the website was online in August 2015.[4]

The criterion followed by OLAC is that the census processes can only be more democratic if there is greater participation in both the production and the use of data by the academy and the public (Eyraud, 2018) – in this sense, the blog language is Spanish and Portuguese promoting dialogue between academia, research institutes, independent data users, and the public. Using the two major Latin American languages is essential for empowering local actors, epistemic decolonization, and creating cultural capital (Swaan, 2010). As some scholars argue about other collaborative platforms (Wright, 2011), OLAC highlights the importance of peer exchange among users to acquire a specific census literacy/numeracy. Following ideas of equality, open access, participation, and deliberation in a domination-free environment, we build bridges between frontiers of producers and consumers of information, potentially making all readers contributors. Therefore, although focusing exclusively on censuses in Latin American countries, OLAC's scope is also global since the experiences of other countries have been incorporated into the analyses, particularly those where the Latin American community is relevant, such as the United States and Spain.

Emerging scientific and technological developments allow new questions about census processes and assess whether they present threats or offer opportunities for users or civil society. One of the group's evaluations based on these practices is that the recent census experience in Latin America (the 1960s and onwards) showed that these operations do not have the permanent resources to carry them out on their own and often need the support not only of national governments but also of international organizations, which influence, as in other social and economic areas, sovereign decisions of national states.

OLAC's proposal differs from international organizations, such as CELADE, The Census Project,[5] The African Census Analysis Project,[6] or databases such as IPUMS-I[7] or specialized journals (*Journal of Official Statistics* or *Statistical Journal of the IAOS*) – to name a few programs with similar objectives.

First, since it is a project without funding (only the individual contribution of its members), without concrete institutional residence (has no affiliation with any university, research center, NGO, private company, or professional association), and second, it has an audience that tries not to constrain the expert, and finally, anyone can participate.

Most of these proposed dialogues are supported by the interests of its members and followers of the space. OLAC is characterized by bringing people with some common intellectual interest to read and criticize documents (scientific and non-scientific). Furthermore, this way, the group's proposal is given based on the experience of its members (who worked or have investigated the censuses in their respective countries), their different specialties, and academic training.

In addition to capitalizing on news and compiling documentation accessible to any user, the website promotes technical debates on demographic issues based on census data, providing a terrain that maintains a balance of institutional, academic, and individual opinions. The OLAC team's evaluation was that although there were specific spaces for exchange, such as professional networks, both in statistics and demography, for example, the ProData Network of the *Latin American Population Association* (ALAP), or information being made available through the website of CELADE, little space was open to sharing ideas about censuses on a relatively regular basis, both for experts and the public. We sought to frame the conversations concerning censuses' purely methodological aspects by focusing on the institutional and organizational conditions in which those methodological decisions were made.

Including experts *and* the public as the target audience of our network was a pillar of the project. It started from the premise that population censuses have accumulated learning experiences that are not always socialized. Hence, there were three main criteria through which the OLAC project executed its objectives: open access, friendly dissemination, and academic collaboration. The website was meant to be an online forum for facilitating collaboration and wide dissemination of content. It has been updated relatively systematically every month by the team members, and at the same time, it has been free to access any content. In addition, a resources section was created where different forms of information related to censuses in Latin America are compiled, from questionnaires to manuals, reports, and data access procedures.

Thanks to the enormous increases in the computing power available and the evolution of ideas about demographic dynamics, the analysis of population processes in Latin America has become increasingly accessible and independent of researchers' institutional affiliation. Due to improvements in civil registration and vital statistics systems, the collection of demographic data in censuses, and the growing technological possibility of data exploitation, it is now possible to circulate access with a regional perspective without neglecting the theoretical and political debate behind the information built.

OLAC has observed census in Latin America since 2015, investigating methodologies, registering external experts' opinions, analyzing official statistics unions' roles, identifying regional convergences, and promoting public debates with actors from inside and outside statistical offices. The Observatory also deployed analyses and investigations, paying attention to scientific evaluations of coverage, under-reporting, selection of questions, costs, benefits, and estimations' accuracy. This task did not imply indifference to the technical details of the operations. Still, it allowed us to criticize the census process through an integrative lens regarding the complete production chain.

Two conceptual criteria complement this vision. First, the knowledge of conducting censuses is usually standardized and formalized in academic and bureaucratic centers in the Global North (Santos, 2018), a consequence of historical experiences regarding the operation and the infrastructures in traditional centers of knowledge production. While contributions from the Global North are positive and help the Latin American region avoid determined mistakes, the Observatory aims to add a perspective from the region, understanding the problems generated in countries with different realities compared to high-income countries. This is particularly important regarding statistical infrastructures, public budgeting, social forces participating in guarding the production of public statistics such as the census, and the differentiated set of needs that societies have in the Global South. If we define census as the tool to obtain critical population data about society, and if societies go through distinctive development processes, it is clear to reconsider the content and scope of national public statistics. At first view, we could take this as self-evident; however, when we analyze the sources and references of scientific production of censuses in the region, it usually is the case that a conceptual variation must complement the realities of our countries.

6.4.2   *Implementation and achievements: articles, resources, and sections*

The plan developed by the group has outlined a series of sections on its page, catering to various visitors. Different users have consulted the resource sections, students, NSO workers, and international organizations of the region, as inferred from communication with visitors who interacted directly with the site editors or left public comments. The OLAC page has received, from its inception until mid-2022, more than 150,000 visits. The visits have grown yearly; in the first half of 2022, the blog received 25,500 visits from 19,000 visitors. The ten countries where the most visits originated were Mexico, Argentina, Colombia, Ecuador, the United States, Brazil, Peru, Spain, and Chile. The webpage contains the census questionnaires and, in addition to technical discussions on census issues, the Observatory has served as a repository of technical documents regarding censuses in the region.

One of the Observatory's main activities is producing short articles written by the group's researchers and collaborators. Although there is no formal editorial committee and the authors are responsible for their texts, all published articles need approval from the active members of the Observatory to ensure the technical quality of posted content.

The articles (more than 150 in total) published on the site in the five years since its foundation has dealt with several topics related to population and housing censuses, mainly in Latin America, and their intersection with sociodemographic aspects, policies, and public statistics (Athias, 2016; Campos, 2016c, 2016e; Sacco, 2016b, 2019c, 2022; Villacís, 2015, 2019c, 2021a). The contents published on the OLAC page have highlighted population and housing censuses as the primary source of demographic and socioeconomic information in the region's countries due to administrative records' weaknesses and household surveys' limitations in terms of providing small area information (Borges, 2016a, 2016b; Campos, 2015, 2016d; Sacco, 2016b, 2019c, 2022). The importance of censuses to measure and characterize minorities and marginalized populations has also been pointed out (Campos, 2016a). Censuses remain the only source of information for small areas on various topics, as household surveys do not produce information at this geographic scale (Urdinola, 2018). Administrative records – not only in Latin America – are often inaccurate and incomplete, and different systems are not integrated (Borges, 2016a, 2016b; Campos, 2015, 2016d; Sacco, 2016c).

Some articles describe how to access the documentation on principles, recommendations, and manuals to carry out the population and housing censuses. Other reports have discussed using census data in sociodemographic analysis in the region for research on fertility and reproductive behavior, migration, mortality, and occupation (Campos, 2016b; Marria & Campos, 2016; Minamiguchi, 2019; Sacco, 2016a; Sacco, & Borges, 2018; Sacco, & Fanta, 2017; Urdinola, 2017b, 2020b). In this sense and connected to the Observatory's objective of promoting the use of census data, the articles have also offered an extensive discussion on accessing census data in Latin American countries, examining the different existing modalities (Nathan & Sacco, 2016; Sacco, 2017; Vázquez, 2015).

Concerning technical and operational issues, the quality and coverage of Latin American censuses has been widely discussed on OLAC's website. Conceptual aspects of the collection of different topics have also been debated, such as what is the best way to ask about age (Nathan, Sacco & Borges, 2017), the possibilities of measuring inequality and poverty (Sacco, 2016a; Urdinola, 2018), issues related to population health (Borges, 2017a, 2020; Urdinola, 2020a; Villacís, 2020b, 2020c, 2021b), and the relevance of incorporating the measurement of mortality (Queiroz, & Sacco, 2018), conceptual aspects of quantifying migration (Campos, 2016b; Nathan, 2018), information regarding occupations (Sacco, 2016a, 2016c) and the year of birth of the first child (Borges, 2015; Nathan, 2015). Another conceptual aspect that received attention was discussing the advantages and disadvantages of conducting a *de facto* or *de jure* census in the region's countries, the subject of debate in four articles. The set of articles published on the OLAC website clarifies a substantial discrepancy in the quality and coverage of the Latin American censuses, both between countries and over time (Villacís, 2017b, 2018a, 2018b; Williams, 2019). At the same time, despite different conceptual aspects that reflect the particularities of each country, there has been a standardization of census data collection processes within the Latin American region, primarily due to international recommendations and the exchange of experiences between countries.

A great variety of articles has been written in recent years to fulfill the objective of producing critical material for census processes and data, with an emphasis on the design of the 2020 round censuses (Borges, 2015, 2017b; Sacco, 2019a; Villacís, 2020a, Urdinola, 2021). For example, pieces of investigation have analyzed the organizational heterogeneity of the region, revealing a "census disparity" (Villacís, 2019a), meaning the differences in the statistical capacity between countries due to their census experiences. This has implications not only in descriptive terms but also has helped to interpret reasons behind organizational and operative achievements or operational failures. The disruption in the planning and execution of censuses in the region is a reality familiar to several countries, represented by the irregularity in the periodicity of the application of censuses, their postponement, and cancellation, which has been a recurring theme of the articles a transversal issue in many government areas in the Global South (Merry, 2019). For example, the non-completion of population counts, particularly in Brazil (Borges, 2021) and Mexico in the last decade, has been discussed, as well as the delay in carrying out various censuses, such as the postponement of censuses in Colombia (Sandoval, 2018; Urdinola, 2016a, 2016b, 2017a), no definitions about the date of carrying out the censuses of Argentina (Sacco, 2019b), El Salvador (Tresoldi, 2018), Guatemala, among others. More recently, some articles discussed the impact of the COVID-19 pandemic in the next census round (Borges, Urdinola, Villacís, & Sacco, 2020).

To contribute to the debate on planning and executing the 2030 census in the region, *OLAC* has paid particular attention to evaluating the recent experiences in Latin America and other areas to consider lessons learned. It has been identified that the recurrent economic, political, and institutional crises in the region are factors that have affected the quality of censuses. Furthermore, the structural aspects related to the organization of NSOs have also played an essential role in their quality. The 2010 census round in the region had some drawbacks, aggravated by a shortage of human and financial resources, a lack of autonomy for statistical institutes, and difficulties coordinating the census within the national and international statistical systems. The prominent examples of failure are the censuses of Chile and Paraguay, both carried out in 2012 (Neupert, 2017). These censuses were by far the worst in their countries regarding quality and coverage. Experiences from developed countries have also been discussed. The contrasts between the event of the last Canadian census and the difficulties faced by the most recent census in Australia differed in terms of communication, financing, and the momentum of social cohesion experienced by the countries (Villacís, 2016b, 2017a). The planning and processes that led to the unsatisfactory results of these two experiences were discussed in various articles published by the Observatory. The 2018 Colombian Census recommendations have also been addressed (Borges, 2019a, 2019b; Villacís, 2019b). One of the lessons learned is that long-term investments to develop solid national statistical systems, in which statistical bodies play a critical role, are fundamental to ensuring the quality of censuses. This involves, among other things, providing financial and human resources for them. Furthermore, censuses must be adequately planned, and the NSOs need to have guarantees that they can conduct their technical project. In addition, a permanent technological update plan is recommended, and careful with implementing new methodologies or collection systems.

Another topic of discussion present in at least five articles published on the blog is technological and methodological innovations in censuses, such as self-completed internet questionnaires, use of administrative records, and incorporation of Big Data in official statistics (Borges, 2016b; Campos, 2016d). Several potentialities of Big Data have been identified. Still, the articles also indicate the need for a census combined with traditional surveys, where demographic censuses are essential, which implies that these are complementary sources (Villacís, 2016a, 2021c).

6.5   The 2020 census round in Latin America

Increasingly, population censuses, and official statistics, in general, are forced to differentiate themselves from the emergence of modern technologies that disseminate an enormous amount of information today (MacFeely, 2016). Furthermore, three factors weigh in the 2020 census round and make it a particular challenge for Latin American countries. First, the advent of the Covid-19 pandemic disrupts the operational plans of each statistical office in the region. The second issue is each country's local institutional capabilities and competence to adapt to organizational challenges. The third element is the regional and local economic conditions, which directly influence the comfort or adjustment that each country needs to go through. Together, these factors determine that we face one of the most complex and challenging scenarios in the recent history of census statistical operations in Latin America. The coverage, quality, and efficiency results are still early to analyze, but we would like to outline three scenarios on the possible impact of these challenges.

In the best scenario, we will have statistical offices with an economic and adaptive aptitude to organize (or continue to organize) a census under the impact of the pandemic; or due to their execution schedules, they were only slightly impacted by its effects. In an intermediate scenario, we will have statistical offices that, although they have the institutional capacity and tools to face crises, have nonetheless been affected by the economic imbalances before and after the pandemic. This situation makes them vulnerable to changes in their schedules and methodological planning. The third case, the most pessimistic scenario, involves countries with little institutional strength and high vulnerability to economic shocks and pandemic impacts. In these cases, census taking might be completely abandoned, or data quality could be severely impaired.

It is not easy and not very responsible to attempt to predict which way those Latin American countries that have yet to implement their censuses will take with certainty. However, *OLAC* has analyzed the institutional logic behind census taking and proposes three factors as relevant when identifying related challenges: heterogeneous institutional preconditions, the external shocks from the pandemic, and the economic crisis; together, they are likely to impair the quality, opportunity, and efficiency of the region's census operations. This implies that the intellectual, organizational, and interinstitutional efforts should focus on solving the problems resulting from missing census data for at least ten years. In this direction, we would like to raise some urgent questions: Will it be necessary to rethink the role of population projections? What steps should be taken to avoid the naive reliance upon administrative registry systems plagued by problems of under-coverage and over-coverage simultaneously? What intra- and inter-regional collaboration mechanisms are needed to support the looming statistical deficit supposedly induced by COVID-19 and its impact on public policies lacking new or high-quality data?

Population censuses result from institutional and governmental conflicts and struggles that must be visible and understood – accepting censuses as an isolated analytical tool or as a mere producer of social data neglects the responsibility and need to understand these devices as entities inserted in complex social and institutional realities. The analyst who ignores this distinction is in danger of using information that, in extreme cases, has no practical utility or functionality. An analyst can take advantage of the information produced only once the dynamics that lead to running a census have been considered. Using census data gathered under sensible planning schedules, with budgets managed according to needs and questionnaires elaborated through processes valuing public participation and scientific contributions, is not comparable to data from an erratically performed census deprived of resources and without a participatory approach.

For these reasons, OLAC is at the forefront of those aware that the upcoming decade requires even more work and commitment to understand and alert census stakeholders about the probable data quality conditions. Failure to do so would naturalize the alarming absence and deficiency of census information and allow decisions to be made based on questionable primary data sources. This problematization is particularly necessary due to an unmistakable trend in which governments value data-based policies, but without questioning too much the origin of that data or, when process automation requires high efficiency of information management, underestimate the necessity to understand the origin of that information.

Finally, OLAC's challenge lies in didactics. Censuses produce countries' crucial, demographic, social, and economic statistics. Observing censuses' management, organization, deficiencies, and strengths require new didactic concepts. For example, there is an increasing demand for technical indicators in social stratification, more complexity in treating gender and race, and more dimensions in the analysis of poverty (income, basic needs, consumption, and subjective poverty, among others) that users demand to be quantified. In this context, describing the scope and limitations of censuses requires increasingly complex didactic processes, which are not easily implemented in a digital environment.

For this reason, it is essential that the Observatory improves and redesigns the channels and formats of its analyses to facilitate the communication of messages that lie within its competence and turn the communication process into a two-way mechanism where we ensure that the information is received and understood. Of course, these challenges summon the need to build bridges, form alliances, and generate responsible academic and intellectual communities. This chapter intends to make a step in this direction.

Notes

1 Authors in alphabetical order, all authors collaborated equally in all sections of the chapter.
2 We use Observatory in the same sense as the acronym OLAC.
3 We are using the label modern the label modern as some other authors in the literature, for example Baffour, King, and Valente (2013), Ventresca (1995), Whitby (2020).
4 https://observatoriocensal.org/.
5 https://thecensusproject.org/.
6 http://www.acap.upenn.edu/.
7 https://international.ipums.org.

References

Abbott, O., Tinsley, B., Milner, S., Taylor, A. C., & Archer, R. (2020). Population statistics without a census or register. *Statistical Journal of the IAOS, 36*, 97–105. doi:10.3233/SJI-190593

Alonso, W., & Starr, P. (Eds.) (1987). *The politics of numbers*. New York, NY: Russell Sage Foundation.

Aragão, R., & Linsi, L. (2020). Many shades of wrong: What governments do when they manipulate statistics. *Review of International Political Economy, 29*(1), 88–113. doi:10.1080/09692290.2020.1769704

Athias, L. (2016). Recomendações internacionais sobre estatísticas sociais e como importantes institutos de estatística divulgam seus temas, com destaque para a área social. *OLAC*. Retrieved from https://observatoriocensal.org/2016/02/02/recomendacoes-internacionais-sobre-estatisticas-sociais-e-como-importantes-institutos-de-estatisticas-divulgam-seus-temas-com-destaque-para-a-area-social/

Baffour, B., King, T., & Valente, P. (2013). The modern census: Evolution, examples, and evaluation. *International Statistical Review/Revue Internationale de Statistique, 81*(3), 407–425.

Borges, G. (2015). Avaliação dos censos demográficos brasileiros. *OLAC*. Retrieved from https://observatoriocensal.org/2015/11/24/avaliacao-dos-censos-demograficos-brasileiros/

Borges, G. (2016a). ¿Se acabaron los conteos intercensales de población en Latinoamérica? ¿Qué hacer ahora? *OLAC*. Retrieved from https://observatoriocensal.org/2016/01/20/se-acabaron-los-conteos-intercensales-de-poblacion-en-latinoamerica-que-hacer-ahora/

Borges, G. (2016b). Estatísticas oficiais na era do Big Data: ainda precisamos de censos demográficos? *OLAC*. Retrieved from https://observatoriocensal.org/2016/07/25/estatisticas-oficiais-na-era-do-big-data-ainda-precisamos-de-censos-demograficos/

Borges, G. (2017a). A investigação da saúde nos censos demográficos do Brasil*. *OLAC*. Retrieved from https://observatoriocensal.org/2017/01/31/a-investigacao-da-saude-nos-censos-demograficos-do-brasil/

Borges, G. (2017b). Where are our kids? The undercount of children in the Latin American censuses*. *OLAC*. Retrieved from https://observatoriocensal.org/2017/12/19/where-are-our-kids-the-undercount-of-children-in-the-latin-american-censuses/

Borges, G. (2019a). Lições aprendidas para a garantia de um censo de qualidade. *OLAC*. Retrieved from https://observatoriocensal.org/2019/04/28/licoes-aprendidas-para-a-garantia-de-um-censo-de-qualidade/

Borges, G. (2019b). O que podemos aprender com as avaliações recentes do Censo 2018 da Colômbia? *OLAC*. Retrieved from https://observatoriocensal.org/2019/07/16/o-que-podemos-aprender-com-as-avaliacoes-recentes-do-censo-2018-da-colombia/

Borges, G. (2020). Demografia e epidemia no Brasil. *OLAC*. https://observatoriocensal.org/2020/03/24/demografia-e-epidemia-no-brasil/

Borges, G. (2021). Do we need a census in the middle of a pandemic? *OLAC*. Retrieved from https://observatoriocensal.org/2021/04/16/do-we-need-a-census-in-the-middle-of-a-pandemic/

Borges, G., Urdinola, P., Villacís, B., & Sacco, N. (2020). El impacto del COVID-19 en la ronda censal de América Latina y el Caribe. *OLAC*. Retrieved from https://observatoriocensal.org/2020/03/26/el-impacto-del-covid-19-en-la-ronda-censal-de-america-latina-y-el-caribe/

Bourdieu, P. (2014). *On the State: Lectures at the College de France, 1989–1992*. Cambridge, England: Polity Press.

Burawoy, M. (2005). Por una sociología pública. *Política y sociedad, 42*(1), 197–226.

Campos, M. (2015). Census limits: A reference to marginal populations. *OLAC*. Retrieved from https://observatoriocensal.org/2015/11/22/census-limits-a-reference-to-marginal-populations-or-individuals-departed-from-the-average-population/

Campos, M. (2016a). Rights groups say male/female option insufficient in Census. *OLAC*. Retrieved from https://observatoriocensal.org/2016/01/12/rights-groups-say-malefemale-options-insufficient-in-2016-census/

Campos, M. (2016b). International out-migration measures based on information from former residents. *OLAC*. Retrieved from https://observatoriocensal.org/2016/01/01/international-out-migration-measures-based-on-information-from-former-residents/

Campos, M. (2016c). Os censos e a construção da realidade. *OLAC*. Retrieved from https://observatoriocensal.org/2016/04/15/os-censos-e-a-construcao-da-realidade/

Campos, M. (2016d). Censos e "Big Data": fontes concorrentes ou complementares? *OLAC*. Retrieved from https://observatoriocensal.org/2016/11/17/censos-e-big-data-fontes-concorrentes-ou-complementares/

Campos, M. (2016e). Responding to the census: Should it be mandatory? *OLAC*. Retrieved from https://observatoriocensal.org/2016/01/25/responding-to-the-census-should-it-be-mandatory/

Coleman, D. (2013). The twilight of the census. *Population and Development Review, 38*, 334–351. Retrieved from http://www.jstor.org.ezaccess.libraries.psu.edu/stable/23655303

Connelly, M. J. (2008). *Fatal misconception: The struggle to control world population.* Cambridge, MA: Belknap Press of Harvard University Press.

Davis, K. E., Fisher, A., Kingsbury, B., & Merry, S. E. (2015). *Governance by indicators global power through quantification and rankings.* Oxford, England: Oxford University Press.

Desrosières, A. (2002). *The politics of large numbers: A history of statistical reasoning.* Cambridge, MA: Harvard University Press.

Desrosières, A. (2012). Words and numbers: For a sociology of the statistical argument. In A. R. Saetnan, H. M. Lomell, & S. Hammer (Eds.), *The mutual construction of statistics and society* (pp. 41–63). New York, NY: Routledge.

Dygaszewicz, J. (2020). Transition from traditional census to combined and registered-based census. *Statistical Journal of the IAOS, 36*, 165–175. doi:10.3233/SJI-190566

Emigh, R. J., Riley, D., & Ahmed, P. (2016). *Changes in censuses from imperialist to welfare states: How societies and states count.* New York, NY: Palgrave Macmillan.

Eyraud, C. (2018). Stakeholder involvement in the statistical value chain: Bridging the gap between citizens and official statistics. In Eurostat (Ed.), *Power from statistics: Data, information and knowledge. Outlook report* (pp. 103–106). Luxembourg: Publications Office of the European Union.

Grajalez, C. G., Magnello, E., Woods, R., & Champkin, J. (2013). Great moments in statistics. *Significance, 10*(6), 21–28. doi:10.1111/j.1740-9713.2013.00706.x

Ho, J.-M. (2019). *Social statisticalization: Number, state, science* (Dissertation). Cornell University, New York.

Loveman, M. (2014). *National colors: Racial classification and the state in Latin America.* Oxford: Oxford University Press.

MacFeely, S. (2016). The continuing evolution of official statistics: Some challenges and opportunities. *Journal of Official Statistics, 32*(4), 789. doi:10.1515/jos-2016-0041

Marria, I., & Campos, M. (2016). Zika virus and demography. *OLAC*. Retrieved from https://observatoriocensal.org/2016/02/22/zika-virus-and-demography/

Merry, S. E. (2019). The sustainable development goals confront the infrastructure of measurement. *Global Policy, 10*(S1), 146–148. doi:10.1111/1758-5899.12606

Minamiguchi, M. (2019). Evolução dos municípios brasileiros mais populosos. *OLAC*.

Nathan, M. (2015). ¿Vale la pena preguntar la edad de la madre al nacimiento del primer hijo en los censos? *OLAC*. Retrieved from https://observatoriocensal.org/2015/11/24/vale-la-pena-preguntar-la-edad-de-madre-al-nacimiento-del-primer-hijo-en-los-censos/

Nathan, M. (2018). Población y residencia habitual: ¿un problema de los censos? *OLAC*. Retrieved from https://observatoriocensal.org/2018/04/02/poblacion-y-residencia-habitual-un-problema-de-los-censos/#:~:text=Este%20problema%20suele%20agudizarse%20en, como%20en%20el%20de%20destino

Nathan, M., & Sacco, N. (2016). Modos de acceso a los datos censales en los países de América Latina. *OLAC*. Retrieved from https://observatoriocensal.org/2016/06/17/modos-de-acceso-a-los-datos-censales-en-los-paises-de-america-latina/

Nathan, M., Sacco, N., & Borges, G. (2017). ¿Cuál es la mejor manera de preguntar la edad en un censo de población? *OLAC*. Retrieved from https://observatoriocensal.org/2017/04/24/cual-es-la-mejor-manera-de-preguntar-la-edad-en-un-censo-de-poblacion/

Neupert, R. (2017). La planificación en el Censo de Chile. *OLAC*. Retrieved from https://observatoriocensal.org/2017/09/13/la-planificacion-en-el-censo-de-chile/

Otero, H. (2006). *Estadística y nación: una historia conceptual del pensamiento censal de la Argentina moderna, 1869–1914*. Ciudad de Buenos Aires, Argentina: Prometeo Libros.

Prévost, J.-G. (2019). Politics and policies of statistical independence. In M. J. Prutsch (Ed.), *Science, numbers and politics* (pp. 153–180). Cham, Switzerland: Springer International Publishing.

Prévost, J.-G. (2020). Independent official statistics: Origins and challenges. *Canadian Public Administration, 63*(2), 318–323. doi:10.1111/capa.12377

Prevost, J. G., & Beaud, J. P. (2015). *Statistics, public debate and the state, 1800–1945: A social, political and intellectual history of numbers*. New York, NY: Taylor & Francis.

Prewitt, K. (2010). What is political interference in federal statistics? *The ANNALS of the American Academy of Political and Social Science, 631*(1), 225–238. doi:10.1177/0002716210373737

Queiroz, B., & Sacco, N. (2018). ¿Es relevante incorporar la medición de la mortalidad en los censos de América Latina y el Caribe? *OLAC*.

Sacco, N. (2016a). Desigualdad social y censos de población: en búsqueda de una relación relegada. *OLAC*. Retrieved from https://observatoriocensal.org/2016/02/24/desigualdad-social-y-censos-de-poblacion-en-busqueda-de-una-relacion-relegada/

Sacco, N. (2016b). Sobre las irregularidades detectadas en el Censo 2010 de Argentina. *OLAC*. Retrieved from https://observatoriocensal.org/2016/08/17/sobre-las-irregularidades-detectadas-en-el-censo-2010-de-argentina/

Sacco, N. (2016c). 5 cosas que hay que saber acerca de los clasificadores de ocupación: el caso de los cuatro últimos censos de población de la Argentina. *OLAC*. Retrieved from https://observatoriocensal.org/2016/02/12/5-cosas-que-usted-necesita-saber-acerca-de-los-clasificadores-de-ocupacion-el-caso-de-los-cuatro-ultimos-censos-de-poblacion-de-la-argentina/

Sacco, N. (2017). ¿Cómo hacer un censo de población? *OLAC*. Retrieved from https://observatoriocensal.org/2017/08/24/como-hacer-un-censo-de-poblacion/

Sacco, N. (2019a). Brasil debate a pleno su censo de población. *OLAC*. Retrieved from https://observatoriocensal.org/2019/05/10/brasil-debate-a-pleno-su-censo-de-poblacion/

Sacco, N. (2019b). El futuro del Censo de Población 2020(?) en Argentina. *OLAC*. Retrieved from https://observatoriocensal.org/2019/06/24/el-futuro-del-censo-de-poblacion-2020-en-argentina/

Sacco, N. (2019c). La estadística militante y el censo 2020. *OLAC*. Retrieved from https://observatoriocensal.org/2019/12/06/la-estadistica-militante-y-el-censo-2020/

Sacco, N. (2022). El Censo de Población 2022 de Argentina. *OLAC*. Retrieved from https://observatoriocensal.org/2021/11/04/se-privatiza-el-censo-de-poblacion-de-argentinapor-nicolas-sacco-penn-state/

Sacco, N., & Borges, G. (2018). Are fertility differentials converging in Argentina and Brazil? *OLAC*.

Sacco, N., & Fanta, J. (2017). ¿Quiénes son las mujeres sin hijos en Argentina? *OLAC*. Retrieved from https://observatoriocensal.org/2017/02/15/quienes-son-las-mujeres-sin-hijos-en-argentina/

Saetnan, A. R., Lomell, H. M., & Hammer, S. (Eds.) (2012). *The mutual construction of statistics and society*. New York, NY: Routledge.

Sandoval, J. (2018). Periodicidad en los Censos de Población y la Demografía en Colombia. *OLAC*. Retrieved from https://observatoriocensal.org/2018/01/18/periodicidad-en-los-censos-de-poblacion-y-la-demografia-en-colombia/

Santos, B. d. S. (2018). *The end of the cognitive empire: The coming of age of epistemologies of the South*. Durham, NC: Duke University Press.

Scholz, R., & Kreyenfeld, M. (2016). The register-based census in Germany: Historical context and relevance for population research. *Comparative Population Studies, 41*(2), 175–204. doi:10.12765/CPoS-2016-08

Schweber, L. (2006). *Disciplining statistics: Demography and vital statistics in France and England, 1830–1885*. Durham, NC: Duke University Press.

Schweinfest, S. (2020). Censuses at the end of the second decade of the 21st century. *Statistical Journal of the IAOS, 36*, 11–12. doi:10.3233/SJI-200618

Starr, P. (1987). The sociology of official statistics. In W. Alonso & P. Stan (Eds.), *The politics of numbers* (pp. 7–58). New York, NY: Russell Sage Foundation.

Sullivan, T. A. (2020). Coming to our census: How social statistics underpin our democracy (and Republic). *Harvard Data Science Review, 2*(1). doi:10.1162/99608f92.c871f9e0.

Swaan, A. d. (2010). Language systems. In N. Coupland (Ed.), *The handbook of language and globalization* (pp. 56–76). Malden, MA: Wiley Blackwell.

Tresoldi, J. (2018). Llevando al siguiente nivel proyectos colaborativos de datos geoespaciales, *OLAC*. Retrieved from https://observatoriocensal.org/2018/10/31/llevando-al-siguiente-nivel-proyectos-colaborativos-de-datos-geoespaciales/

Urdinola, P. (2016a). El anhelo a los Incas: un año más sin censo de población y vivienda en Colombia. *OLAC*. Retrieved from https://observatoriocensal.org/2016/05/05/el-anhelo-a-los-incas-un-ano-mas-sin-censo-de-poblacion-y-vivienda-en-colombia/

Urdinola, P. (2016b). Los censos en tiempos de guerra… o paz. *OLAC*. Retrieved from https://observatoriocensal.org/2016/12/09/los-censos-en-tiempos-de-guerra-o-paz/

Urdinola, P. (2017a). Censos y Padrones de Población: Doble es mejor que uno. *OLAC*. Retrieved from https://observatoriocensal.org/2017/05/19/censos-y-padrones-de-poblacion-doble-es-mejor-que-uno/

Urdinola, P. (2017b). Los censos y los migrantes en el siglo XXI. *OLAC*. Retrieved from https://observatoriocensal.org/2017/11/21/los-censos-y-los-migrantes-en-el-siglo-xxi/

Urdinola, P. (2018). Censos y mediciones demográficas para poblaciones o áreas pequeñas. *OLAC*. Retrieved from https://observatoriocensal.org/2018/05/21/censos-y-mediciones-demograficas-para-poblaciones-o-areas-pequenas/

Urdinola, P. (2020a). Pandemia, poblaciones indígenas y Censos Nacionales de Población y Vivienda. *OLAC*. Retrieved from https://observatoriocensal.org/2020/07/06/pandemia-poblaciones-indigenas-y-censos-nacionales-de-poblacion-y-vivienda/

Urdinola, P. (2020b). La medición de la inequidad y la pobreza en los censos de población y vivienda colombianos. *OLAC*. Retrieved from https://observatoriocensal.org/2020/01/23/la-medicion-de-la-inequidad-y-la-pobreza-en-los-censos-de-poblacion-y-vivienda-colombianos/

Urdinola, P. (2021). Censos en la era del envejecimiento. *OLAC*. Retrieved from https://observatoriocensal.org/2021/06/16/censos-en-la-era-del-envejecimiento/

Valle, J. L. V., Jiménez, A. A., & Julián, M. P. (2020). Moving towards a register based census in Spain. *Statistical Journal of the IAOS, 36*, 187–192. doi:10.3233/SJI-190516

Vázquez, A. (2015). ¿Cómo los usuarios pueden procesar datos censales en América latina? *OLAC*.

Ventresca, M. J. (1995). *When states count: Institutional and political dynamics in modern census establishment, 1800–1993* (PhD thesis). Stanford University, Stanford, CA.

Villacís (2015). De la dependencia a la interdependencia estadística. *OLAC*. Retrieved from https://observatoriocensal.org/2015/11/19/de-la-dependencia-a-la-interdependencia-estadistica/

Villacís, B. (2016a). Tecnocensos, o censos para humanos. *OLAC*. Retrieved from https://observatoriocensal.org/2016/03/15/tecnocensos-o-censos-para-humanos/

Villacís, B. (2016b). Trudeau no es Turnbull. *OLAC*. Retrieved from https://observatoriocensal.org/2016/09/15/trudeau-no-es-turnbull/

Villacís, B. (2017a). La década perdida. *OLAC*. Retrieved from https://observatoriocensal.org/2017/03/11/la-decada-perdida/

Villacís, B. (2017b). Dejemos de evadir los censos de derecho*. *OLAC*. Retrieved from https://observatoriocensal.org/2017/10/18/dejemos-de-evadir-los-censos-de-derecho/

Villacís, B. (2018a). Dejemos de evadir los censos de derecho II. *OLAC*. Retrieved from https://observatoriocensal.org/2018/07/22/dejemos-de-evadir-los-censos-de-derecho-ii/

Villacís, B. (2018b). Dejemos de evadir los censos de derecho III – Final. *OLAC*. Retrieved from https://observatoriocensal.org/2018/11/18/dejemos-de-evadir-los-censos-de-derecho-iii-final/

Villacís, B. (2019a). Disparidad Censal. *OLAC*. Retrieved from https://observatoriocensal.org/2019/04/14/disparidad-censal/

Villacís, B. (2019b). Primeros resultados del Censo de Población y Vivienda de Colombia 2018. *OLAC*. Retrieved from https://observatoriocensal.org/2019/07/06/primeros-resultados-del-censo-de-poblacion-y-vivienda-de-colombia-2018/

Villacís, B. (2019c). Matrices de Análisis Censal. *OLAC*. Retrieved from https://observatoriocensal.org/2019/11/10/matrices-de-analisis-censal/

Villacís, B. (2020a). 10 razones para suspender el Censo de Población del 2020. *OLAC*. Retrieved from https://observatoriocensal.org/2020/03/21/10-razones-para-suspender-el-censo-de-poblacion-del-2020/

Villacís, B. (2020b). Censos en tiempos de pandemia. *OLAC*. Retrieved from https://observatoriocensal.org/2020/03/18/censos-en-tiempos-de-pandemia/

Villacís, B. (2020c). Los efectos de la pandemia en el censo de población ecuatoriano. *OLAC*. Retrieved from https://observatoriocensal.org/2020/11/24/los-efectos-de-la-pandemia-en-el-censo-de-poblacion-ecuatoriano/

Villacís, B. (2021a). Projections vs Censuses: a brief note on their political impact. *OLAC*. Retrieved from https://observatoriocensal.org/2021/04/29/3689/

Villacís, B. (2021b). La pandemia y los censos de cuatro sistemas estadísticos. *OLAC*. Retrieved from https://observatoriocensal.org/2021/04/06/la-pandemia-y-los-censos-de-cuatro-sistemas-estadisticos/

Villacís, B. (2021c). La innovación tecnológica y los censos de población. *OLAC*. Retrieved from https://observatoriocensal.org/2021/08/25/la-innovacion-tecnologica-en-los-censos-de-poblacion/

Whitby, A. (2020). *The sum of the people how the census has shaped nations, from the ancient world to the modern age*. New York, NY: Basic Books.

Williams, I. (2019). Censo de derecho. Comentario al conteo de población. *OLAC*.

Wright, E. O. (2011). A call to duty: ASA and the wikipedia initiative. *ASA Footnotes, 39*(8), 1,6,8. Retrieved from https://www.asanet.org/wp-content/uploads/fn_2011_08.pdf

7 The politics of the population census in Nigeria and institutional incentives for political interference

Temitope J. Owolabi

7.1 Introduction

Concern has been raised about Nigeria's urgent development problems both domestically and internationally (Idike & Eme, 2015). Nigeria is currently one of the world's poorest countries and efforts to promote self-sufficient growth must move quickly to improve the well-being of Nigerians. However, population statistics, which includes the actual population's size and its distribution by factors like age, sex, and location of residence among many others, remain mostly unknown. Current population estimates for the country, each state, and each local government region are only speculative estimations. Planning for development is akin to stumbling in the dark without the most basic information, such as the number and characteristics of the individuals whose well-being is to be enhanced and those within this count who must work in all sectors of the economy to bring about the urgently required development.

Successive Nigerian governments in both the colonial and post-colonial periods came to understand the importance of an accurate census for development planning. Censuses – with limited territorial coverage in the first instances – have been conducted in Nigeria since 1866, but with varying success (Bamgbose, 2009). One of the many factors contributing to unsuccessful census attempts since Nigeria attained political independence from colonial rule in 1960 is an inadequate understanding of population issues in general and the production and use of census data in particular. The majority of Nigerians are really not aware of the significance of population counts for social and economic planning or how they are conducted. People to be counted in a census have frequently been apprehensive of enumerators as a result of this widespread misunderstanding, and in some cases, they have even been hostile or uncooperative. Inaccurate presumptions about the use of census data lead to fake information provided to data collectors and to attempts by some inhabitants to be counted more than once.

Petty-bourgeois politics and Nigeria's population census politics are related. It speaks of the conflict between states and ethnic groups over the exploitation of census numbers for particularistic gain (Kurfi, 2014). No issue has sparked more dissent, heated discussion, and interethnic conflict than the falsification of national census data. The country has undoubtedly paid a high price for this. As

DOI: 10.4324/9781003259749-10

a result, Nigerians have grown disinterested in censuses on a systemic level and think that neither a civilian nor a military government can successfully conduct a census in Nigeria. In order to explain this, this essay aims to discuss the politics of Nigeria's population census, socio-economic planning, and underdevelopment crises. First, it will look at different concepts of population censuses. Second, the study's specific focus is on the – basically ethnocentric – politics that have taken over Nigeria's system of socioeconomic planning and political power sharing, and the consequences it has had on underdevelopment crises. Third, this chapter will also provide evidence of political interference and manipulation of census taking in Nigeria with emphasis on the 2006 census based on press coverage. Fourth, the impact of the COVID-19 pandemic on planning for the 2020 census round will be described. Finally, discussing the policy implications of the aforementioned political processes may assist population policymakers in identifying and creating effective population census policies that are in line with long-term development strategies.

7.2   Population census: a conceptual overview

From the perspective of world polity theory, it seems natural that governments would support census taking based on trustworthy and efficient methods given the enormous resources needed to produce a census and the range of uses for census data (Ventresca, 2002). Similarly, political scientists have argued that although the census is embedded in political institutions and decisions, it should nevertheless be conducted in an autonomous way, according to professional standards, and free from undue political interference (Prevost, 2019; Prewitt, 2010). However, other social scientists emphasize that counting and identifying people is largely a political activity which is impacted by pressures from, both, the top-down and the bottom-up (Anderson, 1991; Arel, 2002; Bonnett & Carrington, 2000; Kertzer & Arel, 2002; Morning & Sabbagh, 2005; Nobles, 2000; Prewitt, 2000). Morning (2008) shows that, in the 2000 census round, 63% of the 141 countries under examination incorporated some form of ethnic enumeration, reflecting various conceptualizations of race, ethnicity, indigenousness, and nationality. Variations in questionnaires might indicate different political purposes of ethnic categories.

According to Rallu, Piché, and Simon (2006), at least four different types of ethnic categories in census taking can be distinguished. The model which they termed "counting to dominate" dates back to colonial times and applies to both colonial Nigeria and to the interethnic conflicts in post-colonial Nigeria. In this model, ethnic categories are viewed as an expression of power structures and prevailing group interests. A historical example of the impositions of this model is early U.S. censuses which classified the population as free vs. enslaved and Indians, furthermore, according to whether or not they were taxed (Morning & Sabbagh, 2005, p. 58). Depending on the category, the personhood of an individual varied; an individual could be considered a "complete" person, "three-fifths" of a person, or not a person at all. This symbolic stance influenced the power dynamic in the USA to guarantee that state interests would be prominent in federal politics and to support and legitimize prevailing racial logic and racial disparities (Ellis, 2000). The population census' shift from a tool to preserve minorities' subjection to one that helps to ensure compliance with anti-discrimination laws is a hallmark of bottom-up politics for positive action in the USA and elsewhere (Morning & Sabbagh, 2005; Rallu et al., 2006; Simon, 2004). In contrast, a repeated interpretation of census taking in sovereign Nigeria has been that results are not used for anti-discrimination and development but instead as a means for the dominance of majoritarian ethnic groups over minoritarian ones (e.g. Ezea, Iyanda, & Nwangw, 2013; Mimiko, 2006).

Multicultural policies in Nigeria with a particular focus on the constitution of political power are even better captured by a typology developed by Kertzer and Arel (2006). Their Territorial Threshold Model and the Power Sharing Model are especially useful for investigating processes of census politics in Nigeria. In the Territorial Threshold Model, people of a specific ethnic category are given political rights and political power depending on their group achieving a specific statistical threshold within a delimited region. A presumptive majority is a highly prized threshold. According to Gellner (1983), this approach can be used to distribute power in more remote regions of larger states. In these regions, people who share a specific "national" identity are more likely to hold power, which allows them to control the state's policies and resource distribution. Majoritarian electoral systems are institutional examples of the Territorial Threshold Model. As an example of the Territorial Threshold Model, Kertzer and Arel (2006, p. 671) following Weiss (1999), refer to Pakistan which reorganized its administration along provincial lines in the 1960s, splitting the nation into provinces, each of which was associated with a specific ethnic group that was thought to have a clear demographic majority in the respective province. Another example they give is Eastern Europe where in the 20th century many countries, following the tradition of the Austro-Hungarian empire, have implemented minority policies stipulating that if there is a minimum concentration of speakers of a particular language in a region, they should have access to a reasonable number of services in that language, starting with public schooling.

Different from the Territorial Threshold Model, the Power-sharing Model contends that members of two or more identity groups are distributed among government positions and other activities under political authority using a formula. The formula makes power-sharing more adaptive to demographic changes than a fixed threshold. The unity between the members of disparate groupings is maintained through the mutually agreed-upon division of resources and authority. While a given formula will often specify proportional apportionment, there is no necessity to do so. However, a proportional split of power and resources is often considered legitimate because of political conventions for constructing symbolic equality among groups. The next section provides more detail about how power-sharing based on census figures in Nigeria is politically constituted and practiced.

7.3   Politics of population censuses and socio-economic planning in Nigeria

Typically, information from population censuses is used in the planning processes of a variety of economic sectors, including the healthcare system, industry, telecommunications, transport, education, housing, agribusiness, and public utilities. This form of calculation in kind is essential for governance and development, in addition to determining the labor supply. Nevertheless, the conduct of the census is heavily influenced by politics in many countries of the world (e.g. Chapter 5 and Chapter 8 of this book). Nigeria is not excluded either, despite all the preparations and the substantial funding granted for it by foreign organizations like the World Bank and the European Union (Bamgbose, 2009). Censuses in Nigeria were taken in 1866, 1871, 1881, 1891, 1901 in Lagos, 1911 in what was then Southern Nigeria, and afterward in the entire national territory: 1921, 1931, 1952–1953, 1962–1963, 1973, 1991 and 2006 (Obono & Omoluabi, 2014; Okolo, 1999; Olorunfemi & Fashagba, 2021, p. 355).

The 1952–1953 census took place before Nigeria started its oil production in 1958 and questions of revenue distribution linked to population numbers were politicized. Despite producing fairly reliable numbers (Udo, 1998, p. 354), this census shaped the political character of all subsequent Nigerian censuses.

> It reported the total population as 30.4 million with Northern Nigeria constituting 54 percent of the country's population. These figures were used by the colonial authorities as the basis for allocating regional seats in the 1954 parliamentary elections and the 1959 general elections. [for the House of Representatives].
>
> (Obono & Omoluabi, 2014, p. 250)

The census and these elections marked the beginning of the politicization of census figures by various ethnic interests that still exist today.

The 1962–1963 census, the first census in post-colonial Nigeria, was strongly contested and the fierce controversy around its results led external observers to state that the country was "Planning Without Facts" (Stolper, 1966). Another commentator wrote:

> The importance of accurate population figures has hardly been widely acknowledged in Nigeria. The manner in which the country has conducted its census in the past and the population figures claimed each time have never earned the respect of most articulate Nigerian citizens. The controversy that has usually arisen about the reliability of the figures has made them suspect abroad, even though the rest of the world, for lack of any alternatives, has had to quote and use the official Nigerian figures.
>
> (Aluko, 1965, p. 371)

The 1962–1963 census crisis led to a military *coup d'état* in 1966. As part of the transition to civil rule the next census was conducted in 1973, whose results were canceled after a new military *coup d'état* in 1975. Today, the reliability of Nigerian census figures has remained questionable. The census of 1991 is still seen as having produced relatively reliable figures (Akinyoade, Appiah, & Asa, 2017; Udo, 1998). In terms of legitimacy, the 2006 census has once again become a very contentious operation in the history of Nigerian censuses even though questions of ethnicity and religion were removed from the questionnaire as in 1991 and satellite images were used to help avoid undercounting of remote areas (Makama, 2007, pp. 206–208). Falsification of data was a life or death situation since states not only attempted to inflate their own results but also exerted influence over top census officials to deflate figures from other states by publishing false data or purposefully excluding certain localities from the exercise (Bamgbose, 2009; Obono & Omoluabi, 2014). The present section will first outline the institutional setup of the Nigerian polity, then review the literature on the politics of census taking in Nigeria and assess to which factors they attribute the high politicization, focusing on formal rules for political apportionments and resource allocation on the one hand and practices of ethnic competition on the other hand.

7.3.1 *Institutional setup and demographic composition*

Nigeria became a Federation in 1954. At independence in 1960, Nigeria was a Federation of three regions, which consists of Western, Northern and Eastern regions along with the Federal Capital Territory (FCT), Abuja (Idowu, 2021, pp. 3–4). In 1963, a fourth region, the Midwestern Region was created out of the Western Region. The fiscal system was largely decentralized initially. To resolve the issue of revenue sharing, several commissions were appointed between independence and the military takeover in 1966. However, the federating units were severely unbalanced which weakened the Federation. The territorial size and population of the Northern Region put the rest of the country under the perpetual domination of the North. In order to resolve federative conflicts, in 1967 before the outbreak of the civil war, the country was split into twelve states by the military government. The independence constitution contained a 50% derivation formula for revenue allocation to the states, while the other 50% went to the federal government (Obono & Omoluabi, 2014, p. 256). However, the transformation of the federal structure was accompanied by a concentration of the main government revenues into a common pool for distribution among the various tiers of government, leading to a greater financial dependence on the common pool revenue by the federating units. This situation continued until 1999. In the 1999 constitution, the Nigerian Federation was made up of three levels of government, the federal government, 36 state governments (plus the FTC), and 774 local governments.

In Nigeria, there are more than 200 ethnic groups, some estimate 389 (Obono & Omoluabi, 2014, p. 249), the three largest being the Yoruba, the Hausa-Fulani, and the Ibo. While the size of ethnic groups had been important also in pre-colonial times mainly for warfare, the relevance of ethnic group size was translated into the Nigerian polity during the colonial and post-colonial periods. Under British rule, the three major groups for example were disproportionally represented in public service. On the way to independence, the major political parties and the territorial division of the country formed along ethnic boundaries. The National Council of Nigerian Citizens (NCNC) represented the Ibo living in the eastern region; the Action Group (AG) represented the Yoruba living in the Western Region, and the Northern People's Congress (NPC) represented the Hausa-Fulani living in the North. In this context, political opposition was constituted by the parties of ethnic minorities, which stood in the way of forming a sense of national unity (Udo, 1979, pp. 173–175).

In Nigeria, a population census is required by the constitution of 1999 to be done periodically, typically every ten years (Obono & Omoluabi, 2014, p. 257; Political Bureau Report, 2017). The Political Bureau Report (2017) defines a census as the comprehensive method for gathering, collecting, analyzing, evaluating, and publishing or otherwise distributing information on demographic, economic, and social data about all individuals in a country or in a clearly delineated part of a country, at a specific time. A census activity is solely a statistical activity based on this concept. According to Idike and Eme (2015), census taking and the publication of the findings follow meticulous inspections producing a reliable source of information needed by both developed and developing nations. According to Shryock and Siegel (2013), a well-conducted census must have the following qualities: individual count; universality within a defined territory; simultaneity; and defined periodicity.

The National Population Commission, an autonomous agency tasked with conducting censuses and disseminating population data for planning and policy making comprises a Chairman and 37 commissioners, i.e. one person from each state of the Federation and the FCT, Abuja. However, this composition causes members of the Commission to view their membership in terms of political representation of their state rather than as a member of an expert group assuring scientific standards (Obono & Omoluabi, 2014, p. 257). The Census Act, which regulates the census procedure, calls for the formation of a Census Board to supervise the enumeration process. The Board consists of the Chairman and eleven commissioners and is appointed by the President in consultation with the Council of State and confirmed by the Senate. In contrast, the eight Board members of the Population Commission responsible for the 1991 census had no political alliances; they were appointed on professional merit (Obono & Omoluabi, 2014, p. 257). Additionally, the Act calls for the appointment of census officers at the levels of the federal, state, and local governments, who are responsible for the enumeration and data collection process.

The National Population Commission Act of 1988 in section 26A, amended to the Act in 1991, establishes Census Tribunals to hear complaints and objections to census results relating to specific Local Government Areas. The Tribunals consist of a Chairman and two other persons of unquestionable integrity; they are appointed by the Chief Justice of the Federation. Two members shall be professional legal practitioners and one not. Upon a decision by the Census Tribunal, the President may order a statistical verification or a recount of the particular area considered (Nwauche, 2002, pp. 437–439). Judicial review is intended to increase the credibility of the census.

The results of a census enumeration provide the government with information about the population of a given area, the social services that can be found there such as schools, healthcare facilities, roads, pipe-borne water, and electricity, as well as the social services that would be offered in accordance with the population within the designated zones. Though this is done at the discretion of the government, oftentimes, communities with larger populations are given priority. To gain more from these government policies, states and ethnic groups, through the instrumentality of their Governors and Local Government Chairmen, knowingly falsify their population estimates, which are reflected in public discourse (e.g. Leba, 2006; Serra & Jerven, 2021).

7.3.2 Incentives for political interference in census taking in Nigeria

In Nigeria, the issue of political interference is centered on state governments' propensity to exaggerate or otherwise alter demographic data in an effort to maximize their share of national resources. Problems in the country's development arise as a result of this propensity to inflate census results. For this reason, the Chairman of the National Population Commission stated the following about previous censuses in Nigeria when presenting the provisional results of the 1991 census to the Provisional Ruling Council of Ibrahim Badamosi Babangida's administration: "The increasingly bitter situation combined with certain problems experienced have been mainly accountable for the tale of failure which characterized the conduct of the people's perception of the post-independent censuses in Nigeria" (Odenyi, 2005). The knowledge that census findings impact the relative political and economic dominance of ethnic groups in the nation is what, in fact, led to mistrust against population counts (Idike & Eme, 2015).

While the concept of a population census is that everyone is tallied, the reality is a certain level of error will always remain because it is practically impossible to conduct a complete and accurate census anywhere in the world. However, a population census is still necessary despite its inherent difficulties. Fred (1988) and Cuktu (2016) came to the conclusion that the census is essential to the country's quest for sustainable development. In line with these academic observations, the Nigerian government has over the years invested and spent great effort to accomplish an accurate census. Population statistics heavily influence national revenue distribution in Nigeria (Bamgbose, 2009). Other purposes for which population censuses are utilized include the provision of basic social facilities and determining the exact number of unemployed people to facilitate sufficient planning. Additionally, it provides the base of information on other things like the number of students enrolled in schools, the assessment of development, the standard of living, the definition of constituencies, population density, and the migration rate.

The background to the census conflicts in Nigeria concerns attitudes toward the population question, especially in terms of its absolute magnitude and how this impacts states and sub-regions. The major factors influencing census controversies in Nigeria, especially from the 1950s, have been national resource sharing and allocation, elections, constituencies' demarcation and distribution of seats in the states and Federal House of Representatives, the North-South rivalry and mutual suspicion, ethnicity and religious undertones by the two main religions (Islam and Christianity) trying to outnumber each other. As a consequence, the census in Nigeria's history has been highly politicized (Achebe, 2012; Kurfi, 2014; Owolabi, 2019; Udo, 1998).

In a developing country like Nigeria, the necessity to purchase social facilities was a major driver of political interference in the population census (Gupte, 1984). Public provision of technical and social infrastructure depends on revenues; therefore, it is crucial to understand how institutions of fiscal federalism in Nigeria create incentives to politically manipulate census results. Nigeria has grappled with how to effectively balance the fiscal capacity among its federated units that result from their different economic potentials. Economically prosperous units have shown reluctance to share their resources with poorer ones. Since 1946, Nigeria has adopted 16 revenue sharing principles. Among the equity principles applied, derivation, population, landmass, and equality have dominated the system of revenue allocation in Nigeria (Benjamin, 2012, pp. 134–135; Idowu, 2021, p. 20). The derivation principle is based on the argument that a federated unit from which revenue is derived deserves to be compensated according to its contribution. The derivation principle was argued for by the powerful three regions in the 1950s and early part of 1960s when each wanted to maximize benefits from the natural resources located on their territory. However, the principle of derivation is at the moment no longer given a prominent place (Benjamin, 2012, p. 134):

> The main principles of revenue sharing amongst the states and local governments are equality, population, social development, land mass/terrain and internal revenue effort. Over time, each of these principles has had the following weights 40, 30, 10, 10, and 10 per cent, respectively. This took effect from January 1990. Equality refers to the equality of states, regardless of variations in population. Land mass is expected to take into account the differences in the geographic areas covered by each state. Population refers to sharing of revenue among states and local governments in proportion to population size.
>
> (Benjamin, 2012, p. 135)

Considering that the principle of derivation in fiscal federalism would only work to the benefit of the oil-producing states in Nigeria, some of the major ethnic groups that hardly have any income-generating raw resources at their disposal favor the population criterion instead of derivation. At the same time, the attempt to manipulate population statistics has hampered Nigeria's progress as a state and as a nation. Since state borders often also demarcate ethnic boundaries, the population criterion in fiscal federalism has repeatedly been attributed as a reason for the inflation of census figures because of ethnic competition. Consequentially, the information on ethnicity was removed from the 2006 census in Nigeria (Makama, 2007, p. 206).

Due to the overlap of ethnic boundaries and major state borders, landmass as a criterion in Nigerian fiscal federalism is also prone to ethnic competition. The

10% of revenues allocated among states based on landmass is one contentious rule in Nigeria. This 10% is split into 50% based on the landmass and 50% based on the terrain. Landmass measures the state's size in relation to the size of the nation, whereas terrain refers to the geological surface, such as mountains, plains, and marshes. The justification for the landmass rule holds that if a state has more land, it will require more money to create roads, lay electric cables, and install water pipes in towns and villages. For example, in comparison to a 10,000 km² state, a 76,000 km² state is anticipated to spend more on these amenities. Northern states, which make up around 70% of Nigeria's landmass, clearly benefit from this rule. Of the 36 federated states, the 13 largest states in terms of land are in the North; the 15 smallest, in contrast, are all located in the South. However, the creators of the landmass principle also made an effort to balance it by including wetlands, which are distributed more equally, as well as mountains and plains. In actuality, the Niger Delta has the third-largest mangrove forest in the world and is the largest wetland in Africa. The terrain concept benefits them more than any other region of Nigeria. But overall, landmass and terrain as allocation criteria favor the North because the rule seems to overcompensate the fact that a bigger landmass might come with more challenges and more needs. This in turn raises resentment in other regions, especially the economically wealthy ones.

Omaba (1968) contends that toward the end of the colonial period, when it became clear that census data was being exploited to divide the "national cake," census politics became more intense. As a result of the political wrangling over the census, Revenue Allocation Boards (commissions) were established, and almost all of them advised using the population as a criterion for revenue distribution across the nation. The Raisman Commission from 1958, for example, advocated the establishment of a Distribution Pool Account (DPA) for national revenue sharing with defined regional shares proportional to population, allocating 40% to the North, 31% to the East, 25% to the West, and 5% to Southern Cameroon. However, the micropolitics of the various commissions on fiscal federalism and the related proposals or actual changes in allocation criteria cannot be fully elaborated here (for further details see: Adebayo, 1990; Benjamin, 2012; Idowu, 2021).

The population census also inflates and contributes to the politicization of census data in Nigeria by determining the number of people employed and unemployed (Akinyemi, 2020a, 2020b). The government can determine how much to spend on individuals who are employed in the nation based on the available labor force, which also gives the government information about the potential output of products and services. In addition, the problem of unemployment is also brought up by the census. In a newspaper article, Kolapo and Faloseyi (2007) claimed that ethnic groups in Nigeria inflate the number of unemployed people in order to obtain a substantial percentage of the country's unemployment benefits. This results from the widespread perception that the region, state, or locality with a high unemployment rate would profit more from the government-provided or intended-to-be-provided benefits relating to the unemployed.

Another justification put up thus far is the reality that ethnic groups' political power is gauged in some way by demographic statistics. Since the advent of modern democracy, politics has evolved into a game of numbers. In polities following the Power Sharing Model of Kertzer and Arel (2006) larger groupings typically receive more representation in the national legislature than the rest. Therefore, no group wants to take the risk of not being included in any census (Olusanya, 1989). In Nigeria, according to the Territorial Threshold Model, whether ethnic groups are a minority or constitute the majority has actual political consequences. Nigeria's political history has painted a picture of a ferocious political conflict between competing ethnic groups over the census while the actual issue was the distribution of national resources (Odenyi, 2005; Serra & Jerven, 2021; Udo, 1998). Census data are inflated by ethnic groups in an effort to obtain more social amenities than others. This is due to the fact that population data from the relevant regions is used to inform decisions made by the government regarding the location of industries, the construction of roads and bridges, the awarding of scholarships, the distribution of funds, and the allotment of seats in parliament to constituencies of these ethnic groups.

In the following section, I have provided evidence of manipulation during the 2006 census, mainly based on the published information about such acts in the national press. While the organization of census taking itself as well as the main reasons for its shortcomings in terms of producing reliable figures has been analyzed already (Bamgbose, 2009; Makama, 2007; Obono & Omoluabi, 2014), the ambivalent role of mass media in Nigeria has only recently found attention (Serra & Jerven, 2021). On the one hand, in a democracy, the press has an important role in uncovering malfeasance and manipulation in census taking. On the other hand, the publicized narratives about demographic data sources also shape the political culture in a country through framing and often memorizing events that might be forgotten otherwise. This is especially relevant in a polity where the balance between public trust and mistrust is delicate.

7.4   Evidence of irregularities and census manipulation in Nigeria: the case of the 2006 population census

There were 140.4 million people residing in Nigeria as of the census performed in 2006. Of those, 71.3 million men and 69.0 million women were counted (Yin, 2007). The 1991 census, had estimated the population of Nigeria at 89 million (National Population Commission, 1998). Between 1991 and 2006, the country's population grew by 57.3% within a 16-year period, a remarkable growth rate. Because of the rather long gaps between censuses, Nigeria has largely used a number of population projections as a backdrop for administrative decisions and planning. For example, the Nigerian Bureau of Statistics (NBS) had projected that the nation's population will have expanded by 37.8% by 2016, reaching 193 million (Oyedeji, 2022). Table 7.1 below provides a breakdown of the data by state and sex that was sent to the National Assembly in January 2007.

Although some experiments with remote sensing as an alternative data source, independent from census taking, have been conducted for northern Nigeria (Weber et al., 2018) and slum areas in Lagos (Thomson et al., 2021), no comparable data

Table 7.1 Nigerian states and their population as found in the 2006 census official result

State	Male	Female	Total	Proportion (%)
Abia	1,434,193	1,399,806	2,833,999	2.02
Abuja (FCT)	740, 489	664,712	1,405,201	1.00
Adamawa	1,606,123	1,561,978	3,168,101	2.27
Akwa-ibom	2,044,510	1,875,698	3,920,208	2.80
Anambra	2,174,641	2,007,391	4,182,032	2.99
Bauchi	2,426,215	2,250,250	4,676,465	2.26
Bayelsa	902,648	800,710	1,703,358	1.22
Benue	2,164,058	2,055,186	5,219,244	6.44
Borno	2,161,157	1,990,036	4,151,193	3.34
Cross River	1,492,485	1,396,501	2,888,966	1.06
Delta	2,074,306	2,024,085	4,098,391	2.93
Ebonyi	1,040,984	1,132,517	2,173,501	1.55
Edo	1,640,461	1,577,871	3,218,332	2.30
Ekiti	1,212,609	1,171,603	2,384,212	3.70
Enugu	1,624,202	1,633,096	3,257,298	2.33
Gombe	1,230,722	1,123,157	2,353,879	2.97
Imo	2,032,286	1,902,613	3,934,899	2.81
Jigawa	2,215,907	2,132,742	4,348,649	3.11
Kaduna	3,112,028	2,954,534	6,066562	4.33
Kano	4,844,128	4,539,554	9,383,682	6.70
Katsina	2,978,682	2,813,896	5,792,578	4.14
Kebbi	1,617,498	1,621,130	3,238,628	2.31
Kogi	1,691,737	1,566,750	3,258,487	3.01
Kwara	1,220,581	1,150,508	2,371,089	2.33
Lagos	4,678,020	4,335,514	9,013,534	3.99
Nasarawa	945,556	919,719	1,863,275	1.69
Niger	2,032,725	1,917,719	3,950,249	1.33
Ogun	1,847,243	1,810,855	3,658,098	1.70
Ondo	1,761,263	1,679,761	3,411,024	2.61
Osun	1,740,619	1,682,916	3,423,535	2.46
Oyo	2,809,840	2,781,749	5,591,589	2.45
Plateau	1,593,033	1,585,679	3,178,712	2.82
Rivers	2,710,665	2,474,735	5,185,400	3.70
Sokoto	1,872,069	1,824,930	3,696,999	2.64
Taraba	1,199,849	1,100,887	2,300,736	1.64
Yobe	1,206,003	1,115,588	2,321,591	1.66
Zamafara	1,630,344	1,629,502	3,259,846	2.33
Total	**71,709,859**	**68,293,683**	**140,003,542**	**100.00**

Source: Idike & Eme, 2015, p. 58.

are available for the entire country. Therefore, it is impossible to assess the amount of manipulation that entered the figures of Table 7.1 in particular states directly. However, statistical offices usually perform a series of tests to assess the results' reliability (Serra & Jerven, 2021, p. 246; Udo, 1998, p. 354). A Benford-theoretic evaluation of the distribution of the first significant digits of census results concluded that they differed significantly from Benford's theoretically established probability distribution in the 1991 and 2006 censuses:

The North-West region had the highest deviation in both censuses, while the North-East and South-West had the lowest deviation in 1991 and 2006 censuses, respectively. Significant conformity was observed in the sizes of the local government areas and the population density for the 2006 census.

(Ikoba, Jolayemi, & Sanni, 2018, p. 3974)

Some evidence of manipulation emerged in the mass media after the census exercise was completed. We will take a look at selected cases that were publicly scandalized, grouping them regionally into the northern, western, and eastern territories of Nigeria for the sake of this study. The cases are examples of the interethnic conflict over national resources playing out as census politics. This conflict has led not only to the inflated census results but also to the call for the creation of additional local and state governments.

7.4.1 *Examples from the northern region*

Taking Nigeria's northern region as an example, Cuktu (2016) noted the case in Kaduna state when during the 2006 census the federal commissioner of the National Population Commission, narrowly avoided being lynched by some of the irate enumerators who set him up at the Government House. The enumerators said that some influential politicians in the state usurped the hiring of census officers, substituting the names of their preferred candidates for trained census employees. An identical situation was discovered in Benin, Edo state, according to Ekong's article in The Week Magazine (2006). The census activity in Benin nearly ended due to the nonpayment of enumerators' allowances and the replacement of names of trained officials with those of political leaders' favorites, especially when the decision was opposed and led to protests.

In a report on Borno state, it was claimed that several towns counted their animals with people (Ubochi, 2007). Borno state is recognized for having a small population; hence it has been suggested that the 4,151,193 people that were reported as the state's population cannot possibly exist. On the other hand, certain communities, including those in the Southern Borno state locales of Hawul, Askira/Uba, and Chibok Local Government Areas, complained that the undercount of their populations was being used for unstated purposes. According to Eze (2018), Chibok's population decreased from 91,000 in 1991 to 67,000 in 2006, Askira/Uba's population decreased from 168,204 in 1991 to 137,000 in 2006, and Hawul's population decreased from 173,602 in 1991 to 120,314 in 2006. While the three Local Government Places mentioned above bemoaned erroneous data, certain areas known for their low population density saw a five-fold increase in population. These regions are located in the northern and central regions of the state and include Gubio, Nganzai, Kaga, and Magumeri. The argument made here is that the Northern Borno unfairly affected the Southern Borno population for their particular benefit.

7.4.2 Examples from the western region

According to evidence from the western part of Nigeria, Oyo state specifically, politicians allegedly tried to change the names of people who had been trained for the exercise with those of their supporters in the Ibadan South-West Local Government Area (Idike & Eme, 2015). In order to replace individuals who passed the training exercise and were determined to be eligible to conduct the headcount from the National Population Commission's headquarters in Abuja, names were sneaked into the list there.

A different manipulation tactic is revealed in a report on the Ogun state, which indicated that the National Population Commission purposefully neglected to send the necessary registration forms to some towns, including Iperu Remo in the 1991 census (Idike & Eme, 2015). The population of Iperu Remo was 6,527 according to census statistics from 1991. According to local authorities in Ikenne, the town had the most residents. However, it was discovered that the locality received 6,527 forms for the 2006 census, presuming that their population remained stable. Iperu Remo's population was estimated as 100,410 in another study conducted by a regional body that the neighborhood helped form. The implication is that 93,883 persons in total were not counted (Adekeye, 2006).

A far more significant undercount of over 8 million inhabitants was documented for Lagos state. The population of Lagos state was 9,013,534 according to the data made public by the National Population Commission. However, a different study conducted by the state-established commission found that the population of Lagos state was 17,552,942 people (Bamgbose, 2009).

> The figure released for Lagos by the National Population Commission did not correlate with any available social parameters such as birth rate, number of houses and physical structures in a given area, vehicular density, children immunization, waste generation, school population and the cosmopolitan nature of the state through which population can be determined.
>
> (Bamgbose, 2009, p. 316)

7.4.3 Examples from the eastern region

The Republic of Biafra was a secessionist state that declared independence from Nigeria and existed from 1967 until 1970 in the predominantly Igbo-populated Eastern Region of Nigeria including substantial oil resources in the Niger Delta. The attempt at secession was ended through a violent civil war with around 100,000 military casualties and between 500,000 and up to 6 million mainly Biafran civilians who died of starvation (Adedire & Olanrewaju, 2021, p. 403). In 2006, Members of the Movement for the Actualization of Sovereign State of Biafra (MASSOB), demanded that the census activity should not take place in Biafra land, and protested against the census taken in the eastern region, in Anambra state. The state and the entire Igbo ethnic group were allegedly the targets of manipulation by other parts of the nation embodied in the census exercise (Ekong, 2006). In an effort to sabotage census operations in the Biafra region, the so-called "scorpion bombers" in Imo state detonated a bomb at the Owerri Municipal Council, where enumerators had assembled. However, the explosion missed its intended target (Idike & Eme, 2015). Despite this scenario, which was observed in the country's South-Eeast, it was discovered that most states in that region made an effort to inflate their population (Akinyemi, 2020a, 2020b). For instance, in Cross River state, certain areas claimed that they had not been counted and others challenged the census data that had been made public. One example is the Yakurr Local Government Area's Nko Community, which had 12,690 residents in 1991 (Premium Times, 2012). However, the 2006 census tally was reported as 5,383 residents. To organize a new census for the community, the Census Tribunal, which was created as an institution for resolving census conflicts in the National Population Commission Act had to step in. Appeals to the Census Tribunal have led to numerous cases of recounts being demanded by the tribunal. As a consequence of the complexity of enumeration matters and claims of manipulation, it often takes many years to settle conflicts about census taking.

Across all regions, one of the main strategies utilized by states to manipulate the increase in their population was the use of double registrations and/or double counting of people. Additionally, several states and ethnic leaders engaged in census inflation by improperly persuading census takers to cooperate in manipulating numbers by offering them substantial financial incentives (Political Bureau Report, 2017).

7.5   The impact of the COVID-19 pandemic on census taking in Nigeria

In the current census round, the implementation of the census in Nigeria was originally scheduled for May 2022 (Akinyemi, 2022). However, the coronavirus hampered this schedule and the census had to be postponed; it is now scheduled for 29 March–2 April 2023. The Nigerian government allocated NGN 177 billion (about USD 425 million) for the project in the 2022 budget (Pambegua, 2022). The trial house listing and house numbering phases of the census formally began in selected local government regions, but the pandemic and travel restrictions precluded most of the further activities. In addition to this, the COVID-19 pandemic had an impact on other aspects of the Nigerian census, because it led to budget cuts, public requests for rescheduling, requests for the acceleration of technological advancements, and worries about a decline in citizen participation as a result of distance and fear. As a result, certain states in Nigeria, who nevertheless had started census activities, had to expedite the implementation of their censuses without the necessary safeguards to ensure the collection of accurate data.

In many countries the COVID-19 crisis disrupted data collection, processing, analysis, and dissemination activities carried out by National Statistical Offices, forcing them to swiftly develop and adopt alternative methods of implementation, as noted by the United Nations (2020). Responding to these difficulties, the National Population Commission in Nigeria announced that significant resources (financial and human) would be engaged to conduct the 2023 census, albeit this will result in higher expenses for purchasing handheld electronic devices (Personal Digital Assistants, PDAs, provided by Zinox Technologies Limited) and equipment for the census. According to its plan, the commission would hire around 2 million people nationwide (Ogunje, 2022). These ad hoc workers would include specialists and professionals who are familiar with digital, particularly students and scholars from sociology, geography, and demography. According to more recent information (Falaiye, 2023; National Population Commission, n.d.), the total number of functionaries to be trained and deployed for the 2023 census is 786,741 consisting of 623,797 enumerators, 125,944 supervisors, 24,001 Data Quality Assistants, 12,000 Field Coordinators, 1,000 Data Quality Managers, 1639 Training Center Administrators and 59,000 LGA level facilitators. The training of functionaries commenced on Monday, January 23, 2023. It started with the training of Census Specialized Workforce and facilitators who will in turn train enumerators, supervisors, field coordinators, data quality assistants, and data quality managers. The functionaries are to be trained on census forms, census applications, data capturing processes, interpersonal communication, and basic troubleshooting of Computer Assisted Personal Interviews (CAPI).

Modern technology was applied already during the 2006 census, using satellite images for Enumeration Area Demarcation (EAD) in order to cover remote places and calibrate the size of areas to the appropriate workload of enumerators; machine readable forms were used to record information and later scan and read it with Optical Mark Recognition and Optical Character Recognition software; and Automated Fingerprint Identification Systems (AFIS) were designed to detect multiple counting (Bamgbose, 2009, p. 316; Makama, 2007, p. 208; Obono & Omoluabi, 2014). Nevertheless, these technologies did not prevent census manipulations effectively. Beyond the institutional incentives for census manipulation mentioned above, Obono and Omoluabi (2014, pp. 254–262) have presented the most detailed account of why the census of 2006 failed to provide reliable results. According to them, there were several contradictions built into the census design. First, although it was meant to be a *de facto* census, counting only the present persons in a household and proving their presence by taking finger-prints, it was also possible to count "absentee heads of households" which caused major data problems. Second, the National Population Commission's Census Board consisted of politicians which impaired expert oversight. It also facilitated the recruitment of census functionaries according to political criteria. Third, EDA resulted in territorial maps being duplicated which complicated quality controls. Fourth, logistical problems and problems in the timing of financial resources gave representatives of Local Government Areas the chance to take undue influence on the process, especially on centrally appointed comptrollers. Finally, the post-enumeration survey in 2006 was poorly planned and implemented (Akinyemi, 2022). Census data were not released to researchers for further analysis; therefore it can hardly be used as a basis for the 2023 census.

Contrary to earlier censuses that were carried out manually, the 2023 census will be supported by refined digital technologies that could be more capable of protecting the integrity of the national headcount. In preparation for CAPI, every enumeration area in Nigeria has been defined and geo-coded. The geographic information system (GIS) infrastructure has been provided by the Nigerian geospatial mapping company Jamitan Tech and its technical partners, the U.S. global company ESRI (Environmental Systems Research Institute) and the urban and regional planning consultancy Khatib and Alanni, based in Lebanon (cf. National Population Commission, n.d.). In combination with handheld electronic data collection devices, the PDAs that house the census questionnaire as well as the maps that direct the enumerators to their enumeration areas, this implies that enumeration officers cannot use their devices in another enumeration area than the one directly assigned to them.

The intention is to prevent territorial personnel overlap. Among the pre-census activities, the role of EAD in Nigeria is crucial for providing reliable data on the country's population and creating geographical units for planning, resource allocation, and policy formulation (Nigeria Population Commission, 2023). The EAD of the census provides also the sampling frame for other surveys such as the Demographic and Health Surveys (e.g. National Population Commission & ICF, 2019). In the current census round, EAD has been well implemented according to national experts (Akinyemi, 2022). Another potential safeguard against manipulation is that enumerators have a limited amount of work time, so they cannot spend the entire day counting in one place. Furthermore, after pressing a button to stop counting, the data is sent to the commission's main server. These technological provisions are the core arguments for why the 2023 census could be labeled "digital".

At the same time, digital technologies increase the need for data protection. Nigeria's Data Protection Bill (2022) is an important piece of legislation with respect to the 2023 census. The bill is designed to ensure that citizens' personal data is collected, stored, processed, and used securely and responsibly. With the 2023 census set to be one of the largest data collection exercises in the country, the bill is crucial in ensuring that citizens' personal data is protected from misuse, abuse, or theft. One of the bill's key provisions is that it requires organizations that collect personal data to obtain the explicit consent of the individuals whose data is being collected. The provision intends to ensure that individuals are aware of what their data is being used for and have the ability to control who has access to it (Part VI, Section 27). The bill supports trust in the census by requiring organizations to implement appropriate security measures to protect the personal data they collect (Part VIII, Section 40).

Despite these technical and legal improvements, Nigeria's current census round can still be politicized, just as the former ones have been. Firstly, there is the structural reason, that the results of the census will significantly impact the allocation of resources and the distribution of political power. Secondly, the current political climate in Nigeria is highly charged, and there is much tension between different groups. This makes it even more likely that politicians will try to manipulate the census results in order to secure a political advantage, but it is still too early to make any definitive conclusions.

7.6   Conclusions and policy implications

Every planning effort revolves around the people; without a population census, there can be no real development planning. A population census is the process of gathering, compiling, analyzing, publishing, or disseminating demographic, economic, and social data relevant at a particular time to all people in a country or in clearly defined regions of the country. Typically, it is implemented once every ten years. According to this definition, a census is only a statistical exercise. Since Nigeria is one of the poorest countries in the world and has a higher birth rate than death rate, particularly in its rural areas, efforts to promote self-sufficient growth must move forward quickly to secure and enhance the welfare of Nigerians, particularly those in rural areas. The population's composition by age, gender, and place of residence, among other factors, are fundamental planning elements for development, but they are still inadequately known. There are only educated guesses as to the nation's total population in each state and local government district.

It is widely acknowledged that effective and efficient planning and administration of development policies in Nigeria is impeded by the absence of population data of sufficient quantity and quality. Aspects of Nigeria's unique political and economic history have prevented the development of a shared understanding of the nation's goals and aspirations. Nigeria currently possesses the manpower and financial means to conduct an effective census. Additionally, more individuals today are literate and are aware of the increased need for population census data, especially to support planning for growth and sustainable development for all Nigerians. It is possible that in the future, some parts of the country may even reject an accurate census simply out of mistrust because no one is aware of the true population size of each region, state, and local government area.

For Nigeria to overcome its census taking difficulties, a system for gathering essential statistics must first be established. Vital events (births, deaths, marriages, and divorces) and migration change the composition of the population and its dynamics. While census data can be used to acquire vital statistics, census data can be updated in years without a census by using vital statistics derived from civil registers. These still require improvement in Nigeria. Although the National Population Commission is tasked with enhancing the national vital statistics system, it has been criticized that efforts for developing complementary data sources were insufficient (Akinyemi, 2022). The National Population Commission actually disposes of local offices in every state and in the 774 Local Government Areas. Typically, birth certificates are issued shortly after a child is born but can still be given to those below 18 years old in one of these offices (Wangare & Simwa, 2022).

A step toward the depoliticization of the census could be to strengthen the institutional autonomy of the National Population Commission (Udo, 1998, pp. 364–367). One element would be to alter the composition of the National Population

Commission, appointing members not according to federated states but according to expertise (Nwauche, 2002). Furthermore, revising the revenue allocation method used to distribute money from the federal tax account to states and local government councils is crucial. Because population size is heavily emphasized in intergovernmental fiscal relations, almost all political territories of the country have historically inflated census statistics due to these financial incentives. While population size is often used as an allocation criterion in systems of fiscal federalism (Bartl, 2015; OECD, 2021), the Nigerian case might be more politicized than others because of ethnic cleavages that map quite neatly onto territorial borders. In reality, the allocation of revenue is an economic element that influences the success of public policy measures. Instead of making lump sum allocations based on population size, revenue allocation could become more targeted toward particular economic and social initiatives meant to raise the standard of living for the people in question. This should be facilitated by transparent administration and strong governance in running state affairs.

The census process is a time-consuming and complicated task requiring significant resources and manpower. However, having rescheduled the census to 2023 means that it will take place in the same year as the national elections. The government may find it challenging to conduct a census and organize elections at the same time. This overlap could result in a shortage of resources, manpower, and political will, leading to the inefficient conduct of both the census and the elections. Moreover, the political tensions that have often accompanied elections in Nigeria may disrupt the census process. Political parties could use the census as a tool for propaganda and manipulation, which may negatively affect the credibility and accuracy of the data collected. The future will show which of these potential developments and outcomes will actually unfold.

References

Achebe, C. (2012). *There was a country: A personal history of Biafra*. London, England: Penguin Press.

Adebayo, A. G. (1990). The 'Ibadan School' and the handling of federal finance in Nigeria. *The Journal of Modern African Studies, 28*(2), 245–264. doi:10.1017/S0022278X00054446

Adedire, S. A., & Olanrewaju, J. S. (2021). Military intervention in Nigerian politics. In R. Ajayi & J. Y. Fashagba (Eds.), *Nigerian politics* (pp. 395–405). Cham, Switzerland: Springer.

Adekeye, F. (2006, April 10). Another peculiar censu. *The Week Magazine, 23*(13), 6.

Akinyemi, A. I. (2020a, December 9). Nigeria's census has always been tricky: Why this must change. *The Conversation*. Retrieved from https://theconversation.com/nigerias-census-has-always-been-tricky-why-this-must-change-150391

Akinyemi, A. I. (2020b, December 11). The tricky, politicized history of Nigeria's census and how to undo its long-term harm. *Quartz*. Retrieved from https://qz.com/africa/1944964/why-nigeria-census-has-a-difficult-and-politicized-history

Akinyemi, A. I. (2022, March 8). Nigeria's 2022 census is overdue but preparation is in doubt. *The Conversation*. Retrieved from https://theconversation.com/nigerias-2022-census-is-overdue-but-preparation-is-in-doubt-177781

Akinyoade, A., Appiah, E., & Asa, S. (2017). Census-taking in Nigeria: The good, the technical, and the politics of numbers. *African Population Studies, 31*(1), 3383–3394. doi:10.11564/31-1-997

Aluko, S. A. (1965). How many Nigerians? An analysis of Nigeria's census problems, 1901–63. *The Journal of Modern African Studies, 3*(3), 371–392. doi:10.1017/S0022278X00006170

Anderson, B. (1991). *Imagined communities: Reflections on the origins and spread of nationalism*. London, England: Verso Press.

Arel, D. (2002). Demography and politics in the first post-Soviet censuses: Mistrusted state, contested identities. *Population, 57*, 801–827.

Bamgbose, J. A. (2009). Falsification of population census data in a heterogeneous Nigerian state: The fourth republic example. *African Journal of Political Science and International Relations, 3*(8), 311–319.

Bartl, W. (2015). Why do municipalities 'think' in demographic terms? Governing by population numbers in Germany and Poland. In R. Sackmann, W. Bartl, B. Jonda, K. Kopycka, & C. Rademacher (Eds.), *Coping with demographic change. A comparative view on education and local government in Germany and Poland*. European Studies of Population (Vol. 19, pp. 67–94). Dordrecht, The Netherlands: Springer. doi:10.1007/978-3-319-10301-3

Benjamin, S. A. (2012). *Politics of accommodation in Nigeria's federalism (1993–2007)* (PhD thesis). University of Ibadan. Retrieved from http://80.240.30.238/handle/123456789/751

Bonnett, A., & Carrington, B. (2000). Fitting into categories or falling between them? Rethinking ethnic classification. *British Journal of Sociology of Education, 21*, 487–500.

Cuktu, P. B. (2016, August 6). Population and economic development. *Daily Times*, p. 4.

Ekong, A. (2006, April 10). The new uprising. *The Week Magazine*, p. 6.

Ellis, J. J. (2000). *Founding brothers: The revolutionary generation*. New York, NY: Alfred A. Knopf.

Eze, B. U. (2018). Errors and disfigurations in Nigeria's census data: Evidences and implications. *American Journal of Humanities and Social Sciences Research, 2*(9), 124–131.

Ezea, P., Iyanda, C., & Nwangw, C. (2013). Challenges of national population census and sustainable development in Nigeria: A theoretical exposition. *IOSR Journal of Humanities and Social Science (IOSR-JHSS), 18*(1), 50–56.

Falaiye, H. (2023, January 25). NPC commences training for 2023 census. *Punch*. Retrieved from https://punchng.com/npc-commences-training-for-2023-census/

Fred, T. (1988). Changing perspectives of the population in Africa and international responses. *African Affairs, 87*(347), 267–276.

Gellner, E. (1983). *Nations and nationalism*. Ithaca, NY: Cornell University Press.

Gupte, P. (1984). *The crowded earth: People and the politics of population*. New York, NY: W. W. Norton.

Idike, A., & Eme, O. I. (2015). Census politics in Nigeria: An examination of 2006 population census. *Journal of Policy and Development Studies, 9*(3), 47–72.

Idowu, O. O. (2021). *The dynamics of fiscal federalism in Nigeria 1999–2011* (PhD thesis). University of Ilorin. Retrieved from https://uilspace.unilorin.edu.ng/handle/20.500.12484/7516

Ikoba, N. A., Jolayemi, E. T., & Sanni, O. O. M. (2018). Nigeria's recent population censuses: A Benford-theoretic evaluation. *African Population Studies, 32*(1), 3974–3981.

Kertzer, D. I., & Arel, D. (2002). *Census and identity: The politics of race, ethnicity, and language in national censuses*. Cambridge, England: Cambridge University Press.

Kertzer, D. I., & Arel, D. (2006). Population composition as an object of political struggle. In R. E. Goodin & C. Tilly (Eds.), *The Oxford handbook of contextual political analysis* (pp. 664–677). Oxford, England: Oxford University Press. doi:10.1093/oxfordhb/9780199270439.003.0036

Kolapo, Y., & Faloseyi, M. (2007, February 6). Lagos and the fallacies in national census figures. *The Punch*, pp. 3, 8.

Kurfi, A. (2014). *My life and times (An autobiography)*. Ibadan, Nigeria: Spectrum Books.

Leba, L. (2006, April 10). Nigeria: 2006 Census: Matters arising. *Vanguard*. Retrieved from https://allafrica.com/stories/200604100764.html

Makama, S. D. (2007). Report of Nigeria's national population commission on the 2006 census. *Population and Development Review, 33*(1), 206–210.

Mimiko, F. (2006). Census in Nigeria: The politics and the imperative of depoliticization. *African and Asian Studies, 5*(1), 1–22. doi:10.1163/156920906775768273

Morning, A. (2008). Ethnic classification in global perspective: A cross-national survey of the 2000 census round. *Population Research and Policy Review, 27*(2), 239–272.

Morning, A., & Sabbagh, D. (2005). From sword to plowshare: Using race for discrimination and antidiscrimination in the United States. *International Social Science Journal, 57*(183), 57–73.

National Population Commission. (1998). *1991 population census of the Federal Republic of Nigeria*. Abuja, Nigeria: NPC.

National Population Commission. (2023). *2023 Census*. Retrieved from National Population Commission website: https://nationalpopulation.gov.ng/2023-census.html

National Population Commission. (n.d.). [Various Facebook posts]. Retrieved from https://www.facebook.com/natpopcom

National Population Commission, & ICF. (2019). *Nigeria demographic and health survey 2018*. Retrieved from https://www.dhsprogram.com/pubs/pdf/FR359/FR359.pdf

Nigeria's Data Protection Bill. (2022). *The data protection bill, 2022*. Retrieved from the National Population Commission website: https://ndpc.gov.ng/Files/Nigeria_Data_Protection_Bill.pdf

Nobles, M. (2000). *Shades of citizenship: Race and the census in modern politics*. Stanford, CA: Stanford University Press.

Nwauche, E. S. (2002). The 1999 Nigerian constitution: Accuracy and acceptability of the census results. *Verfassung und Recht in Übersee, 35*(3), 431–441.

Obono, O., & Omoluabi, E. (2014). Technical and political aspects of the 2006 Nigerian population and housing census. *African Population Studies, 27*(2), 249–262. doi:10.11564/27-2-472

Odenyi, N. B. (2005). *Population census and national development* (Master's thesis). Department of Political Science, University of Nigeria, Nsukka.

OECD. (2021). *Fiscal federalism 2022*. Paris, France: OECD Publishing. doi:10.1787/201c75b6-en

Ogunje, V. (2022). NPC to Hire 2m Nigerians for 2023 census. *This Day*. Retrieved from https://www.thisdaylive.com/index.php/2022/11/10/npc-to-hire-2m-nigerians-for-2023-census/

Okolo, A. (1999). The Nigerian census: Problems and prospects. *The American Statistician, 53*(4), 321–325. doi:10.1080/00031305.1999.10474483

Olorunfemi, J. F., & Fashagba, I. (2021). Population census administration in Nigeria. In R. Ajayi & J. Y. Fashagba (Eds.), *Nigerian politics* (pp. 353–368). Cham, Switzerland: Springer.

Olusanya, P. O. (1989). Population and development planning in Nigeria. In T. N. Tamuno & J. A. Atanda (Eds.), *Nigeria since independence: The first twenty-five years* (pp. 319–326). Ibadan, Nigeria: Heinemann.

Omaba, R. N. (1968). The role of government in population census project in Africa. In J. C. Caldwell & C. Okonjo (Eds.), *The population of tropical Africa* (pp. 40–46). London, England: Longmans.

Owolabi, T. (2019). Population equilibrium and development issues in Nigeria. *International Journal of African and Asian Studies, 58,* 32–36.

Oyedeji, O. (2022). Nigeria gears for 2022 population and housing census, approves N177 billion budget. *Dataphyte.* Retrieved from https://www.dataphyte.com/latest-reports/governance/nigeria-gears-for-2022-population-and-housing-census-approves-n177-billion-budget/

Pambegua, I. (2022, July 21). Nigerian census and its challenges. *Punch.* Retrieved from https://punchng.com/nigerian-census-and-its-challenges/

Premium Times. (2012, October 8). Census tribunal orders fresh census in Cross River community. *Premium Times.* Retrieved from https://www.premiumtimesng.com/news/102953-census-tribunal-orders-fresh-census-in-cross-river-community.html

Prévost, J.-G. (2019). Politics and policies of statistical independence. In M. J. Prutsch (Ed.), *Science, numbers, and politics* (pp. 153–180). Cham, Switzerland: Springer.

Prewitt, K. (2000). The U.S. decennial census: Political questions, scientific answers. *Population and Development Review, 26,* 1–16.

Prewitt, K. (2010). What is political interference in federal statistics? *The ANNALS of the American Academy of Political and Social Science, 631*(1), 225–238. doi:10.1177/0002716210373737

Rallu, J.-L., Piché, V., & Simon, P. (2006). Demography and ethnicity: An ambiguous relationship. In G. Casellu, J. Vallin, & G. Wunsch (Eds.), *Demography: Analysis and synthesis* (Vol. 6, pp. 415–516). Paris, France: Institut National d'Etudes Demographiques.

Serra, G., & Jerven, M. (2021). Contested numbers: Census controversies and the press in 1960s Nigeria. *Journal of African History, 62*(2), 235–253. doi:10.1017/S0021853721000438

Shryock, H. S., & Siegel, J. S. (2013). *The methods and materials of demography* (Vols. 1 and 2). Washington, DC: U.S. Department of Commerce, Bureau of the Census.

Simon, P. (2004). *Comparative study on the collection of data to measure the extent and impact of discrimination within the United States, Canada, Australia, Great Britain and the Netherlands.* Luxembourg: Office for Official Publications of the European Communities.

Stolper, W. F. (1966). *Planning without facts* (N. G. Carter, Trans.). Boston, MA: Harvard University Press. doi:10.4159/harvard.9780674594203

Thomson, D. R., Gaughan, A. E., Stevens, F. R., Yetman, G., Elias, P., & Chen, R. (2021). Evaluating the accuracy of gridded population estimates in slums: A case study in Nigeria and Kenya. *Urban Science, 5*(2), 48. doi:10.3390/urbansci5020048

Ubochi, T. C. (2007, May 7). "2006 census" The political imperative. *The Nigerian World,* March, 12.

Udo, R. K. (1979). Population and politics in Sub-Saharan Africa. In R. K. Udo (Ed.), *Population education source book for sub-Saharan Africa* (pp. 172–181). Nairobi, Kenya: Heinemann.

Udo, R. K. (1998). Geography and population censuses in Nigeria. In O. Areola (Ed.), *Fifty years of geography in Nigeria: The Ibadan story. Essays in commemoration of the golden jubilee of the University of Ibadan, 1948–1998* (pp. 348–372). Ibadan, Nigeria: Ibadan University Press.

United Nations. (2020). Impact of COVID-19 on 2020 round population and housing censuses. Luxembourg. Retrieved from https://unstats.un.org/unsd/demographic-social/census/COVID-19/

Ventresca, M. J. (2002). *Global policy fields: Conflicts and settlements in the emergence of organized international attention to official statistics, 1853–1947* (Institute for Policy Research Working Papers No. 02–45). Evanston, IL: Northwestern University. Retrieved from https://www.ipr.northwestern.edu/documents/working-papers/2002/IPR-WP-02-45.pdf

Wangare, J., & Simwa, A. (2022, October 20). National Population Commission birth certificate obtaining guide. *Legit.* Retrieved from https://www.legit.ng/ask-legit/guides/1151650-national-population-commission-birth-certificate-obtaining-guide/

Weber, E. M., Seaman, V. Y., Stewart, R. N., Bird, T. J., Tatem, A. J., McKee, J. J., . . . Reith, A. E. (2018). Census-independent population mapping in northern Nigeria. *Remote Sensing of Environment, 204*, 786–798.

Weiss, A. M. (1999). Much ado about counting: The conflict over holding a census in Pakistan. *Asian Survey, 3*, 679–693.

Yin, S. (2007, April 18). Objections surface over Nigerian census results. *PRB.* Retrieved from https://www.prb.org/resources/objections-surface-over-nigerian-census-results/

8 Censuses in Ukraine

Not trusted and not needed?

Tetyana Tyshchuk and Ilona Sologoub

8.1 Introduction

Ukraine gained its independence from the Soviet Union (USSR) in 1991. Its first and only census was conducted in 2001. At that time, its organisation was quite modern. Almost 250,000 people were involved in the implementation of the census. Of them, 186,000 (approximately one person per 100 households) were visiting people's homes as enumerators, while people could also provide information about themselves via a hotline. Data processing after the census took about a year, and a special website for the dissemination of results was created (Ukrstat, 2003–2004). Up until the start of the full-scale war on 24 February 2022, the State Statistics Service of Ukraine (Ukrstat) provided estimates of population quantity based on the results of that census. Namely, to the numbers provided by the 2001 census, it added the difference between births and deaths and the number of registered migrants (all derived from administrative records). Therefore, such estimates did not include unregistered internal and external migration, which resulted in errors that accumulated over time.

Characteristics of the population other than its quantity (e.g., household composition, employment status, views and opinions) have been derived from sample surveys. The two largest surveys are Ukrstat's quarterly Household Budget Survey and Labour Force Survey (each has about 10,000 respondents in each wave). Polling companies conduct a variety of other surveys – their samples are usually 1500–2000 people. All these samples are constructed using 2001 census data.

The 2001 census was initially planned for 1999 (Cabinet of Ministers of Ukraine [CMU], 1995) but was then rescheduled to 2001 (CMU, 1998) due to the dire financial situation of the country because of the 1998 economic crisis. In 2008, the government planned the next census for 2011 (CMU, 2008), and in 2010 it even organised a trial census in one of the rayons (districts).[1] Later, the census was postponed to 2012, then to 2013, later to 2016, and finally to 2020. The statistical agency implemented a trial census again at the end of 2019 and was preparing to implement a full-scale census in 2020. However, in late 2019 the government decided that it had more important things to finance, and redistributed funds planned for the census to other needs. Moreover, it presented a population estimate resting on the data from public registries and mobile phone providers (Sarioglo et al.,

DOI: 10.4324/9781003259749-11

2019) as a census (we, therefore, refer to it as a quasi-census), further undermining the sense of urgency for a proper census (we discuss this issue in detail below).

In 2020, the outbreak of the COVID-19 pandemic made the implementation of a census impossible. In 2021, the government was determined to finally implement the census in 2023 (CMU, 2020), and Ukrstat was preparing for the organisation of the census (e.g. developing necessary changes to normative documents so that users could fill in the questionnaire online).

The full-scale Russian war on Ukraine has obviously cancelled this schedule. Moreover, the war forced many people to relocate within Ukraine (about 7 million) and abroad (over 5 million).[2] After the war ends, some people may stay where they relocated, some may return home, and some may choose another region of Ukraine as their permanent place of living. Future reconstruction of Ukraine will require up-to-date population data to understand in which places the need for houses, schools, kindergartens, etc. is the most acute. However, the situation will be fluent for some time since people who relocated once may be more willing to relocate further looking for a better combination of earnings/cost of living/social infrastructure. Thus, there is no sense in implementing a census right after the war, when many people will still be settling down, but people should be incentivised to declare their actual place of living.

In February 2022 the government (CMU, 2022) simplified the procedure for declaring or registering one's place of living and updated the procedure for data exchange between government agencies (CMU, 2022). In August 2022, the Ukrainian parliament approved a law that allows launching of a single registry of addresses and a single registry of buildings and infrastructure (such as bridges).[3] The Ministry of Digital Transformation will set up the registries and local governments will enter the data via special software. This will help local governments to better plan the provision of public services and encourage citizens to declare their place of living to obtain those services. Hopefully, the new procedure will be fully implemented in practice.

Against this background, we address the following research questions: Why did the government pay so little attention to the need to know how many people live in Ukraine, where, and how they live between 2011 and 2021? After interviewing public officials, we see three main reasons for that. First, the government rarely (if at all) uses the data to develop policies or to evaluate their impact, second – neither people nor the government (as representatives of the people) trust official statistics (we elaborate on this issue in part 2), and third – officials do not understand that sample surveys rely on census data as a framework (hence, they believe, it is possible to replace a census with sample surveys).

Prior to the full-scale war, Ukrstat was caught in a "trust trap" with respect to the census. If it aggressively advocated for the need to implement a census based on arguments about inaccurate population data, it would further undermine trust of people and policymakers in official statistics. If it did not, the government would prioritise spending on other things, which it did since 2008. Huge (mostly unregistered) migration movements and fatalities caused by the war made advocation for a census much easier since it is obvious that the accuracy of population data has deteriorated. It is important that any "quick" population counts implemented after the war are *not* called a census so that a proper census is implemented when it becomes feasible. An honest and open dialogue between the government, statistical agency, scientists, and society is needed to restore the trust of people in statistics and to engage them in population counts and eventually the census.

The chapter is structured as follows: In the next section, we look at censuses organised on Ukrainian territory by the USSR. This analysis allows to clearly see the reasons for the low trust in statistics by Ukrainians. In the USSR, statistics was manipulated for "political necessity" – in line with the general government policy of ruling by lies and repression. We still hear the echo of these policies today. Section 8.3 describes the development of population records in Ukraine. Section 8.4 discusses why a typical democratic government would need a census and why these reasons only weakly apply to Ukraine. Section 8.5 provides the reasons for the continuous postponement of censuses after 2001. Section 8.6 describes the results of the semi-structured interviews with central and local government officials describing their views on the census, and Section 8.7 concludes.

8.2   A brief history of censuses in Ukraine

Until the end of World War II, the current territory of Ukraine was distributed between different states (an independent Ukrainian state existed only for a brief period of 1918–1921, afterward it was occupied by Russia and Poland). Therefore, before the 1950s the data on the population and its characteristics are rather fragmented (Table 8.1). Thus, in our discussion, we concentrate on the censuses implemented by the Russian Empire and USSR since these covered a major part of today's Ukrainian territory.[4] Table 8.1 shows that only one census (in 2001) was implemented by the Ukrainian government rather than some occupation power.

As discussed in the introduction to this book (Chapter 1), censuses have never been free from politics. Apart from a pure population count, governments usually relate other goals to a census. This is evident when we look at questionnaires of censuses implemented on the territory of Ukraine. Census questionnaires are worth analysing because they reveal the primary purposes pursued by the implementing government, as well as provide an idea about the social structure and social relations. Specifically, in Table 8.4 in the Annex, we see that in the 1920 census, the government was mostly interested in the workforce: questions on occupational specialisation and employment of a person were much more detailed than in subsequent censuses. This is understandable, because World War I and the subsequent Russian-Ukrainian war (1918–1921) exhausted labour resources, so the government wanted to know how many people it could put to work.

Both the 1897 and 1920 censuses asked about disabilities and whether a person could work despite of them. In contrast, the 1959 census did not ask about disabilities, although in the aftermath of World War II this question would have been more than legitimate.[5] Moreover, the 1949 census, which should have naturally followed the 1939 one, was cancelled to mask huge population losses of the Soviet Union in World War II.

Table 8.1 Censuses organised on the territory of Ukraine at different times between 1784 and 2001

Years of censuses and countries that organised them	Number of 25 oblasts (regions) of independent Ukraine that participated in censuses*
Austrian-Hungarian empire: 1784	1
Austrian-Hungarian empire: 1854	2
Austrian-Hungarian empire: 1869, 1880, 1890, 1900, 1910	3
Russian Empire: 1897	20
Poland: 1921, 1931	5
Czechoslovakia: 1921, 1930	1
Romania: 1930	1 and a part of another oblast
USSR: 1920**	14
USSR: 1926, 1937, 1939	18
Hungary, 1941	1
USSR, 1959, 1970, 1979, 1989	25
Ukraine, 2001	25

Source: compiled by authors based upon Ukrstat (2004).

Notes: * Ukraine has 24 oblasts and the Autonomous Republic of Crimea. The cities of Kyiv and Sevastopol are separate administrative units but were covered by censuses within Kyiv oblast and Crimea respectively. ** The USSR census of 1920 was taken during the Ukrainian War of Independence. Its organisers estimated that only up to 72% of the population within Ukrainian borders that existed before September 1939 (i.e. those territories that entered the USSR in 1922) participated in the census (Steshenko, 2019).

The 1949 census was not the only one being cancelled. The fate of the 1937 census and its organisers is even more tragic. Initially, it was scheduled for 1933 but was rescheduled later three times to hide the consequences of the 1932–1933 artificial famine that killed about 4 million Ukrainians, the Holodomor. Historians have attributed the Holodomor to the failed Stalinist collectivisation policy; as of November 2023, it has been recognised as genocide by 32 countries and the European Parliament. Until the 1990s, discussions about and even mentions of the famine were prohibited, but since 1991 Ukrainian researchers have joined the international scientific community working on this subject and collected a substantial body of evidence – documents and stories of eyewitnesses – that allow assessing the scale of this man-made deadly famine (Institute of History of the National Academy of Sciences of Ukraine n.d.).

During 1913–1930, the population of Ukraine was growing at an average pace of 276,000 people per year, despite the war losses of 1914–1921. Ukrainian demographer Mykhailo Ptukha forecasted that by 1937 Ukraine would have had 34 million people (Revenko, 2001). Other forecasts provided the number of about 31 million people. However, fragmented data found in archives in 1989 allow inferring that the 1937 census accounted for just 28.2 million people. This is a 4% decline compared to 29.4 million people reported by the 1926 census, which is generally considered a well-organised one (although perhaps some

250,000 children were undercounted). For the entire USSR, the 1937 census showed 8 million fewer people than expected.[6]

When the first results of the 1937 census were calculated, it was labelled "defective", its results were classified, and the statisticians who organised it were either executed or jailed.[7] Another reason to destroy the results of this census, besides the revelation of the population losses, was that, unlike other Soviet censuses, it contained the question of religion. It showed that despite the aggressive anti-religious campaign of the 1920s–1930s, during which priests were killed and churches turned into warehouses, more than half of people older than 16 defined themselves as religious. Stalin himself approved the questionnaire for the 1937 census. Among others, he removed the question on the place of birth – because this question would have revealed the scale of forced deportation or semi-voluntary migration within the USSR. During the 1920s–1940s, millions of people from the European part of the USSR were deported to Siberia and the Far East or fled their homes to avoid prosecution, while people from Russia moved into houses of deported Chechens and Crimean Tatars, Kazakhs, and Ukrainians who died of famine. The place of birth question returned to the census only in 1989. Other reasons for the absence of the place of birth question were that many people did not know their place of birth since they grew up in orphanages because their parents were killed during wars, executed, or died in the GULAG. But perhaps more importantly, the place of birth question would have contradicted the Soviet concept that propagated the entire USSR as one motherland and was implemented together with a "single Soviet nation" policy, which implied erasing national identities of people living in the USSR.

Having destroyed the results of the 1937 census, as well as the people who had organised it, the Soviet government scheduled the next census for 1939. Unlike the 1937 census in which fieldwork lasted for just one day, the fieldwork for the 1939 census lasted seven days in cities and ten days in villages, and statisticians did their best to account for every person (Steshenko, 2019). They also did not try to avoid double counting. For example, they counted both permanent (*de jure*) and actual (*de facto*) population. Despite that, they had to increase the final population number by 2.9 million to match the total number of 170 million USSR people announced by Stalin at the Communist Party meeting before census results had been processed. The regional distribution of the population was distorted by that census too. The Soviet government wanted to "hide" military personnel and prisoners, thus they ordered to "redistribute" 4.7 million people between regions other than those in which these people actually resided. Naturally, a large part of these "extra" people was recorded in Ukraine and Kazakhstan – places that had suffered most from the 1932 to 1933 famine.

Ukrainian researchers estimate that without three famines (1921–1923, 1932–1933, 1946–1947) the Ukrainian population in 2013 would have been 52.3 to 53.7 million while in fact, it was 45.4 million (Hladun, 2013; see also Figure 8.1). Besides the famines, about 140,000 people living in Ukraine were killed during the mass repressions of 1937–1938 (Hladun & Rudnytskyi, 2009). Obviously, an honest census would have revealed that the repressive policies of the Soviet government in the 1920s and 1930s led to huge population losses.

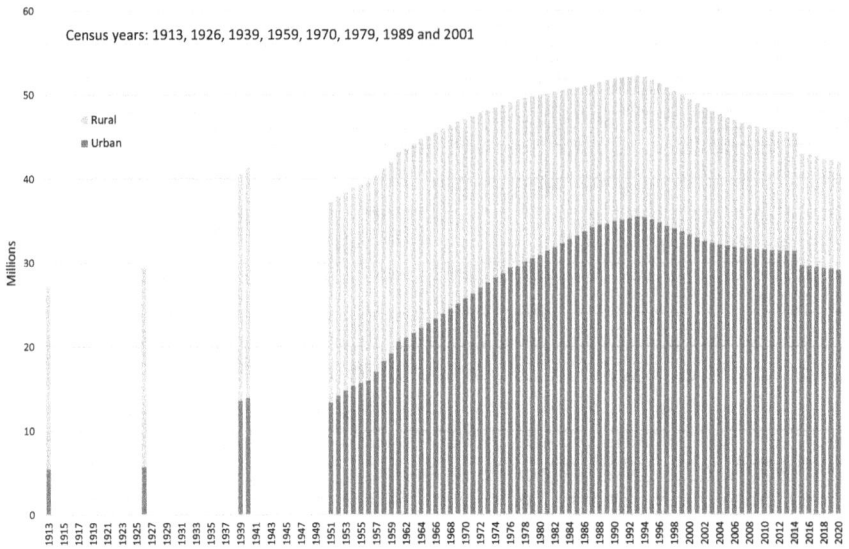

Figure 8.1 Population of Ukraine according to censuses and intercensal estimates, million people.

Source: Ukrstat (different years). Note: 1913 and 1926 are based on census data that do not include Western oblasts. 1939 includes census data on the territories that were a part of the USSR and an estimate for Western oblasts that were annexed by the USSR in 1939. Annual data started in 1951. According to official data, Ukraine's population reached its 1939 level in 1957 - birth rates were quite high at that time. Since 2015, Ukrstat does not include occupied Crimea in its estimates. However, it does include occupied territories of Donetsk and Luhansk oblasts. Since the start of the full-scale war, Ukrstat has not published population data

8.2.1   Accounting for ethnic and language issues

Even more politicised than the population size were the questions on its ethnic and language composition (Arel, 2002). These issues inevitably cause heated social debates and often call for policy actions (Duchene & Humbert, 2018), but in the USSR, which acted *de facto* as an empire and tried (often successfully) to erase non-Russian languages and nations from its territory, these questions were particularly acute.

Silver (2016) provides a comprehensive discussion of ethnic and language questions in the Soviet censuses. The devil hides in many details – from defining stand-alone ethnic groups (e.g. until 1989, Crimean Tatars (Qirimli)[8] were recorded together with Tatars, as a single nationality [Anderson & Silver, 1989]) over vague instructions to enumerators on how to record different nationalities, to scarce, if any, publication of ethnic data. Botev (2002) notes a contradiction in the Soviet ethnic policy: whereas on the one hand, the state proclaimed the merging of all ethnic groups into a unified "Soviet people", on the other hand, the nationality of a person was indicated in his/her passport, and there existed explicit (or implicit but widely known) ethnic discrimination, e.g., ethnic quotas. Thus, Jews, for example, had difficulties entering some higher education institutions,[9] while people from Asian republics had privileges entering other (mainly agricultural) universities.

However, the true goal of the Soviet (and Russian) ethnic policy was Russification (Anderson & Silver, 1983; Kuznetsova, 2022; Silver, 1974). This was done with three methods: deportations and discrimination of non-Russians as well as squeezing out of their national languages and cultures. It was not unusual for the USSR to forcefully relocate entire nations. Examples are the deportation of Crimean Tatars to Central Asian republics in 1944,[10] the relocation of Ukrainians from Western regions of Ukraine either to the East of Ukraine or to Siberia (1947 and the following years), or an attempt to relocate Jews to the Amur region near the Chinese border in 1934. The deported population was replaced by Russians, and this happens today as well: the Ukrainian government estimates that between 600,000 and 1 million people moved from Russia to Crimea since its occupation in 2014. From March to September 2022, Russians deported over 2 million Ukrainians from the occupied Ukrainian territories to Russia. Among them, there are nearly 300,000 children. Ukrainian children in Russia are taken away from their parents and put into orphanages or adopted by Russians. Very few people manage to return to Ukraine.

The second method of Russification was discrimination, which led to the semi-voluntary identification of people of other nationalities as Russians. Sometimes people even changed their names – either to avoid discrimination, or Soviet officials changed their names for them (e.g. the great-grandfather of one of the authors of this chapter had a Russian ending *-ov* added to his Ukrainian surname, and this was the rule rather than an exception). After the demise of the USSR, an opposite process started. Thus, Stebelsky (2009) estimates that about 85% of the increase in the share of Ukrainians and the respective decline in the share of Russians in Ukraine between 1989 and 2001 is due to reidentification rather than other factors such as migration or birth and death rates.

Finally, perhaps the most effective method of Russification was the promotion of the Russian language: measures ranged from higher pay to school teachers of Russian and more hours devoted to studying Russian at schools to forming the perception of other languages and cultures as inferior to Russian. For example, in Ukraine, those who spoke Ukrainian in a city would often hear humiliating phrases such as "Are you from a village?", "Speak human language!". Urbanisation became a powerful Russification tool (Silver, 1974). Until 1974 rural dwellers did not have passports and required a special permit even for a trip to the nearest town. No wonder those who managed to escape into a city tried to assimilate as fast as possible, including switching to the Russian language.[11]

Language adds another layer of complexity to the ethnic question in censuses. As noted by Silver (2016), although the Russian Empire's 1897 census and all Soviet censuses asked for a "mother tongue", no clear definition of this term was provided to census takers or readers of census results. Thus, some census takers asked for the mother tongue – in which case they most likely got an answer supporting the respondent's nationality, even if a respondent did not know that language very well. Some census takers asked "What language do you speak at home?", in which case they probably recorded the language which a person used most frequently.[12] But this answer may not reflect the "mother tongue", i.e. the language spoken to a person by her parents in her childhood, because kindergartens and schools were a powerful Russification tool. There, children of parents who spoke another language quickly became more comfortable with Russian. The 1970 and 1979 censuses, along with "mother tongue", asked whether a person spoke "another language of the Soviet Union" (an implicit assumption was that people would name Russian as this "another language"). Probably this was used to measure the success of the Russification policy and the extent to which the use of other languages could be narrowed since "everyone understands anyway" (what is printed in newspapers or said on TV in Russian). The language question was perceived as highly political not only by the government but also by the people, which can be illustrated by the fact that, between 1970 and 1979, the share of Estonians who answered that they knew Russian dropped from 29% to 24% (Silver 1986). This result is improbable given the extent of Russian language promotion in the USSR. However, people increasingly answered that they did not understand Russian to support their native languages.

8.2.2   Economic and educational issues in censuses

Other questions in censuses provide an understanding of the long-term societal changes along different dimensions. (Annex, Table 8.4). A question on literacy was present until 1959 but not afterward – since by that time the literacy rate reached almost 100%. At the same time, the question on education in 1959 became more detailed, reflecting a modern classification of education levels, and has remained unchanged since then.

Perhaps the most interesting is to compare the questions on employment and social status which show how the socio-economic structure of society and government policies towards society changed. The 1920 and 1926 censuses asked whether a person was an employer, i.e. whether he/she was an owner of a small business or a farm and employed workers other than family members. USSR believed that a person could work only for the state. Those who "exploited" (i.e. hired) other people were repressed – either killed or deported to remote parts of the USSR, and their possessions were confiscated. Naturally, Ukrainians opposed such policies – in 1929 and 1930 there were thousands of revolts. Thus, Stalin organised the Holodomor to "solve the Ukrainian question" and implement the collectivisation policy. By 1939 all the businesses were owned by the state and the creation of collective farms was completed. Thus, the "employer" category disappeared from census questionnaires – the state was the only employer for everyone.

In the 1939 census, the employment question included a "non-working element" category. This reflects two guiding ideas of the Soviet state: first, that everyone should work, and second, that a person can be called "an element". The latter was a part of a broader dehumanisation policy that devalued the human rights of individuals. Thus, the Declaration of Workers and Exploited Peoples, adopted by Bolsheviks in 1918 and later included in the USSR Constitution, proclaimed that "exploiters" could not participate in any government bodies and introduced a general obligation to work (Declaration, 1918). In 1961, the Soviet government started punishing unemployed people (calling them "deadbeats") by forcefully relocating them outside of major cities (Decree of the Soviet government of 04.05.1961). At the same time, the definition of a "deadbeat" was quite wide – it could be applied to people who, for example, sold vegetables or poultry grown at their garden plots or to poets, writers, and creative artists who did not belong to writers' or artists' unions. In other words, since entrepreneurship was banned, people whom we would now call entrepreneurs or freelancers were prosecuted. Although in 1975 this decree was cancelled, the government's fight against "deadbeats" continued until the onset of *perestroika* in 1985.

The 1989 census admits that people could work in a cooperative or as employees for an individual. This reflects some softening of labour laws – cooperatives were allowed in 1986, and other forms of entrepreneurship were allowed in 1990.

Not only the questionnaires or the timing of censuses were highly politicised in the USSR but also was the publication of census results (Tables 8.2 and 8.3). Clem (2016a), Anderson and Silver (1989), and other researchers describe, how hard it is to extract meaningful, let alone comparable data from official publications of USSR censuses. Soviet scientists, in their papers, often refer to census data that were never published, which suggests that these scientists had to obtain special permissions to work with this data. This secrecy can be explained in part by military interests: the Soviet government wanted to hide the number of people that could be drafted into the army in the next war. In part, it is also attributable to a general tension between party ideology about the future society to be created and the professional ethos of statisticians manifest in the census (Mespoulet, 2022). Ethnic data, for example, were not published to support the formation of a "single Soviet nation".

Table 8.2 Dates of publications of censuses that included at least parts of Ukraine

Census year	Year of publication of results
1897	1905–1915 (publication was not completed due to World War I and the following revolution)
1920	1922–1923 brief version, 1928 full version
1926	1928–1933
1937	1991, 2007 (partially). As discussed in this text, census results were classified and implementers repressed. Only in 1989, some pieces of information from this census were found in archives and later published
1939	1940 brief version, 1992 full version
1959	1962–1963
1970	1972–1973
1979	1984 brief version, 1989–1990 full version*
1989	1990 brief version, later Ukrstat published some results concerning Ukraine
2001	2002–2004

Note: * Toft (2014) thinks that the publication of the 1979 census results was delayed five years because the Soviet government saw a decline in the share of the European population in these results. She believes that the high growth of the Muslim population was the major factor behind the USSR's demise. However, we are sceptical about this – it is hardly possible to define which of the many factors that together caused the USSR's downfall was the major one.

8.3   Development of the modern system of population records in Ukraine

Population registers are a source of population data alternative to the traditional census. In the present section, we will describe the development of the modern system of population registers in Ukraine (based on Hladun & Rudnytskyi, 2009), because they are a crucial resource for the possibility to switch to a register-based census in the future. We also describe their relationship with personal authentication technologies of the state, such as birth certificates, identification (ID) cards, and passports.

Before the revolution of 1917, births, deaths, and marriages were recorded in church books. The new Communist government tried to replace the traditional religion with the "new religion" – communism. Thus, in early 1918, it introduced books for recording civic status by local governments or party organs, and by 1919, population records had been completely transferred to civil authorities.[13] The USSR system of population records was primarily used to control people. Therefore, the State Security Service (first called the Extraordinary Commission, then the People's Commissariat of Internal Affairs [NKVD], then the State Security Committee, better known as KGB) took control over this system from its very inception.

Nevertheless, statistical authorities created in Ukraine from 1919 to 1921 had rather full information on the population and used advanced at the time methods of data collection. They used administrative data (population records on birth, deaths, and registered migration) to estimate the number of people and used censuses to update population registers and to learn about other characteristics of the population that were not contained in population registers, such as employment or social status, ethnicity and language, as discussed above.

From the second half of the 1920s to 1932, statistical authorities published rather detailed annual reports on the Ukrainian population. In 1922, after Ukraine formally entered the Soviet Union, the Ukrainian statistical agency was subordinated to the Soviet one, and in 1930 it was merged with Derzhplan (State Planning). Derzhplan was a government agency responsible for central planning, i.e., it prescribed what, where and how much should be produced. After independence, it became Ukraine's Ministry of Economy. Although today the ministry's functions are very different from Soviet-style planning, until the fall of 2019 Ukrstat was still subordinate to it. Now, Ukrstat is subordinate to the Cabinet of Ministers. Unfortunately, the recently adopted new law on statistics[14] does not provide the mechanisms for ensuring Ukrstat's political independence, but at least it foresees the professional independence of Ukrstat (see Kupina, 2020 for a discussion of the draft law). It also provides the obligation of other government agencies to provide administrative data to Ukrstat, collect the data according to Ukrstat methodology, and foresees the provision of microdata for research purposes. Adoption of the law is just the first step in the modernisation of Ukraine's statistical system. Implementation of the law will take some time and will require effort from both the Ukrainian government and Ukrstat (in particular, Ukrstat will need to develop procedures for providing microdata).

Since 1925, *propiska* (the mandatory registration of the place of living) was introduced in cities: anyone arriving there needed to register at the police within 24 hours. Nevertheless, it was not uncommon for Soviet people, especially in the second half of the 20th century, to live without *propiska,* although one could not formally obtain public services such as healthcare without it. Yet, informally, bribery and personal relations were effective tools for people's survival under the repressive state.

To better control the population, in December 1932, the USSR introduced passports, and in 1934 – birth certificates. Since 1937 it became mandatory to include a photo in a passport, and a copy of that photo was stored at the police. Passports were necessary for travelling and for getting a job or public services. However, villagers (those who worked at the collective farms – *kolkhozniki*) were not issued passports until 1974 in order to keep them in villages as a very cheap workforce for the inefficient Soviet agriculture. Many of them tried hard to get a passport and escape from the *kolkhoz* slavery; this was possible by going to a technical school or a university, joining the army service, or one of the USSR's "grand projects". Because of severe disparities between rural and urban areas and between large and small cities, it was possible to use place of living restrictions as a means of punishment. Thus, certain categories of people (e.g. dissidents) were prohibited from living in major cities.

In 1992, Ukraine replaced Soviet passports with its own passports. Neither Ukrainian passports nor the related administrative records contain information on nationality; therefore, the census is the only way to learn about the ethnic composition of the population. In 2003, Ukraine replaced the obligatory *propiska* system with the voluntary registration of citizens. Since that time, the government has worked on improving the registration system to make it more convenient for people. In 2012, Ukraine created a central population registry, the Single Demographic Registry.[15] However, the registry is still incomplete, for only people who obtained some document since that time (e.g., a passport for travelling abroad) have been included there. More generally, Ukraine has many registries but no single registry that would include *all* its people (this is discussed below in more detail).

8.4   Why are censuses important and how this applies to Ukraine

From the discussion above it is clear that censuses in Ukraine too often fell victim to political necessity. Can it be so that, after 2001, the census fell victim to a *lack* of political necessity? Before discussing this question, we consider, why censuses are important and whether these reasons resonate with the Ukrainian government.

Firstly, censuses are essential for democracy. In the USA, for example, political representation is based on the population data gathered through the census as constitutionally required (Sullivan, 2020). In most countries, voting is tied to the place of living, so it is essential to have information not only on the number of people but on their actual place of residence. Thus, a government willing to protect voting rights must be interested in census data or an equivalent source of information. Unfortunately, this is not the case in Ukraine.

In Ukraine, the electoral legislation was changed practically before every election, because the ruling parties in parliament tried to ensure a high representation for themselves in the next parliament (VoxUkraine Editorial Board, 2019). Electoral fraud has been quite common.[16] One type of electoral fraud is screwing up registries of voters – so that when a voter arrives at a polling station, he/she cannot find him/herself on the list. This artificially reduces the number of voters who showed up and (theoretically) allows electoral commissions to fill in the unused ballots. If many such cases are reported, this undermines trust in the entire electoral procedure.[17] Citizens and NGOs rather than governments have been protecting voting rights; the most well-known case is the Orange Revolution of 2004. Naturally, if the government doesn't care about voting rights protection, it will neither need a proper voter registry nor a census.

Second, census data is used for the distribution of budget funds (e.g. Eniayejuni & Agoyi, 2011). In Ukraine, subvention for financing secondary education as well as some other subsidies are distributed to communities based on the number of recipients of certain public services. A study (see Table 8.3) of several dozen communities showed a substantial discrepancy between the number of people estimated by Ukrstat and the number of people as counted by local governments (Tyshchuk, 2019). Community heads have an incentive to exaggerate the number of dwellers to receive more subsidies (although we have no evidence that they actually do this). Therefore, the position of the central government was to rely on official Ukrstat data until the census was implemented. Now, because of the war, it will have to rely more on local government data and encourage citizens to register or declare their actual place of living. Is the government actually interested in the fine-tuning of subsidies and subventions? Theoretically, it should be. During the last decades, different Ukrainian governments declared switching public policy to means-tested social support (targeted at households and thus reliant on census data). Despite that, the social support system remains poorly targeted and very complicated (Ierusalymov & Marchak, 2021). For more than ten years the national government (with the help of international donors) unsuccessfully tried to create a single registry of social support recipients. The actual creation of this database will be one of the tasks for reconstructing the country after the war.

Thirdly, census population data is needed for the construction of representative samples to implement surveys. In principle, policymakers should be interested in household surveys or labour force surveys to estimate the potential impact of policies on macro indicators (e.g. unemployment) and on different social groups. To construct proper representative samples, data on the entire target population is needed, which typically comes from a census. Moreover, only a census would allow us to "see" isolated or small social groups that are "invisible" to sample surveys. Another group of powerful stakeholders that should be interested in good quality surveys are politicians since they closely watch opinion polls. But they are probably satisfied with the data that polling companies currently provide; indeed, their forecasts of election results are usually quite close to actual numbers.

Fourth, census data is needed for calculating "per capita" indicators used in international comparisons (Baffour, King, & Valente, 2013). Thus, if a census shows that Ukraine has fewer people than current estimates suggest, its per capita GDP and other most common development indicators will improve. But this is probably not a sufficient incentive for Ukrainian governments to strongly push for a census.

Table 8.3 Difference in communities' number of inhabitants according to data from the local registers and central government data, 2019

Community	Difference	Inhabitants, thousands
Khorostkivska	−19.3%	5.4
Smolinska	−14.2%	9.3
Tavrychanska	−12.0%	20.2
Pomichnianska	−11.1%	11.8
Stanyslavska	−10.4%	17.8
Novoraiska	−9.8%	22.6
Mykolaivska	−8.4%	9.5
Vysokopilska	−8.6%	10.6
Novoraiska	−5.4%	7.4
Voznesenska	−5.9%	11.9
Pischanska	−2.1%	7.8
Kompaniivska	−1.1%	18.3
Tlumatska	−1.2%	6
Zolochyvska	−1.4%	12.7
Arbuzynska	−1.4%	8.5
Dzvyniatska	0.0%	7.7
Velykopanivska	0.0%	9
Lanchynska	0.0%	8.4
Petrykivska	0.0%	6.4
Piadytska	0.0%	35.9
Pidvolichyska	0.0%	4.8
Zborivska	0.0%	19.4
Malynivska	−0.2%	25.2
Lanovetska	−0.3%	8
Shevchenkivska	1.4%	10.1
Borschivska	3.5%	4.3
Zolotnykivska	6.7%	9.6
Dmytrivska	6.7%	13.2

Source: Ukrstat's oblast departments. Data collected from different publications by the authors. Data are based on the intercensal estimates and are used for the distribution of subsidies; communities' data stem from their registers or from the portal of communities maintained by the Ministry of Regional Development: https://gromada.info/gromada; The data for Table 8.3 were compiled by the programme DOBRE USAID. The communities in Table 8.3 were studied within the framework of the sub-programme "Financial Management Assessment" of the DOBRE programme. Results of this programme can be retrieved at https://www.slideshare.net/Decentralizationgovua/dobre-results-of-financial-management-assessment.

Note: Negative numbers mean that the population estimates in the local registers are bigger than the central government data.

Finally, businesses should be interested in knowing the number and composition of people in the smallest administrative units. This information would be useful for decisions on the allocation of production facilities, targeting places where there is workforce or the market, or both (Martins, Yusuf, & Swanson, 2012). However, processing the census results takes quite a long time. So, businesses would rather get the needed information (even imprecise) from a survey or from local authorities.

8.5   Excuses for postponing the census since 2011

As we see from the above discussion, in theory, both government (especially local governments) and businesses should be interested in the census data. However, in practice, obtaining this data is low on their priority list. The most frequent explanation for continuous rescheduling of a census was lack of money. Other explanations are the war (which was limited to certain areas before February 2022), elections, overestimated capacity to conduct a register-based census, and the COVID-19 pandemic. Obviously, the full-scale invasion cancelled the census planned for 2023 and now its timing is uncertain.

At the end of 2019, when the new government, which had come into office in September 2019, reviewed the draft budget for 2020, it excluded the money planned for the census from it (about USD 185 million[18] or 0.1% of Ukraine's 2019 GDP). At that time, policymakers initiated a series of discussions with the expert community on the need to spend money on a census. The authors of this chapter were present at some of these discussions. Experts and researchers tried to explain the need for a traditional census. Their arguments were: (1) the problem with selective registries; (2) the need to have information on the population structure to make sample surveys more precise; (3) the need for socio-demographic household information that can be gathered only via a census; (4) the lack of registries suitable for a registry-based census.[19] However, the minister who was responsible for statistics at that time argued that (1) he interviewed about 100 government officials and only a few of them saw a need for more than just the population size; and (2) a quarter of the households did not open their door during the pilot census at the end of 2019; therefore, the data collected in a personal census interview would be incomplete. Although a proper communication campaign and a mixed-mode approach, allowing people to fill in the questionnaire online, would mitigate the second risk, the government prioritised spending on other things, as happened many times before.

Moreover, claiming to spend zero public money, the government commissioned an estimate of the population size (Sarioglo et al., 2019). The experts commissioned with that estimate combined the data from a few registries with the data from mobile phone comhpanies and estimated that the total number of people living in the government-controlled area of Ukraine as of December 1st, 2019 was 37.3 million (this estimate is consistent with Ukrstat estimates). Then they applied existing age, sex, and regional information provided by Ukrstat based on the 2001 census to this number and produced estimates of regional, gender, and age distributions. Of course, that was *not* a census. However, it was presented as such, and many people from the government wrongly believed that this estimate was indeed a census. In the next section, we elaborate on this.

8.6   Do Ukrainian officials understand the necessity of the census? Analysis of semi-structured interviews

To find out reasons *why* Ukrainian policymakers have not been very interested in a census, we implemented 20 semi-structured interviews with civil servants from central and local government offices involved in the policy-making process.

Respondents from the ministries were middle-level managers – heads of depart-ments, directorates, or divisions – with 3–20 years of experience in the government. Respondents from local authorities were deputy heads of oblast administrations and communities. The interviews were implemented in May and June 2021. The questionnaire and the organisational affiliation of respondents can be found in the Annex. Interviews were audio-recorded and transcribed, and then interpreted and summarised manually. Based on the interview results, we can identify three main reasons why policymakers believe that a census is not important.

1 Policymakers rarely use data to predict or assess policy outcomes

Our respondents explained that they often use data to monitor the situation in the area of their responsibility. They use data on labour market, industrial production, agricultural output, service sector, consumer and producer price in-dexes, and population data by regions and communities. However, they rarely forecast probable policy outcomes during the process of policy formulation or try to evaluate policy impact using data or models after implementation. Thus, they do not perceive a need for detailed micro-level data that a census or repre-sentative surveys based on a census would provide.

Policymakers at, both, central and local levels use data to identify the prob-lems that policies should address. However, decisions about the need for cer-tain interventions are rarely accompanied by an in-depth analysis of their likely impact. Most of our respondents never were in a situation when poor quality of intercensal data limited their capacity for policy impact assessment – because they don't perform policy impact assessments. The reasons for that are, both, a lack of data-analysis skills and a lack of incentives to develop quality manage-ment regarding policies. This indicates more general problems with the quality of government in Ukraine.

Highly precise population data, which can be provided only by a census, is needed to develop complex policy solutions, such as reform of the pension sys-tem, healthcare services, or education infrastructure networks. When developing these types of reforms, Ukrainian governments have usually relied on technical assistance from international organisations, such as the World Bank. However, these organisations do not have the necessary bargaining power to insist on a census, since the final decision is a national political competency anyway.

From our experience with technical assistance projects and interviews with public officials, the low interest of policymakers in high-quality population data can be explained by several reasons. Firstly, the analytical capacity of govern-ment agencies is quite low: although they are required by law to do a Regulatory Impact Analysis,[20] quite often, they either don't do it or they do it just formally.[21] Secondly, even when people with the necessary skills for such an analysis come into government, they cannot get individual-level data, because Ukrstat has not set up a defined procedure for providing such data (the new law on statistics ad-dresses this question but its implementation will take time).

From 2015 to 2019, the government with support from the European Union (EU) implemented a civil service reform. One of its aims was the development of the capacity for policy analysis within government agencies. A series of training programmes were implemented to equip civil servants with policy assessment skills. However, the situation has not changed much. Thus, since Ukrainian ministries do not use the data, they have little interest in collecting them. During our interviews, the problem of poor quality of population data was only mentioned by representatives of the Ministry of Regional Development and by local authorities (the subsidy problem discussed in Section 8.4).

The other side of the coin of government officials' disinterest in censuses or the proper functioning of a unified civil registration and vital statistics system (Silva, Snow, Andreev, Mitra, & Abo-Omar, 2019) is that they have developed their own databases for their specific tasks or spheres of responsibility. For example, the Ministry of Health does not rely on Ukrstat data. In 2017 they launched their own electronic registry of declarations signed by Ukrainians with their family doctors; in early 2021 the registry contained information about 31 million people. The Ministry of Occupied Territories runs a registry of internally displaced people (IDPs). The Ministry of Social Policy has established a registry of people who obtain subsidies for communal services. The Tax Administration and the Pension Fund have their registries of taxpayers and pensioners respectively. Hladun (2021) reports that there are 32 different electronic registries that contain some information about people and/or their property. Unfortunately, one cannot merge the data from these registries in such a way as to ensure that each person living in Ukraine *enters* the resulting dataset and enters *only once* since there is no unique identifier for persons present in every registry. Only people who obtained some documents since 2012 have a unique identifier used in the Single Demographic Registry. Therefore, Ukrainian researchers argue for the implementation of at least one traditional census before a transition to registry-based censuses. Moreover, such a transition is impossible without the resolution of many issues related to personal data protection which today is rather poor.

2   Policymakers do not believe that a census will provide data of adequate quality

The 2001 census was implemented by the Ukrainian statistical agency according to international standards and its data are quite reliable. Since administrative data used to update population statistics are incomplete, with time the quality of these statistics deteriorated, and the size of the error is unknown (in 2001 the discrepancy between census results and previous population estimates was about 1 million people). However, when the issue of population quality data is communicated to policymakers, they start questioning the ability of the national statistical agency to collect reliable data in principle. Even more harm is done when policymakers share their doubts with citizens. The situation with the 2019 quasi-census is a bright illustration of this.

Ukrstat was preparing to implement the 2020 census with the traditional method of personal interviews. In December 2019 it ran a pilot census in two rayons (one urban and one rural). Simultaneously, the government set an ambitious goal to implement a register-based quasi-census instead. Experts pointed to the risks of this approach: low quality or incomplete data in the registers and the impossibility of merging data from different registers. However, the government did not address these risks during their population estimate. The detailed methodology for this quasi-census was neither published nor discussed in advance. Moreover, this project was implemented by a group of experts outside of the national statistical agency, although some Ukrstat representatives were also involved.

The results of the quasi-census were presented in early 2020. Although Ukrstat and experts explained that this was *not* a census, the minister who organised it communicated that it *was*. Moreover, in his communication, he discredited the 2001 census calling it "the big fake" (Dubilet, 2020). He inferred that if during the trial census in 2019, a quarter of respondents did not open the door to census takers, they did not open their doors in 2001 either, and claimed that the data was "invented" by census takers. Despite its flawed logic, this message was effectively delivered: almost all of our respondents said that in the case of a traditional census, Ukrstat would not be able to collect data of appropriate quality, since many people would not open their doors to enumerators.

3 Policymakers do not fully understand that up-to-date census data affect the quality of sample surveys

Ukrstat, as well as polling companies, use census data as a sampling frame, i.e. ideally a complete list of the target population, to construct representative samples for their surveys. Having census data that are more than 20 years old negatively affects the results of these surveys: in particular, problems of coverage and weighting errors are larger than they could be. Unfortunately, many people do not understand the difference between the target population and a sample aiming to represent it in an unbiased way. Over two thirds of our respondents believe that it makes sense to replace the traditional census with estimates based on sample surveys, and that this procedure would provide better results than a traditional census. Perhaps this belief is also a consequence of the quasi-census implemented in 2019.

8.7   Conclusions

The only proper census in independent Ukraine was implemented in 2001. After that, Ukrstat provided population estimates using the administrative data which do not account for the unregistered migration. By 2022 the error contained in these figures may have reached millions of people. The Russian invasion of 24 February 2022 forced about a third of the population to flee their homes making a traditional census both impossible and useless.

For 20 years Ukrainian governments did not conduct a census, finding some more urgent needs to finance. Government agencies, except for Ukrstat, did not advocate for it because they don't use the data for policy development or evaluation. They also do not trust Ukrstat's ability to collect high-quality data and believe that sample surveys can substitute for a census. In its turn, Ukrstat did not push for a census too much for two reasons: in order not to undermine the trust in statistics even more (if existing population statistics could not be trusted why would census data be better in the view of laymen?) and because it is not truly independent from the government.

Moreover, in 2019 the government initiated a discussion that discredited a traditional census in general and the 2001 census in particular. They argued that (1) census data would not be reliable anyway since many people would not open the door to enumerators; (2) it was possible to estimate the number of people using registers and the data from mobile operators; (3) thus there was no need to spend so much money on a census. This suggests that the fundamental problem with the census is the quality of governance in Ukraine. Thus, Ukrainian governments were too often busy with winning the next elections rather than with protecting voter rights; and they focused on developing quick and popular solutions rather than designing data-based and efficient policies.

The damage done to Ukraine by Russia's war – both in terms of lives and in terms of physical infrastructure – is enormous. After the war, a long and costly reconstruction will be needed. There is an understanding of both Ukrainians and Ukraine's international partners that the reconstruction should include not only rebuilding material infrastructure but also rebuilding or creating strong institutions and processes (e.g., Becker et al., 2022). Thus, during the reconstruction, Ukraine will have a chance to finally introduce proper policy procedures. Since the reconstruction process will be coupled with the EU accession process (Ukraine became an EU candidate in June 2022), Ukraine could benefit from EU technical assistance and conditionality to build proper institutions and develop its public service.

Given the huge demographic shifts caused by the war, the necessity of a census becomes obvious. At the same time, resources to implement a proper census will be unavailable in the first years after the war. However, considering that migration is likely to be quite intense during these few years while people settle down, we recommend using administrative data for population estimates during or right after the war; the new procedure for registering or declaring the place of living from February 2022 should help with this. Implementation of a traditional census makes sense within three to five years after the war ends to collect the data on households and their living conditions which, among others, will be needed for targeted and inclusive social policies. This data will also provide the basis for sample surveys.

Afterward, Ukraine can launch a process for switching to a register-based census. In implementing the traditional census, the Ukrainian government would appreciate the financial support of international organisations (e.g., the UN), while for introducing the register-based census it will need intense technical assistance. Switching to a register-based census (initially planned for 2030) would require substantial investments into the further development of the Single Demographic Register and its integration with other registries (first of all, the registry of addresses). This will also require sophisticated technical solutions for data protection, access to the registry data, data processing, etc. Thus, although a register-based census has lower operational costs, its initial (fixed) cost is high. International organisations can help cover this cost and provide expertise for the organisation of the registry-based census.

It is important that the government does not repeat its 2019 mistake of devaluing the census as a policy instrument. There should be an honest and professional dialogue between the stakeholders of a census so that problems (e.g., poor quality of registries) are solved rather than swept under the carpet.

Notes

1 Ukraine is divided into 27 administrative units – Autonomous Republic of Crimea, 24 oblasts, and the cities of Kyiv and Sevastopol. Each oblast is divided into rayons. At the end of 2020, as a conclusion of the decentralization reform launched in 2014, Ukraine reduced the number of rayons from 490 to 136.
2 According to the UNHCR data, most Ukrainian refugees settled in Poland, Germany and Czech Republic. Reported huge numbers of refugees in Russia are doubtful – these may be people who were deported (i.e. involuntarily moved) to Russia from occupied territories of Ukraine (UNHCR, 2022).
3 https://zakon.rada.gov.ua/laws/show/2486–IX#Text
4 The book edited by Clem (2016b) provides a very detailed discussion of organization, methodology and caveats of Russian and Soviet censuses since 1897. Here we make a much more concise discussion.
5 Treatment of veterans with disabilities by the Soviet government was inhuman. Often, they were forcefully moved outside of cities to remote places like Balaam islands, where they died with little care provided.
6 To "compensate" for the population decline, in 1936 Stalin banned abortions, and this ban was in place until 1955.
7 Recently, the Security Service of Ukraine published archive documents related to the 1937 census (Center for Research on the Liberation Movement, n.d.).
8 Bocale (2016) provides a good overview of the history and current situation in Crimea.
9 Antisemitism was an official or implicit policy of the USSR during most of the time of its existence.
10 When Russian Empire occupied Crimea in the late 18th century, it deported or forced to leave many Qirimli and Greeks who lived there (Snyder, 2022).
11 Matviyishyn and Michalski (2018) discuss how Soviet Russification policy is felt in Ukraine even today.
12 In contemporary Ukraine, according to different surveys, the share of people who defined Ukrainian as their mother tongue was constantly 20–30% higher than the share of those who responded that they spoke Ukrainian in the everyday life. Many people who identify themselves as Ukrainians but prefer to speak Russian is one of the most pronounced results of Russification policy. Some people have been switching from Russian to Ukrainian in their everyday communication. This process was accelerated by 2014 and especially 2022 Russian attacks.
13 In 1921–1923, during the first famine caused by collectivization, the Soviet government took away possessions from churches (e.g., ritual tableware, crosses, gold from cupolas and others). Many churches at that time were closed and priests were executed (Syrota, 2013).
14 https://zakon.rada.gov.ua/laws/show/2524-IX#Text
15 Registry host is the State Migration Service, a government agency subordinate to the Ministry if Internal Affairs. State Migration Service deals with all the issues related to citizenship (e.g. it issues IDs and passports) and immigrants.
16 Many NGOs provide reports on electoral violations, for example, NGO Opora (n.d.). One can also look at OSCE reports on Ukrainian elections (e.g. OSCE, 2013).
17 Maintaining a voter registry is quite complicated too, so genuine mistakes are possible. The registry is kept by the Central Electoral Commission. It operates via regional

branches, and it relies on the data of local governments for the majority of voters. But it also receives data from heads of military units and penitentiary institutions on people serving there, from the Ministry of Foreign Affairs on the voters who live abroad, from courts on people who were recognized as unable to take responsible decisions, from Civil Acts registration centers on people who died, from social support services on homeless people and people unable to walk. To update his/her information (e.g. a change in the place of living), a voter must submit the respective documents to the local administration or local branch of the electoral commission.

18 The 2001 census costed USD 20 million (0.03% of Ukraine's 2001 GDP), but at that time, salaries in Ukraine were much lower and enumerators used paper rather than tablets

19 See, for example, the interview of Dr. Libanova, the the head of the Institute for Demography and Social Studies from the National Academy of Sciences: https://zn.ua/ukr/personalities/ella-libanova-te-scho-proviv-dubilet-ce-prosto-ne-perepis-337785_.html

20 Regulatory Impact Analysis was introduced as early as 2000 by the Presidential Decree #89 (President of Ukraine, 2000), and then reintroduced a few times by the Cabinet of Ministers.

21 The understanding of the government role as a market regulator and provider of public goods is very vague among the Ukrainian government. As one of the officials put it, "I need to write a regulatory act, I don't need to meet with business [to assess the impact of that act]".

References

Anderson, B., & Silver, B. (1983). Estimating russification of ethnic identity among non-Russians in the USSR. *Demography, 20*(4), 461–489.

Anderson, B., & Silver, B. (1989). *Demographic sources of the changing ethnic composition of the Soviet Union.* Prepared for presentation at the annual meeting of the American Association for the Advancement of Slavic Studies at the University of Michigan, 2–5 November 1989. Retrieved from https://www.ucis.pitt.edu/nceeer/1989-000-00-Anderson.pdf

Arel, D. (2002). Demography and politics in the first post-soviet censuses: Mistrusted state, contested identities. *Population, 57*(6), 801–827. Retrieved from https://www.cairn-int.info/journal-population-2002-6-page-801.htm

Baffour, B., King, T., & Valente, P. (2013). The modern census: Evolution, examples and evaluation. *International Statistical Review, 81*(3), 407–425. doi:10.1111/insr.12036

Becker, T., Eichengreen, B., Gorodnichenko, Y., Guriev, S., Johnson, S., Mylovanov, T., . . . Di Weder Mauro, B. (2022). *A blueprint for the reconstruction of Ukraine* (Rapid Response Economics No. 1). London, England: Centre for Economic Policy Research. Retrieved from https://cepr.org/system/files/2022-08/BlueprintReconstructionUkraine.pdf

Bocale, P. (2016). Trends and issues in language policy and language education in Crimea. *Canadian Slavonic Papers, 58*(1), 1–20. doi:10.1080/00085006.2015.1130253

Botev, N. (2002). The ethnic composition of families in Russia in 1989: Insights into the soviet "nationalities policy". *Population and Development Review, 28*(4), 681–706.

Center for Research on the Liberation Movement. (n.d.) *NKVD and Census 1937.* Lviv. Retrieved from https://avr.org.ua/?idUpCat=1847&locale=en

Clem, R. S. (2016a). On the use of Russian and soviet censuses for research. In R. S. Clem (Ed.), *Research guide to the Russian and Soviet censuses* (pp. 17–35). Ithaca, NY: Cornell University Press.

Clem, R. S. (Ed.) (2016b). *Research guide to the Russian and Soviet censuses.* Ithaca, NY: Cornell University Press. doi:10.7591/9781501707087 (Original work published 1986).

CMU. (1995). *Cabinet of Ministers Decree #536 of 20.07.1995. On the organization in 1999 of all-Ukrainian census.* Retrieved from https://zakon.rada.gov.ua/laws/show/536-95-%D0%BF#Text

CMU. (1998). *Cabinet of Ministers Decree #1536 of 28.09.1998. On the organization in 2001 of all-Ukrainian census.* Retrieved from https://zakon.rada.gov.ua/laws/show/1536-98-%D0%BF#Text

CMU. (2000). *Cabinet of Ministers Decree #767 of 06.05.2000. On approval of methodological recommendations on preparing the justification for draft regulatory acts.* Retrieved from https://zakon.rada.gov.ua/laws/show/767-2000-%D0%BF/ed20040311#Text

CMU. (2008). *Cabinet of Ministers Order #581-r of 09.04.2008. On the organization in 2011 of all-Ukrainian census.* Retrieved from https://zakon.rada.gov.ua/laws/show/581-2008-%D1%80/ed20080409#Text

CMU. (2020). *Cabinet of Ministers Order #1542-r of 09.12.2020. On the organization in 2023 of all-Ukrainian census.* Retrieved from https://zakon.rada.gov.ua/laws/show/1542-2020-%D1%80#Text

Declaration. (1918). *Декларация прав трудящегося и эксплуатируемого народа* [Declaration of rights of workers and exploited people]. 03.01.1918. Retrieved from http://www.hist.msu.ru/ER/Etext/DEKRET/declarat.htm

Decree of the Soviet government of 04.05.1961. *Обусиленииборьбыслицами(бездельниками, тунеядцами, паразитами), уклоняющимися от общественно-полезного труда и ведущими антиобщественный паразитический образ жизни.* [On strengthening counteraction to persons that shirk from socially useful work and lead anti-social parasitic lifestyle]. Retrieved from https://bankstoday.net/wp-content/uploads/2018/12/Ob-usilenii-borby-s-litsami-uklonyayushhimisya-ot-obshhestvenno-poleznogo-truda-i-vedushhimi-antiobshhestvennyj-paraziticheskij-obraz-zhizni.pdf

Dubilet, D. (2020). *Інтерв'ю з Дубілетом: 40% зростання ВВП, перепис населення, «Приватбанк» та книги* [Interview with Dubilet: "40% GDP growth, census, Privatbank and books"]. Dmytro Dubilet interview to Radio Free Europe of 24.01.2020. Retrieved from https://www.radiosvoboda.org/a/30417099.html

Duchene, A., & Humvert, P. N. (2018). Surveying languages: The art of governing speakers with numbers. *International Journal of the Sociology of Language, 2018*(252), 1–20. doi:10.1515/ijsl-2018-0012

Eniayejuni, A. T., & Agoyi, M. (2011). A biometrics approach to population census and national identification in Nigeria: A prerequisite for planning and development. *Asian Transactions on Basic & Applied Sciences, 01*(05), 60–67.

Hladun, O. (2013). *Гладун. О.М. Вплив голоду у XX столітті на чисельність населення України* [Impact of XX century famine on population quantity in Ukraine]. In the Materials of the scientific conference "Голод в Україні у першій половині XX століття: причини та наслідки (1921–1923, 1932–1933, 1946–1947)" [The famine in Ukraine in the First Half of XX Century: Reasons and Consequences (1921–1923, 1932–1933, 1946–1947)], Kyiv, Ukraine, Hubersky, L., Kvit, S., Libanova, E., & Smoliy, V. (Eds.). Retrieved from https://idss.org.ua/monografii/Golodomor_2013_20_11.pdf

Hladun, O. (Ed.) (2021). *Ред. Гладун О.М. Електронні реєстри: стан в Україні* [Electronic registries: Status in Ukraine.] Issued by the Institute for Demography and Social Research named after M.V. Ptukha of the National Academy of Sciences of Ukraine. Retrieved from https://bit.ly/3GEbcf0

Hladun, O., & Rudnytskyi, O. (2009). *Статистика населення в Україні в 1920–1930-ті роки* [Population statistics in Ukraine in 1920s–1930s]. *Демографія та процеси відтворення населення* [Demography and the processes of population reproduction.]. Retrieved from https://dse.org.ua/arhcive/12/5.pdf

Ierusalymov, V., & Marchak, D. (2021). *Ієрусалимов, В., та Д. Марчак. Аналіз ефективності законодавчого регулювання надання соціальної допомоги в Україні для основних цільових груп* [Analysis of efficiency of legal regulation of social support provision in Ukraine for the main target groups]. Centre for Public Finance and Governance of the Kyiv School of Economics. Retrieved from https://kse.ua/wp-content/uploads/2021/07/Sotsialni-dopomogi_doslidzhennya.pdf

Institute of History of the National Academy of Sciences of Ukraine. (n.d.) *Цифровий архів Голодомору.* [Holodomor digital archives]. Retrieved from http://holodomor-era.history.org.ua/cgi-bin/holodomor/person.exe?I21DBN=DAH_DT&P21DBN=DAH&S21FMT=dct&T21PRF=GD%3D&T21CNR=1000&C21COM=T

Kupina, E. (2020, May 25). Statistical information in Ukraine: Outdated law, new challenges. *VoxUkraine.* Retrieved from https://voxukraine.org/en/statistical-information-in-ukraine-outdated-law-new-challenges/

Kuznetsova, E. (2022, April 24). Game of Reversi: Why there are Russian speakers in Ukraine. *VoxUkraine.* Retrieved from https://voxukraine.org/en/game-of-reversi-why-there-are-russian-speakers-in-ukraine/

Martins, J. M., Yusuf, F., & Swanson, D. A. (2012). *Consumer demographics and behaviour: Markets are people.* Dordrecht, The Netherlands: Springer. doi:10.1007/978-94-007-1855-5

Matviyishyn, Y., & Michalski, T. (2018). Language differentiation of Ukraine's population. *Journal of Nationalism Memory & Language Politics, 11*(2), 181–197. doi:10.1515/jnmlp-2017-0008

Mespoulet, M. (2022). Creating a socialist society and quantification in the USSR. In A. Mennicken & R. Salais (Eds.), *The new politics of numbers* (pp. 45–70). Cham, Switzerland: Springer.

Opora. (n.d.). *Опора. Вибори* [Opora. Elections]. Retrieved from https://www.oporaua.org/category/vybory?type=report

OSCE. (2013). *OSCE/ODIHR final report on Ukraine parliamentary elections recommends measures to improve transparency, impartiality.* Retrieved from https://www.osce.org/odihr/98571

President of Ukraine. (2000). *Decree #89 "On introduction of the single state regulatory policy in the entrepreneurship sphere."* Retrieved from https://zakon.rada.gov.ua/laws/show/89/2000#Text

Revenko, A. (2001). Ревенко А. Скільки нас було, є та скільки буде напередодні першого всеукраїнського перепису населення [How many of us there were, there are and there will be before the first all-Ukrainian census]. *Mirror Weekly, 30 November 2021.* Retrieved from https://zn.ua/ukr/SOCIUM/skilki_nas_bulo,_e_ta_skilki_bude_naperedodni_pershogo_vseukrayinskogo_perepisu_naselennya.html

Sarioglo, V., Dubilet, D., Verner, I., Patsera, K., Daniuk, M., Chertov, O., Polikarchuk, P. (2019). *Оцінка чисельності наявного населення України* [Estimate of the number of present population of Ukraine as of December 1st, 2019]. Retrieved from https://ukrstat.gov.ua/Noviny/new2020/zmist/novini/OnU_01_12_2019.pdf

Silva, R., Snow, R., Andreev, D., Mitra, R., & Abo-Omar, K. (2019). *Strengthening CRVS systems, overcoming barriers and empowering women and children.* Ottawa, Canada: International Development Research Centre. Retrieved from http://hdl.handle.net/10625/60225

Silver, B. (1974). Social mobilization and the russification of soviet nationalities. *American Political Science Review, 68*(1), 45–66. doi:10.2307/1959741

Silver, B. D. (2016). The ethnic and language dimensions in Russian and soviet censuses. In R. S. Clem (Ed.), *Research guide to the Russian and Soviet censuses* (pp. 70–97). Ithaca, NY: Cornell University Press.

Snyder, T. (2022). *Russia's crimea disconnect.* Retrieved from https://snyder.substack.com/p/russias-crimea-disconnect

Stebelsky, I. (2009). Ethnic self-identification in Ukraine, 1989–2001: Why more Ukrainians and fewer Russians? *Canadian Slavonic Papers/Revue Canadienne des Slavistes, 51*(1), 77–100 (24 pages).

Steshenko, V. (2019). Стешенко В.С., Переписи населення [Population censuses]. In *Encyclopedia of the history of Ukraine.* Institute of History of Ukraine, NAS of Ukraine, Kyiv. Retrieved from http://www.history.org.ua/?termin=Perepysy_naselennia

Sullivan, T. A. (2020). Coming to our census: How social statistics underpin our democracy (and republic). *Harvard Data Science Review, 2*(1). doi:10.1162/99608f92.c871f9e0

Syrota, O. (2013). Сирота О.І. Голод 1921–1923 років в Україні та його руйнівні наслідки для українського народу *[1921–1923 famine in Ukraine and its devastating consequences for Ukrainian people]. In Голод в Україні у першій половині XX століття: причини та наслідки (1921–1923, 1932–1933, 1946–1947)* [The famine in Ukraine in the first half of XX century: reasons and consequences (1921–1923, 1932–1933, 1946–1947)], Hubersky, L., Kvit, S., Libanova, E., & Smoliy, V. (Eds.) Retrieved from https://idss.org.ua/monografii/Golodomor_2013_20_11.pdf

Toft, M. (2014). Death by demography: 1979 as a turning point in the disintegration of the Soviet Union. *International Area Studies Review, 17*(2), 184–204.

Tyshchuk, T. (2019, October 14). *How to count Ukrainians correctly?* Retrieved from https://voxukraine.org/en/how-to-count-ukrainians-correctly/

Ukrstat. (2003–2004). *All-Ukrainian population census 2001 portal.* Retrieved from http://2001.ukrcensus.gov.ua/

Ukrstat. (2004). *First all-national population census: Historical, methodological, social, economic, ethnic aspects.* Kyiv, Ukraine: Institute for Demography and Social Studies, NAS of Ukraine, & State Statistic Committee of Ukraine. Retrieved from http://2001.ukrcensus.gov.ua/d/mono_eng.pdf

Ukrstat. (different years). *Населення України* [Population of Ukraine]. Retrieved from https://ukrstat.gov.ua/druk/publicat/Arhiv_u/13/Arch_nasel_zb.htm

UNHCR. (2022). *Operational Data Portal (ODP): Ukraine refugee situation.* Retrieved from https://data.unhcr.org/en/situations/ukraine/location?secret=unhcrrestricted

VoxUkraine Editorial Board. (2019, May 20). *On behalf of the people. Ukraine requires electoral system reform.* Retrieved from https://voxukraine.org/en/on-behalf-of-the-people-ukraine-requires-electoral-system-reform/

ANNEXES

Questionnaire for policymakers

1　Which data, including official statistics, do you or your colleagues use for policy development, assessment or monitoring?
2　Which statistical data meet your needs and which ones doesn't?
3　Do you think that the data that you use is of appropriate quality? Is the quality of some data poor? Why do you think the quality of this data is poor?
4　In your opinion, why do statistical agencies implement censuses?
5　Which government policies, in your opinion, require census data? How does lack of a census over a long period of time affect the quality of these policies?
6　Do you think the government may use other data sources instead of a census to develop policies?
7　In your opinion, how does the lack of census for a long time affect the quality of other data?
8　In your opinion, how a statistical agency can collect the population data without a census?

Number and organisational affiliation of respondents

- 12 representatives of the central government institutions, including the Ministry of Finance (two respondents), The Ministry of Economy (two respondents), the Ministry of Energy (two respondents), the Ministry of Regional Development (two respondents), the Ministry of Education (one respondent), the Ministry of Health (one respondent), the Ministry of Social Policy (one respondent), the Ministry of Internal Affairs (one respondent).
- Four representatives of oblast administrations from Luhansk (Eastern part of Ukraine), Poltava (Central part), Lviv (Western part), Kherson (Southern part).
- Four representatives of local authorities from the same oblasts.

Table 8.4 Questions of censuses held on the majority of Ukrainian territory

	1897	1920	1923**	1926	1939	1959	1970	1979	1989	2001
Name	Name(s), patronymics and surname	Name, patronymics and surname	Name, patronymics and surname	Name, patronymics and surname	Name, patronymics and surname	Name, patronymics and surname	Name, patronymics and surname	Name, patronymics and surname	Name, patronymics and surname	Name, patronymics and surname
Relation to household head	Relation to both household head and to the head of own family	—	Is a member of a rural household? Gets help from this household?	Family relations – only in urban areas	Relation to family head	Relation to family head	Relation to family head	Relation to household head and family head (separate families identified)	Relation to household head and family head (separate families identified)	Relation to household head and family head
Temporarily absent or present household members	Temporarily absent/present	Temporarily absent/present			If temporarily absent – for how long	If temporarily absent – for how long	If temporarily absent – for how long and why	Temporarily absent/present	Temporarily absent/present	temporarily absent/present
Gender	+	+	+	+	+	+	+	+	+	+
Number of children born	—	—	—	—	—	—	—	For women only*	And how many of them are alive - for women only*	And how many of them are alive
Family status	Married, not married, divorced or widowed	Married, not married, divorced or widowed	+	Married, not married, divorced or widowed	Married or not	Married or not	Married or not	Married, not married, divorced (officially or not) or widowed	Married, not married, divorced (officially or not) or widowed	married, not married, divorced (officially or not) or widowed
Age	Number of years and months from the day of birth	Years from birth or months for children under one year	+	Years from birth or months for children under one year	Years from birth or months for children under one year	Years from birth or months for children under one year	Years from birth or months for children under one year	Year of birth and age (full number of years)	Date of birth, number of full years	Date of birth, number of full years
Place of birth	Exact place	Exact place	—	Exact place	—	—	—	—	Exact place	Exact place

(Continued)

Table 8.4 (Continued)

	1897	1920	1923**	1926	1939	1959	1970	1979	1989	2001
Place of permanent residence	Exact place; and also place of registration for those who should be registered	For how long has been living here permanently? Or lives here temporarily	—	For how long has been living here permanently? Or lives here temporarily	Place of permanent residence and for how long has been absent from there?	Place of permanent residence and for how long has been absent from there?	Place of permanent residence and for how long has been absent from there? For how long has been living here permanently?* If less than two years – when and why moved here?*	Time of uninterrupted living in this place*	Time of uninterrupted living in this place* (if not since birth – then where came from)*	Time of uninterrupted living in this place (if not since birth – then where came from)
Religion	+	—	—	—	—	—	—	—	—	—
Nationality	—	+	+	+	+	+	+	+	+	Ethnic background
Citizenship	—	For foreigners only	—	For foreigners only	+	+	For foreigners only	—	—	+
Mother tongue	+	+	—	+	+	+	+	+	+	+
Other languages	Can speak Russian? For homeless or nomadic people	—	—	—	—	—	Which other USSR languages can freely speak?	Which other USSR languages can freely speak?	—	—

Literacy	Can read?	Can read and write or only read in Russian? Can read and write or only read in other language? Or illiterate?	—	Can read and write or only read in any language? Or illiterate?	Can read and write or only read in any language? Or illiterate?	For people older than nine who don't have primary education: Can read and write or only read in any language? Or illiterate?	—	—	—
Education	Where is/was studying? Completed the studies?	General or specialised school? Completed or not?	If studies, does get a stipend?	—	For those who study – full name of the institution and year of studies; has secondary or higher education	For those who study – full name of the institution and year of studies; Higher, incomplete higher, technical school, secondary school, incomplete secondary (seven years), primary	For those who study – type of institution. Level of education.	For those who study – type of institution. Higher, incomplete higher, technical school, secondary school, incomplete secondary, primary or no primary education	For those who study – type of institution. Higher, incomplete higher, technical school, secondary school, incomplete secondary, primary or no primary education	For those who study – type of institution. Higher, incomplete higher, technical school, secondary school, incomplete secondary, primary or no primary education

(Continued)

Table 8.4 (Continued)

	1897	1920	1923**	1926	1939	1959	1970	1979	1989	2001
Employment	Main and secondary place of work, profession or position that provides means for living	Main and secondary employment that provides means for living – profession, enterprise, position, employment before 1914 and 1917	Main and secondary employment, position, profession	Main and secondary employment, position, profession, address of the place of work	Source of means for living; place of work	Place of work (name of enterprise or kolkhoz) or own subsistence farm; position in that enterprise	Source of means for living; place of work (name of enterprise or kolkhoz). Position in that enterprise*; for pensioners – previous main job*	Source of means for living: work at an enterprise, kolkhoz, own [subsistence] farm or as a craftsman, freelance, pension, stipend, dependent, other. Name of an enterprise* and position there*	Source of means for living: work at an enterprise, kolkhoz, work in a cooperative or for some individuals, freelance, own subsistence farming, pension, stipend, other support from the state or other people. Place of work* and the job done at this place*	Source of means for living: working for an enterprise, for an individual, own business or farm, unpaid employee in family business, subsistence farm, rent from property, pension, stipend, unemployment benefits or other benefits. Main employment: name and address of an enterprise, position there

Agricultural employment	Main employment? Profession? Has employees? Was employed before 1914? Before 1917?	—	—	—	—	—	—
Unemployment	Sources of means for living? Profession of a person who provides means for living. What profession a person considers his/her specialty? Can this person work - at all or in her profession?	Duration of unemployment and the last workplace; profession	—	Duration of unemployment, the last workplace and position in it, previous profession or specialty. What are the sources of means for living? Profession of a person who provides means for living.	—	If there is no employment that provides means for living, what are the sources of means for living?	—

(Continued)

Table 8.4 (Continued)

	1897	1920	1923**	1926	1939	1959	1970	1979	1989	2001
Social status	Social class and rank	Head of an enterprise with employees, family member working for family business, worker, administrative staff, intern	+	Head of an enterprise or farm (in this case has employees, or only family members?), worker, member of cooperative, administrative staff, family member working in family business	Worker, administrative staff, kolkhoz[a] worker, craftsman in a cooperative, an individual farmer, freelancer, priest or "non-working element"	Worker, administrative staff, kolkhoz[a] worker, craftsman in or not in a cooperative, an individual farmer, freelancer, priest	Social group[*]: worker, administrative staff, kolkhoz worker, craftsman, individual farmer, priest	Social group[*]: Worker, administrative staff, kolkhoz worker, craftsman, individual farmer, priest	Social group[*]: worker, administrative staff, kolkhoz worker, craftsman, individual farmer, priest	Social group[*]: worker, administrative staff, kolkhoz worker, freelancer, priest
Army duty	Relation to army duty	Participated in 1914–1917 and 1918–1920 wars?	—			—		—	—	—
Disabilities	Records on household members who are totally blind, dumb, deaf and dumb, incapable to act	Records on severe disabilities or mental illnesses	Is able to work?	Records on severe disabilities or mental illnesses	—		—	—		—

									Living conditions	Processing of census data	Publication of census results

Living conditions / Processing of census data / Publication of census results

Living conditions	—	—	Number of rooms, utilities and rent; description of the house – number of floors, material of walls and roof, number of flats	Description of property and utilities – only in urban areas	New maps and the list of houses were created before the census	—	—	Age and material of the house. Owner^b of the house. Type of premises (individual house, flat, dormitory etc), utilities, number of rooms and square metres	Type of house, number of rooms and area of a flat or a house; land plots and rights for them (own or rented); area of land plots, are you renting them out?
Processing of census data	4+ years	Nine years	—	Two years	Never completed because of the war	Four years	Three years	—	One year – main results
Publication of census results	—	2 general books, 89 books for regions	—	56 books published by 1932	Several very general tables published in newspapers	1 general book and 15 books for each of the republics	20 books	11 books published in Ukraine; aggregated statistics for USSR was never published	Web-site, a number of publications on certain issues

Source: Compiled by authors using Ukrstat (2004).

* this question was asked only of a sample of people (in every fourth household)

** This census was organised only in urban areas and covered about 18% of population. The government expected to get data about rural households from financial records – when the objects subject to taxation would be recorded. However, this never happened since government abandoned the policy of taxation in favour of collectivisation and nationalisation.

a kolkhoz (collective farm) is a form of an agricultural enterprise in the Soviet Union. The ideal model for it was 'commune' or 'kibbutz'. Kolkhozes were imposed onto farmers during 1920s–1930s: individual farmers had to give up all their land except for small plots near their houses, as well as kettle and poultry, to kolkhozes, and then work on that land and grow that kettle and poultry 'for the common good'. Kolkhoz workers would get their pay (a part of product which they have grown) only after they supplied the planned amount of products to the state. Naturally, under such conditions labour productivity in kolkhozes was very low, and farmers were very poor.

b A house in a city/town could be owned by the state, a cooperative, or an NGO (e.g. a trade union), or individual houses could belong to individuals.

III

The politics of socio-technical and methodological innovations

9 Establishing a register-based census in Spain

Challenges and implications

Alberto Veira-Ramos and Walter Bartl

9.1 Introduction

An increasing number of countries in Europe are leaning toward making censuses based entirely or partially on register data. The leading pack has been the Scandinavian countries, followed by Austria and the Baltics. More recently, Spain has decided to join the emerging trend. Consequently, Spain has become the largest country in the world to switch from a traditional to an entirely register-based population census in 2022. How was such a fundamental methodological shift in the cornerstone of public statistics possible? This chapter describes the process of this methodological innovation.

The analytical background of the paper is based on four theoretical approaches that are used as sensitizing concepts. They are useful for highlighting the way in which the process of creating the conditions for carrying out a register-based census in Spain was influenced by national as well as international developments. In particular, municipal registers were an important infrastructural precondition which were enhanced by a series of national "investments in form". Furthermore, the development of statistical standards for member states of the European Union (EU) including register-based censuses proved to be important.

This chapter is based on information contained in official documents publicly available from the National Statistical Office (*Instituto Nacional de Estadística, INE*), and other legal and academic documentation. It lacks references to parliamentary debates or to controversies reflected by the media, because censuses in Spain are governed by decrees and their implementation has not raised major public discussions (Treviño Maruri & Domingo, 2020, p. 120). The documents analyzed allow for the description and explanation of how a series of technical decisions and legal reforms, made since the mid-1990s, led to a shift in the methodology used to implement a census in Spain. Beginning with a brief description of the history of census taking and the origin and development of municipal registers in Spain, this chapter later describes the implementation of censuses since 2001 in several steps. Possible explanations for this shift as well as implications for the future will be reflected upon in the concluding section.

DOI: 10.4324/9781003259749-13

9.2   Four sensitizing concepts

The analytical background of this chapter is based on four theoretical approaches that we combine as sensitizing concepts rather than pitting them against each other in strictly epistemological terms. First, it is based on Emigh, Riley, and Ahmed's (2016a, 2016b) conflict-sensitive approach that focuses on processes of state-society interactions. With regard to census taking, these interactions are understood as rooted in power interests and interpretive sensemaking at the same time. In essence, census taking is typically a state endeavor that formalizes foregoing lay categories. Census taking produces second-order constructions of social reality that might be more or less in line with the first-order constructions of social actors. As such, the knowledge produced is not arbitrary, because the very notion of knowledge implies a certain correspondence between symbols and their referents. In the case of the census, the objects of classification are "interactive kinds" that exist prior to imposed categories and might contest them (Riley, Ahmed, & Emigh, 2021, p. 344). Indeed, the authors see knowledge change as the outcome of interpretive struggles and critique.[1] Empirically, census taking might be based to a varying degree upon interactions with society, which has consequences for the instrumental value of census data. While the mature modern censuses have become more open to social influences since the Second World War, state-society interactions in the United States of America, for example, have been more intense than in Italy, where census taking has basically remained an expert project. This finding does not fit well with the assumption that national statistical offices (NSOs) are more responsive to social demands when elected officials are in control of national statistical offices (Howard, 2021, p. 214). While the Italian NSO is accountable to the national parliament, the U.S. Census Bureau is positioned directly within the hierarchy of the U.S. Department of Commerce (Prévost, 2019, p. 161). The Spanish *Instituto National de Estadística* (INE) in comparison has more formal autonomy than the U.S. Census Bureau, because it was re-constituted as an independent government agency under the Ministry of Economic Affairs by the Statistics Act of 1989 (BOE, 1989, p. 14027).[2] However, the authority of NSOs is not determined by formal regulations but also depends on dramaturgical practices of statisticians (Howard, 2021). Among the institutional features that set the stage for statisticians' professional practices, administrative traditions have been found to be most relevant for the authority of NSOs (Howard, 2021, p. 205).

Spanish public administration is rooted in the Napoleonic tradition, including centralism, priority of the rule of law over managerial qualities and a professional civil service (Alba & Navarro, 2011). While public administration was politically dominated during Franco's authoritarian regime (1939–1976), it is today considered more independent from politics than its counterparts in Italy or Greece (Sotiropoulos, 2018, p. 891). As a consequence of this historical legacy, political and administrative elites in Spain are rather isolated from the larger public and do not put much emphasis on participatory governance. Complementary to that, the political affection of the population is relatively low (Fishman, 2020, pp. 27–28). Despite the decentralization of the Spanish polity during the country's transition to democracy (Colino, 2020), a population survey decades later strikingly showed that a large majority of Spaniards still believed that the central administration was "still responsible for service delivery, including service delivery that has in fact already been transferred to the regional governments" (Alba & Navarro, 2011, p. 798). Against the backdrop of the Napoleonic administrative tradition, we would expect low interaction between the state and society during the transition to a register-based census. However, more recently, an increasing hybridization of traditional Spanish public administration with managerialism was observed (Parrado, 2020, pp. 190–195).

Second, complementing the first perspective, we follow the world polity theory (Kukutai, Thompson, & McMillan, 2015; Ventresca, 1995) in their emphasis on transnational influences on states. Taking social change seriously (as demanded by Emigh and colleagues), we use the world polity approach to highlight transnational influences on the process of quantification itself and on the object of quantification: the population to be counted. In the first respect, international organizations have become standard setters for information gathering that act to a considerable extent independently from the nation-state's interests.[3] Mara Loveman (2014), for example, showed how civil society actors in Latin America coalesced with international organizations for bringing ethnoracial categories to the census. In the case of the EU, it is obvious that the harmonization of national standards of information gathering matters (Desrosières, 2000; Radermacher, 2020, pp. 42–44, 92–93), but this is in itself an arduous process (Grommé, Ruppert, & Ustek-Spilda, 2021; Ruppert & Ustek-Spilda, 2021). In fact, the 1989 statistical act reconstituting the INE was justified by the accession of Spain to the European Communities in 1986 and the ensuing transnational cooperation requirements (BOE, 1989, p. 14027). More recently, international organizations have advocated register-based censuses (UNECE, 2007, 2018).

In the second respect, changes in the population to be counted, world polity theory seems to be of limited use at first sight, because despite postulating the cultural constitution of individuals as legitimate actors of world society (Meyer, 2010), individual agency – for example in migration processes – has largely been neglected. However, it leads us to check which institutional frameworks exist at a transnational level. Up to now, there is no global migration regime in place (Faist, 2010, p. 1670).[4] In comparison to migration, the two major norms of the global refugee regime (asylum and burden-sharing) are institutionalized to a diverging extent: while the norm of asylum is widely accepted, the norm of burden-sharing is weak and largely discretionary (Betts, 2015, p. 363). With the creation of the Schengen area in the EU, migration across national borders of member states was liberalized, while access for individuals from third countries was restricted. The Schengen agreement was accompanied in asylum policy by the highly controversial Dublin Regulation that has not yet been translated to legitimate terms of responsibility sharing among EU states (Doomernik & Glorius, 2022). The fortification of Europe's external border created paradox situations of simultaneous formal inclusion *and* exclusion of irregular immigrants at different levels of governance (Chauvin & Garcés-Mascareñas, 2020). In the case of Spain, becoming a member state of the European Communities meant that the relatively homogenous population that had emerged from the early external isolationism, internal suppression of ethnic minorities, and bilateral agreements for labor emigration of the Franco era (Pardos-Prado, 2020; Parrado, 2020, p. 189) has become more diverse due to legal and irregular immigration of various origins starting from the late 1990s (Pardos-Prado, 2020; Veira, Stanek, & Cachón, 2011). Especially, irregular immigrants might have incentives to remain invisible to state officials and prefer not to be identified or counted (Herzog, 2021). Later, especially during the Great Recession (2008–2014), many young people emigrated to northern EU member states (Lafleur, Stanek, & Veira, 2017). While immigration policy was not a major political concern for almost two decades (Amat i Puigsech & Garcés-Mascareñas, 2019; Pardos-Prado, 2020), official statistics in Spain have been criticized for not representing this growing diversity of the population adequately, for example, with regard to education policy (El-Habib Draoui, Jiménez-Delgado, & Ruiz-Callado, 2019). Critique of this kind leads us to expect growing attention for data on migration in the Spanish statistical agency INE.

Third, while we agree with Emigh and colleagues that state power derived from census taking might be limited by resistance and critique from society, we would like to emphasize a second possible limitation to power from census information, the prerequisite of "investments in form" (Desrosières, 1991; Thévenot, 1984, 1985). From the perspective of French pragmatism, quantification is a production process that requires not only costly negotiations about conventions of classification and equivalence but also investments in material objects and organizational procedures. Hence, quantification is not a merely cognitive endeavor of establishing operational definitions and rules of measurement but requires a corresponding socio-technical basis. Once established, these arrangements can be used for new purposes and serve as shortcuts for facilitating interactive coordination. When applying this concept to the transition to a register-based census, we recognize that such a transition depends on the existence of administrative registers. State technologies to identify persons as individuals date back to the European police state of the late 18th century (Caplan, 2001; Caplan & Torpey, 2001, pp. 8–9). Although the registration of persons as legally recognized entities has been codified in 1966 at a global level in in Article 24 of the United Nations International Covenant on Civil and Political Rights for empowering individuals *vis-à-vis* states (Szreter & Breckenridge, 2012), the respective registers are still lacking in many parts of the world (Silva, Snow, Andreev, Mitra, & Abo-Omar, 2019). However, in countries where local population registers exist, such as in Spain, they need to be linked at the national level, which creates problems in identifying individuals and avoiding duplicate entries. In Spain, this task is especially challenging, because, since the early 19th century, every settlement was allowed to form a municipality. As a consequence, Spain has more than 8,000 municipalities today.

Our fourth argument underpins the third one in a differentiation-theoretical perspective. The growth of public tasks in the world's core states, beginning from the second half of the 19th century (Mann, 1993/2012, pp. 467–472), has led to the organizational differentiation of state actors, both in a horizontal (functional) and in a vertical (multi-level) sense. However, differentiation dynamics are neither linear nor do they automatically increase the efficiency of society, because coordination between conflicting value orientations and organizational interests becomes more difficult. Instead, further differentiation might become limited by scarce resources or a call for mechanisms of reintegration (Schimank, 2005). The growth of tax revenues in the world's core states has basically stagnated since the 1980s (Piketty, 2014, p. 632).[5] Partially in response to scarce resources, the new public management discourse has propagated more efficient procedures of public administration, contracting out of services, and the creation of relatively autonomous state agencies (Pollitt & Bouckaert, 2017). Agencification in particular has also created new coordination problems for public governance (Waluyo, 2022). While differentiation theories emphasize political coordination and steering as mechanisms of reintegration (Schimank, 2005), the growing debate in political science literature about policy coordination, coherence, and integration (Trein, Meyer, & Maggetti, 2019) seems to confirm this assumption. Although official statistics are a medium for societal coordination (Starr, 1987), the ability to link personal information for statistical purposes from a diverse set of administrative sources is itself hampered by organizational differentiation. This is even more so the case, as personal data must be protected against disclosure throughout the linkage procedure (Christen, Ranbaduge, & Schnell, 2020). The introduction of personal identification numbers across different administrative domains has allowed for the interoperability of public and private data repositories, especially in the Nordic welfare states (Alastalo & Helén, 2022); yet their potential introduction was highly contentious elsewhere, as case studies on Germany and the United Kingdom show (Frohman, 2020; Whitley, Martin, & Hosein, 2014). The problem of interoperability is further increased when private date providers are involved. While the production of sample surveys has been contracted out to private providers for decades (Starr & Corson, 1987), the management of Big Data in the public realm is likely to evoke significant tensions because of the different purposes the data were originally collected for and the different conventions applied to their procession (Diaz-Bone & Horvath, 2021). Against the backdrop of the latter strand of literature, the way in which the interoperability of registers was achieved in the 2021 Spanish census is all the more in need of explanation.

9.3   Brief history of Spanish censuses

The first Spanish population counts date back to the 16th century. However, they cannot be considered modern censuses, because they were not aimed at making an exhaustive count of the population; information was not gathered directly from citizens but from local municipal authorities, and data collection was not carried out by professional staff (Ventresca, 1995). Moreover, those "primitive" censuses pursued specific ends such as the enumeration of "*vecinos pecheros*", that is, subjects obliged to pay taxes (Melgar García & Barrionuevo Dolmos, 2009). The first Spanish[6] census containing a modern element aimed at counting the population for its own sake became known as the "*Conde de Aranda Census*" of 1768. It took its name from the aristocrat who was commissioned with its implementation by the King, S. M. Carlos III. Aranda gave precise instructions to bishops for requesting from parish priests the completion of questionnaires aimed at providing an exhaustive count of all the population living within the boundaries of their parishes. In 1787 another census was implemented under the supervision of José Moñino y Redondo, count of Floridablanca (*"Conde de Floridablanca Census"*). This time, population data was gathered by civil servants, instead of clergymen. The next census took place in 1797, with S. M. Carlos IV already on the throne, and was commissioned by one of his ministers, Manuel Godoy.

The first half of the 19th century was characterized by intense political turmoil. Political stability enabled the reestablishment of statistical activity and the creation of the necessary institutions to undertake the tasks associated with the implementation of population censuses such as the General Commission of Statistics of the Kingdom founded in 1856 and the Institute for Geographic and Statistical studies, founded in 1873. The creation of these institutions characterizes the early history of the modern Spanish census as a central state project rather than a societal one. However, municipalities (and later also provinces) as an integral part of the state had a crucial role in coopting the knowledge of the church and other social actors for this endeavor of "bottom-up nation building" (Salas-Vives & Pujadas-Mora, 2021). Modern census taking began with the censuses of 1857 and 1860 (Gozálvez Pérez & Martín-Serrano Rodríguez, 2016; Melón, 1951, p. 211). In 1877, the internationally recommended distinction between *de jure* and *de facto* population was introduced and maintained in the following censuses. It was considered useful for monitoring spatial mobility and migration (Melón, 1951, p. 227). During the second half of the 19th century, five censuses were implemented, in 1857, 1860, 1877, 1887, and 1897. A law passed in 1880 established that from 1900 onwards, censuses would be implemented every ten years.

Since 1945, the INE has been the government agency in charge of organizing and conducting the censuses. From 1950 onwards, Spanish censuses have been referred to as "Population and Housing Censuses", because they collect information on the characteristics of the inhabitants and the dwellings in which they live. However, in Spain, another method of population count besides censuses has been operative since the 19th century.

9.4   The Spanish municipal register

Unlike census data, municipal population registers (*Padrón Municipal*) contain official data with legal implications, including providing the informational base for the Electoral Census. Moreover, entitlement to certain rights and obligations related to services and taxes managed by local municipal authorities apply only to individuals registered as inhabitants in the municipal register. In contrast, data collected by censuses are of merely statistical nature and protected by statistical secrecy.

An early version of municipal registers existed in Spain during the Late Middle Ages (Pino Rebolledo, 1991). The modern register, however, is much more recent.

Article 6 of the Instruction of February 3rd, 1823, ordered the municipalities to conduct a census of their inhabitants in January of each year (Suero Salamanca, 1999). One year later, the Municipal Statute extended this order to gather also information on inhabitants' "qualities" (Presidencia del Directorio Militar, 1924, p. 1225, our translation).

Despite the provisions of the Spanish legislation of the 19th and early 20th century regarding municipal registration and the series of sanctions contemplated for those failing to comply, it was quite difficult to get the population used to registering. The slow but steady increase in registered persons is attributed by García-Pérez (2007) to the evolution of population control and identification instruments. Passports required prior to their issuance that the applicant's residence be accredited by means of the so-called municipal registration ballot (Ministerio del Interior, n.d.). These documents were essential to life in towns and cities because they allowed a person to authenticate her/himself, move from one place to another, get a job, perform public employment, or exercise legal actions and were the basic instrument for interacting with the public administration, including the management of administrative procedures and taxes. All these procedures had as an essential requirement that the person concerned be registered in a municipality, so it is not unreasonable to suppose that they were a decisive stimulus for the normalization of the registration process (García Pérez, 2007).

From 1981 to 1996, the municipal census was carried out every five years, in the years ending in 1 and 6. Its implementation began with the home delivery of the "registration sheets", which had to be filled out by the head of the family. Then, after the data was collected and organized by districts and sections, and approved by the city council, it was exposed to the public so that claims could be presented. Once resolved, these sheets would be sent together with the summaries to the Provincial Statistical Delegations, which were in charge of publishing the resulting figures in the Official Gazette of the Province. In order to keep the data useful and accurate, the register was updated every year with new registrations, de-registrations, and modifications that had occurred.

There have been strong incentives that made municipalities very keen to count every citizen living within their administrative boundaries. With respect to the local treasury, taxes received by a municipality from the government was fixed by its number of inhabitants, so that the greater the number of people registered in a register, the greater the transfer. Likewise, within the prescriptive framework of urban welfare planning that exists in Spain (Caldarice, 2018, p. 21), the population of a municipality determines the minimum mandatory services that a city council must provide. Within that framework, climbing up the hierarchical classification of settlements, based on population thresholds, legitimates means for a somewhat more complex administrative organization.[7] Moreover, population was also a criterion to calculate the number of councilors belonging to a city with a view to local elections. As a result, there were clear incentives for registering citizens, but not for unregistering them. Mistakes in municipal registers were increasingly due to over-coverage; i.e. the failure to unregister inhabitants who had moved away to another municipality (or to another country), rather than under-coverage. From the individual's side, it has already been mentioned that registration assured access to rights and services provided by the local and regional authorities. Hence, there were strong incentives for municipalities and individuals to register but rather weak ones to unregister. Thus, soon enough, authorities realized that duplicity became a major problem of municipal registers, because many individuals who moved away from one municipality ended up registered in at least two different municipalities.

In order to improve the reliability of municipal registers, a series of legal reforms were approved during the 1990s. Reforms, especially from 1992 onwards, took into consideration the new possibilities made available by technological improvements in data processing and storage, as well as the affordable price and usability of computer equipment (Muñoz-Cañavate & Hípola, 2011, p. 77). Such improvements convinced the Spanish government that municipal registers could be permanently updated, thereby ensuring the quality of the accumulated information and that they could be constantly refined. Indeed, a reformulation of the 1985 Regulation of the Bases of Local Government (*Reguladora de las bases del Régimen Local*), approved in 1996, aimed to computerize municipal registers in order to allow for electronic coordination among all registries and to avoid five-year renewals, thus avoiding errors inherent to the individualized management of each registry; at the same time, that would facilitate its permanent update allowing INE yearly updates of official statistics ([BOE], 1996, p. 813).

In the Regulation of the Bases of Local Government, the municipal register was defined as the administrative register listing the residents of a municipality; it constitutes proof of residence in the municipality. Thus, certifications that are issued based on that register have the character of a public and reliable document for all administrative purposes. Data from the municipal register could be transferred to other public administrations that request it without the prior consent of the affected party only when they are necessary and exclusively for matters in which residence or domicile is relevant information. Beyond these assumptions, the municipal register data is confidential. This means that police could not use it to, for instance, track immigrants in irregular situations. Thus, while the 1996 legislation specified that immigrants could not use the municipal register to claim rights related to the recognition of legal residence status or naturalization (BOE, 1996, p. 815), they still could benefit from registering to be entitled to certain municipal or regional services, such as public health care provision or schooling, regardless of their legal status and without fearing being tracked by immigration police. Moreover, irregular immigrants could initiate a procedure to become regular (regularización por arraigo social) after five years (later three years) of living in Spain when they could prove to be socially integrated. One item of evidence that can be presented to authorities to prove integration is a municipal register sheet. The latter two aspects proved to have a remarkable impact on the reliability of the municipal registers when Spain received an inflow of more than 6 million immigrants between 2002 and 2008.[8]

The 1996 reform of the Regulation of the Bases of Local Government also opened up the possibility that municipal register data could be used to produce official statistics subject to statistical secrecy, which became crucial for the transition to a register-based census. Moreover, it also established the standard criteria by which the municipalities have carried out the maintenance, review, and custody of the municipal registers to ensure uniformity in the data. By doing so, municipal registers were intended to serve as a base element for the preparation of population statistics at national level and of the Electoral Census (BOE, 1996, p. 813).

To ensure the achievement of the abovementioned goals, the INE was entrusted with the function of coordinating the different municipal registers. The Registration Council (*Consejo de Empadronamiento*) was created, as a collaborative body in this matter between the state administration and the local units in charge of registries (BOE, 1996, p. 814). The deputy director of INE was appointed to chair the Registration Council, since INE was put in charge of correcting possible errors and avoiding duplications, carrying out the appropriate checks, and informing municipalities of any actions and operations necessary so that the registry data could serve as a basis for the preparation of population statistics at the national level. The resulting data could then be used as the official reference for the elaboration of the Electoral Census. Soon afterward, therefore, the National Register (*Padrón Nacional*) was created, which is hosted by INE. This process has required the close cooperation between municipalities and the INE. Indeed, INE validates information received by municipalities and provides feedback to them after checking data from the civil register on deaths, births, and acquisition of Spanish citizenship. About 250 Consular offices of Spain around the world are also connected to the National Register to provide information on Spanish expatriates in the same manner as municipalities do.

At this point, no plan to change from a traditional census to a register-based census had been adopted or was even remotely in the mind of INE officials, according to the documents consulted. Technological improvements necessary for such tasks were still not available. Municipal-based data would simply serve to build the Electoral Census and to provide statistical information on the population residing in Spain more frequently than that obtained from a traditional census. Moreover, the continuously updated National Register revealed stark discrepancies with results from the previous 1991 census. While the 1991 census counted 38.9 million inhabitants living in 12 million households, the annual population count carried out by about 8,131 Spanish municipalities came out at 39.9 million. Such discrepancies were attributed to the absence of incentives for deregistering from the municipal register after moving away. Until deregistration became automatic, municipal registers were believed to have suffered from over-coverage.

Thanks to the reform of the Regulation of the Bases of Local Government in 1996, it became unnecessary for citizens moving away from a municipality to deregister from the municipality they were about to leave. Instead, the receiving municipality (or the corresponding Consular Office) became in charge of notifying the registration of a new inhabitant to INE. INE then communicates to the municipality of origin within the first ten days of the following month, so that it can proceed to clean up its register. This system of "backward policing" (Coleman, 1990, pp. 431–432, 445) has contributed to reducing many sources of inconsistencies, but not all of them. For instance, foreigners moving away to another country or Spanish nationals moving to another country without reporting to the corresponding Spanish consulate could still be counted as residents of a Spanish municipality. To cope with this situation, foreigners are expected to update their registration every two or five years, depending on the length of their residence permit. Otherwise, they are automatically deregistered.

Maintenance and custody of the municipal register corresponds to local municipal authorities, not to the police, as is the case in other countries. This decoupling of municipal registers from policing the population is a significant deviation of democratic Spain from the Franco period (cf. Galdon Clavell & Ouziel, 2014).

Despite the referred problems with municipal registry data, the availability of the National Register based on these data conditioned the decision-making process on how to implement the censuses of the 21st century. Another element that was still lacking will be described in the following section.

9.5   Investments in the interoperability of administrative data bases

In local contexts, persons can be identified by their name; authentication of the person is usually not a problem, because people know each other anyway. The questions of identification and authentication arise only when chains of interaction extend beyond local situations (cf. Caplan, 2001). Identification means the intertemporal consistency of information about a person, and authentication denotes the relationship between the physical existence of a person and information about her/him. For any registry, the identification of persons is crucial. Today, all people living in Spain are obliged by law to be registered in the municipality where they live. This registration is also a right and therefore should be possible without any legal or administrative restriction for everybody. For each person, the municipal register collects information on address, gender, date, place of birth (country of birth for those born abroad), level of education, nationality, name, and the fiscal identification number of a person (*Número de Identificación Fiscal* [NIF]) used by the national tax agency. For Spanish citizens, the NIF is made up of the number of the national identity card (*Documento nacional de identidad* [DNI]) and a letter. Carrying a DNI was made obligatory under Franco, who held the first ever issued ID card in the country, in 1951 (Galdon Clavell & Ouziel, 2014, p. 139). During his regime, the aspect of control was clearly at the center of the DNI. Although, today, carrying the DNI is formally not obligatory in Spain, it practically is, because citizens can be fined if refusal to show the DNI upon request of a police officer is regarded as obstruction of legal action. Foreigners, according to their legal situation, are registered with their number of residence permits (*Número de identidad de extranjero* [NIE]) plus a letter. Those still waiting to obtain their NIE or in an irregular situation are registered in the municipal register with their passport number.

In 2008, the NIF replaced the former tax identification number CIF (*Código de Identificación Fiscal*) ([BOE], 2007). The NIF is the same number as the DNI plus an alphabetical letter (Esteban Romero, 2016); hence, the linkage between registers from different administrative bodies was possible. Indeed, ever since, all Spanish citizens use their DNI plus a letter in virtually all procedures where they have to identify themselves (with administration, with banks, with phone companies, etc.). The eight-digit DNI number plus the letter (=NIF) is printed on ID cards; hence, in daily life, the NIF is still called DNI. Foreigners also use their eight-digit NIE, preceded by the letter X plus a second letter following the eight-digit number as their fiscal identification number (NIF). Practically, police ID and tax ID have been merged. Rather than being simply a case of mundane imprecision, the common reference to the NIF as DNI indicates that practices of being identified by the state as an individual underwent a cultural process of normalization.

9.6   The Spanish census of 2001

In the 2000 census round, most countries of the world implemented a traditional census based on the comprehensive coverage of the territory (Kukutai et al., 2015, p. 12). Scandinavian countries, in contrast, based their censuses entirely on administrative records. Other countries such as Austria, Switzerland, Belgium, Portugal, or Spain opted for a combined approach, using administrative registers to support the implementation of the census. The Netherlands relied on administrative records and a sample survey to obtain complementary information.

In Spain, when planning the 2001 census, INE officials took into consideration three different options: (1) a traditional census, following Spain's own tradition and the example of the largest countries in Europe, (2) a census based entirely on administrative records, following the example of Scandinavian countries, and (3) the use of data from municipal registers plus a survey based on a sample of individuals, following the example of the Netherlands (INE, 2001, pp. 8–11). While all these options were considered, they were discarded for different reasons.

The implementation of a census based on the National Register plus a survey was discarded because that approach would threaten to misrepresent low populated provinces. Representing low populated provinces accurately had been one of the major achievements of previous censuses, particularly that of 1991. In fact, data from the 1991 census became a reference for private surveying companies or public institutions when estimating the number of interviewees that a national representative sample of individuals should include from each province. Such accomplishment would have been endangered by switching to a methodology similar to that applied in the Netherlands because of the coverage problems of municipal registers and, consequently, the National Register. Due to the importance of the census for providing small-scale data, the implementation of a complete count based on a short questionnaire in combination with a sample survey using a longer questionnaire, as in the United States or Canada, was also discarded (INE, 2001, p. 11).

A traditional census, in principle the safest option, would nonetheless fail to benefit from the creation of the National Register. Improvements in the municipal registers and the digitalization of Spanish administration were features that INE officials deemed to be taken into consideration when designing 21st-century censuses. Yet, following the Scandinavian methodology still required more preparation.

According to INE, the 2001 Spanish census could not be a register-based one because of various difficulties that had remained unresolved, such as legal barriers, a lack of a nationwide identification number for each person across all registers affected (a national tax ID was under development since 1990, but introduced only in 2008; BOE, 2007), a lack of adequate standardization of administrative records and an assumed lack of social acceptance. Furthermore, the National Register counts individuals, but not buildings or households; and linking data between registers and the cadaster would imply a long period of preparation for both data sets. Moreover, information provided by municipal registers was limited to very basic sociodemographic characteristics (sex, age, nationality, education, date and place of birth), thus important variables typically gathered in a census, such as occupation, marital status or the nature of the relationship between members of the same household would be missing.

Despite the lack of resources to design an entirely register-based census for 2001, INE documentation already signaled that officials considered this census as a significant step in that direction. Indeed, the interconnectedness between the census fieldwork and the municipal registers used to build the National Register became an essential element of the objectives of the 2001 census (INE, 2001, pp. 17–19). Deliberations at INE resulted in a fieldwork design that combined the traditional method, that is, an exhaustive inspection of the territory to collect information on all census units (persons, households, houses), with strong support from administrative records. In fact, since the National Register had been created recently, the 2001 census could be used to test the reliability of municipal registers for the future implementation of a census "*à la Scandinave*".

The 2001 census was designed to include any inhabitant residing in Spain plus Spanish citizens working temporarily abroad (including diplomats and their families, sailors, and pilots) and foreigners living regularly in Spain but temporarily abroad. Fieldwork preparation took into account information provided by municipal registers and the cadaster. Combining both data sets contributed to reducing the financial costs significantly and greatly enhanced the data's accuracy. Using municipal register data allowed questionnaires to be sent by mail instead of relying on census agents to drop them into the mailbox. Agents concentrated on gathering the questionnaires and clearing inconsistencies. Thanks to a pilot trial, the vast majority of inhabitants got the questionnaires within the scheduled time, shortening the length of fieldwork. Another advantage of using municipal registers was the implementation of only one field "tour" instead of two. This implied a saving of 36 million euros (INE, 2001, p. 37; our calculation of currency exchange rates).

The availability of certain data from municipal registers freed space from the census questionnaire, allowing for the introduction of new items such as place of work, hours of work, and questions related to the dwelling conditions. The latter could be filled by one person per household and the former were addressed only to persons above 16 years of age, and still on the labor market, which served to save money and coding time.

Each dwelling received the census questionnaire prefilled with the registration data in order to verify its accuracy and to introduce, where appropriate, the relevant variations. Other options, such as computer-assisted telephone interviews (CATI) or computer-assisted web interviews (CAWI) were used to gather data, but those were secondary options, directed only to people hard to reach at home, such as individuals living alone or young couples, both working. This multi-modal approach made Spain the fourth country in the world, after the United States, Singapore, and Switzerland, to make possible the completion of the census questionnaire by internet (INE, 2001, p. 34). Modifications of information affecting register data were sent to municipalities in order for them to verify and make the necessary adjustments. Information from census questionnaires was kept anonymized, in order to avoid violation of statistical secrecy legislation, which was scrutinized by the Spanish Agency for Data Protection.

These achievements in the 2001 census were possible because of the digitalization of the Spanish administration that had taken place during the previous decade and especially the implementation of the automatic deregistration from municipal registers, monitored by INE since 1996. The census of 2001 counted 40.8 million inhabitants living in nearly 14 million households. Results from the National Register had counted 41.8 million; hence there was still an overcount of 1 million in the register data, as was the case after the previous census.

9.7   The Spanish census of 2011

In the 2011 census, the Spanish concept of a census switched from processual to a product definition: What was aimed at was *information* on the entire population, no matter if this information was based on an actual *enumeration*. The preparation of the 2011 Spanish census faced various challenges. Some were anticipated but others were not. After the positive experience with the 2001 census combining cadaster data and municipal register data, there was an established consensus among INE officials on the idea that the country should move toward the implementation of a register-based census. Because of this, INE statisticians were very much concerned with the discrepancies between the results of the 2001 census and the National Register. Legal reforms were needed to grant the INE access to data collected and coded by other governmental branches such as the Spanish Tax Agency (AEAT) or social security. Finally, since the EU requires the housing census to contain georeferences of buildings from 2025 onwards, the experts saw a need for counting buildings, in order to check and potentially correct the information that was available from the cadaster.[9]

A series of events and social changes complicated the preparation of the 2011 census even further. First, between 2002 and 2008, a period of unprecedented economic growth in Spain, close to 6 million immigrants were estimated to have entered the country (Veira et al., 2011), pushing the percentage of foreign-born population upwards from 3.3% in 2001 to 12.2% in 2008. This would be the first time Spanish census agents had to deal with a large number of foreign-born individuals when conducting the interviews. Providing the interconnectedness between census operations and municipal registers, a great deal of success in the precise estimation of the foreign-born population would depend on how accurately local authorities had processed the arrival of immigrants into their municipal registers. Municipal registration offices in Spain are managed independently by the police or any Ministry or governmental body. This peculiarity proved decisive in helping to keep an accurate count of the foreign population living in Spain after the period when most immigration occurred. This is so because foreigners in irregular situations are not at risk of being identified by police or immigration officials when they apply to register.

Second, the 2011 census had to deal with the task of solving the controversy on how many Spaniards had left the country because of the effects of the 2008 financial crisis. The Great Recession of 2008 caused an increase in unemployment rates from 8% in 2007 to 20% by the end of 2010. It is well-documented that unemployment in Spain hits the young particularly hard and that since 2008 an undetermined number of young unemployed Spanish nationals have moved to other European countries looking for a job, or to continue their education (Lafleur et al., 2017). Very often, young Spanish nationals are registered as living in their parental households and they have no incentives to deregister from the municipality even if they live abroad. Moreover, the economic crisis of the construction sector affected the employment of immigrants as well. Many are believed to have moved to other European countries or to have returned to their countries of origin, but with reasonable expectations of returning to Spain. Thus, they were not much inclined to deregister as residents of municipalities where they had been residing and from where they used to benefit from social services. Often, some members of the same family could remain in Spain while others would move, but all would stay as registered inhabitants in order not to lose the entitlements to education or health care provision (Lafleur et al., 2017).

Last, but not least, the financial crisis imposed budgetary constraints on administrative spending, affecting the budget of the 2011 census ([INE], 2011, p. 4; Vega & Argüeso, 2016, p. 6).

Implementing the census according to EU regulations concerning standardization of certain procedures, quality evaluation and compliance with national and European data protection legislation were additional challenges added to a plate already full of tasks. The Regulation of the European Parliament and of the Council of 9th July 2008 on population and housing censuses provides a series of common definitions (Article 2) and general measures for the transmission of data (Article 5) (OJ, 2008). It also includes an Annex with a list of topics that should be covered in Population and Housing Censuses by Member States. Moreover, the Regulation requires reports on the quality of data transmitted from member states to the Commission (i.e. Eurostat). Directions for the quality assessment of data (Article 6) define the following six dimensions of evaluation: relevance, accuracy, timeliness, accessibility, comparability, and coherence. However, the Regulation allows member states to choose their own methodology to generate data from a list of various options from the traditional censuses based on an exhaustive collection of data, up to a census based on information taken exclusively from administrative records. Furthermore, in order to support methodological innovations at the international level, information on how register-based censuses were implemented in Scandinavian countries was made available by the United Nations Economic Commission for Europe (UNECE, 2007).

During the preparation of the 2011 fieldwork, INE officials opted for the model of a census based on administrative records completed with a sample survey. To cope with the abovementioned challenges, preparations included significant improvements on the municipal registers. Additional efforts for the introduction of the georeferencing of buildings were adopted (INE, 2011, pp. 3–5, 13). The diffusion of new technologies among the Spanish population and the use of tablets by census agents facilitated the implementation of the sample survey in a multi-modal approach (mail 52%, CAWI 38%, CAPI 10%; Argüeso & Vega, 2014; INE, 2011, pp. 5, 90). Furthermore, new legislation allowed for the integration of tax and social security data.

The first operation of the implementation process was the creation of a "pre-census file" including information from various administrative sources. This pre-census file took the National Register as the key reference to all other information taken from different statistical sources. Validity controls were made to clean the National Register data set following the "signs-of-life" method (Wallgren & Wallgren, 2014/2022, pp. 131–133). In Spain, it meant the verification of births and deaths bulletins, and the checking of the expiration date of the municipal register for foreigners –who must update their presence every two or five years depending on the circumstances. Further controls were implemented when crosschecking register data with tax agency and social security data. The pre-census file used for the 2011 census, based on information from municipal registers, contained information about only four variables per person (gender, birthdate, birthplace, and nationality). Second, population and dwelling data not available from the pre-census file was obtained from the large multi-modal sample survey mentioned above, aimed at around 12.3% of the population. The survey also served to weight records contained in the "pre-census file" when necessary (Argüeso & Vega, 2014). As a third operation, a plan for the georeferencing of all buildings and their characteristics was also designed. A tour of the territory served to complete and contrast the already available information, listing the dwellings and collecting the missing variables of buildings. Hence, the pre-census file plus the data obtained from the survey and the georeferenced data on buildings provided the final information.

The 2011 census counted 46.8 million inhabitants living in about 18 million households (dwellings). Results from the National Register indicated, again, 1 million inhabitants more. According to Goerlich Gisbert (2007, 2012), municipal registers lead to an overestimation of the population, because of foreign people and individuals from the upper and lower age groups, all of which are more likely not to be deregistered.

9.8   Preparations for the 2021 Spanish census

Once works related to the 2011 census were finished in 2014, the focus of INE statisticians turned to the study of the possibilities of including additional variables in the main 2021 census file, besides those taken from the municipal registers. After two years of preparation, a pilot pre-census file was created based on data from municipal registers in 2016, to which information from other sources was added (Vega Valle, Argüeso Jiménez, & Pérez Julián, 2020).

It must be noted that municipal registers give information on who is living in the same household but do not elucidate the nature of their relationship. Data about relationships between household members was obtained by combining data from the municipal registers with data from the tax agency, from birth and marriage bulletins, from the police database (which contains the names of father and mother of every person), and from previous censuses. Municipal registers do record information on educational attainments, which is not of very good quality, so this time information was obtained from data bases of the Ministry of Education, or from the Unemployment Register or from 2001 to 2011 censuses. In case of any discrepancy between sources, the highest educational level was chosen. This variable was constructed only for those 15 and older. Data on current activity status was obtained from various sources: social security registers, unemployment registers, public aids databases, mutualities registers, register of retired civil servants, registers of students, tax agencies and from the 2001 to 2011 censuses. These included occupation, type of industry, and type of employment. Data about immigrants was obtained from municipal registers or from the 2001 census if arrival to Spain was prior to 1996 (Vega Valle et al., 2020).

Approaches and methods for adding information are specific to the variable concerned; e.g. methods for adding the marital status to the dataset are different from the method for adding information on income. These linkage problems required dealing and negotiating with different institutions and governmental bodies with idiosyncratic characteristics and legal liabilities concerning data privacy as well as specific coding conventions for their data. The results of the efforts to create the pre-census file were deemed satisfactory because the need for imputation on some exceptional cases was similar to what was usual in traditional censuses (Vega & Argüeso, 2016). Moreover, all the variables that are mandatory according to EU regulations were finally obtained from administrative registers, plus other variables that are of interest to users of census data, such as year of arrival in municipality (Vega Valle et al., 2020).

While the quality of municipal registers, and as a consequence, of the National Register had been assessed in previous census rounds by looking at discrepancies between registry data and the census or a sample survey, that procedure was not viable this time; there was neither a traditional census nor a sample survey as an external source of validation. However, a quality assessment was still possible by comparing the National Register data, based on municipal registers, with data from the pilot pre-census file, based on policy-specific register data. Discrepancies between the National Register and the 2016 pilot pre-census file were almost 25,000 individuals, a number that is significantly lower than the discrepancy of around 1 million individuals between the last two censuses and the municipal register data at that time. Discrepancies arose from individuals counted in the municipal registers that should not be counted, given no evidence of their presence according to the signs-of-life approach (around 158,000 in 2017), and individuals not counted that should be counted (around 131,000 in 2017), given evidence of their presence from records on administrative interactions in other sources. Finally, a variable-by-variable analysis was carried out until the end of 2018. Individuals with high percentages of variables imputed or missing values were identified. Most of these individuals turned out to be either foreigners or Spanish nationals living in deprived areas. In the future, these groups could be targeted in a survey, for example, in order to better understand the nature of discrepancies between the National Register and the census file.

There was a set of variables in the 2011 questionnaire that recorded information on daily and seasonal mobility. Such information is very much demanded by census data users because it serves to establish the size of the linked population (*población vinculada*) of each municipality, a term used by INE statisticians and Spanish census data users, to refer to the population that a given municipality needs to provide services for. This includes of course the number of legal residents, but also the people who, residing most of the year in other (perhaps distant) municipalities, tend to spend at least 15 days a year in another one, often because they have a secondary residence. Linked population also includes individuals who are residents of other (mostly neighboring) municipalities but who, because of working or educational reasons, tend to spend much of the day in another municipality. Additional information on the means of transportation used for daily mobility and the time needed for commuting was also collected in the 2011 questionnaire. Information on all these variables by no means could be obtained from administrative data. Hence, pilot studies were conducted to assess the feasibility of using mobile phone companies' data instead. Replacing mobility data from the survey with data provided by the three major mobile phone company operators was considered another promising feature of the register-based census.

9.9   Implementing the first register-based census in Spain 2022

More than 20 years after the creation of the National Register, INE was ready to rely entirely on it to carry out a register-based census. Thanks to having overcome legal barriers to have access to different administrative data and thanks to the development of new IT technologies, Spain became in 2022 the most populated country with a fully register-based census.

Operations began with the creation of an updated version of the 2018 pre-census file, during the first semester of 2020. A final list of additional sources of data was also approved. As in 2011, variables unavailable from administrative records would have to be collected from a survey. However, this time, the survey was much smaller compared to the previous census. The Survey of Essential Characteristics of the Population and Housing 2021 (ECEPOV-2021) was implemented between February and July 2021. It targeted close to 300,000 households, expecting an effective completion of the questionnaire from at least 200,000. Results of the survey are to be released in November 2022, a month after the present chapter is being written. The survey included 12 questions about the household, 13 about the dwelling, 24 questions for each person living in the household, and 12 additional questions for those 16 years of age and older. Some of the information collected by the survey is required by the European Regulation on censuses related to houses and buildings ([OJ], 2017), such as tenure regime, useful area (for the NUTS-3 level) or water supply system, bathrooms and toilets and heating type (for the NUTS-2 level). Additional arguments in favor of implementing a survey were, for example, the fact that administrative records do not offer correct information on unmarried couples or the number of same-sex couples.

Information in the sample survey was obtained from various methods (CAWI, CATI, CAPI, and by mail). Last but not least, thanks to the detailed information collected in the survey, census imputation models, based entirely on administrative records, could be refined. Imputation was based on the Fellegi-Holt (1976) methodology and was applied using software developed by INE programmers and statisticians.

From October 2021 until April 2023, data from the 2020 pre-census file and data from dozens of other administrative records were linked using the DNI number for Spanish nationals and the NIE number for foreigners respectively. In case of conflicting information between administrative records, individuals were counted where they were registered according to the municipal register. There was no intention to modify the residence of individuals after incorporating residence data from other files, except in very particular cases. This deviates to a certain degree from the signs-of-life approach by prioritizing information from the municipal register.[10] The cleaning of data after the merging of administrative records and further post-collection data processing was finished in 2022. Final results were published in June 2023.

Data on daily and seasonal mobility was provided by three main mobile phone operators, each accounting for at least 20% of the Spanish mobile phone market and, combined accounting for almost 79% ([INE], 2020, p. 4). All three operators had rates of coverage of at least 17% per region. A fourth operator, accounting for 13% of the market, was excluded because in some provinces it accounted for less than 5% of the total number of mobile phones. To create the data matrix with information on mobility, the country was divided into 3,214 mobility areas. Each municipality between 5,000 and 50,000 inhabitants constituted a mobility area. Smaller Municipalities were grouped until they added up to at least 5,000 inhabitants. Towns and cities of more than 50,000 inhabitants were split into so-called "sub-city districts" (SCD), providing that each district contained at least 5,000 dwellers. The average size of each mobility area is 15,000 people and 12,000 mobile phones. In the pilot study, data for assessing daily mobility was obtained during a normal week in autumn, without holidays. For seasonal mobility two individual days in summer and Christmas eve were selected. Mobile phone operators were provided with precise instructions on how to create the data matrices accounting for daily weekday mobility, regular weekend mobility, and seasonal mobility. Users are defined to be residents of a certain mobility area when they are observed there for 60 days being present most of the time between midnight and 6:00 am. For each day, operators inform about the number of mobile phones that move out of a given mobility area and about the particular mobility areas where they spend most of the time during the day. Information can be processed to assess: the number of residents in each mobility area, as well as the number of outward and inward commuters. At the aggregate level of mobility areas, this information is believed to provide more accurate data than that gathered in traditional surveys. In contrast, the new method does not result in commuting data at an individual level.

The resident population in Spain as of January 1, 2021, stood at 47,400,798 inhabitants, according to the register-based census. As could be expected, that figure is very similar to the 47,385,107 resulting directly from the population register on that same date.

9.10 Implications and challenges

A major transformation has occurred in the methodology of census taking in Spain. Now, the primary target of the Spanish census is no longer the count of the population, since that task has been *de facto* transferred to municipal registers. The primary objective of the census is to provide an accurate description of the characteristics of the population, taking advantage of the already existing information ([INE], 2019, p. 9).

9.10.1 Advantages

Changes in the methodology of the census have often been controversial. This is very obvious in the case of ethnoracial categories (e.g. Bhagat, 2022; Emigh et al., 2016b; Loveman, 2014; Simon, 2017); but even seemingly more technical methodological changes have been disputed, such as in several censuses in the USA (Hannah, 2001; Sullivan, 2020), in the 2011 census in Germany (Radermacher, 2020, pp. 58–60; Chapter 4 in this book) or in the repeatedly postponed census in Ukraine (Chapter 8 in this book). Yet, in the Spanish case, there has been virtually no contestation from media, civil society, academia, or users of census data in general.

INE officials Jorge Vega and Antonio Argüeso (2016) argued at the European Conference on Quality in Official Statistics, held in Madrid in 2016, that the Spanish census of the current census round would be of much better quality than the previous ones because of various reasons. First, the new system copes better with the increasing non-response and with people who are hard to interview such as foreigners and young people living alone. The lack of response from certain groups varies remarkably within any population, introducing a source of bias, since those who complete the questionnaire are not representative of the whole population. Second, a register-based census reduces the discrepancies between the census and the Labor Force Survey (LFS) when estimating the number of unemployed persons.[11] Third, the additional survey targeted up to 1% of the population and was able to include additional variables, since much space was freed from the standard census questionnaire because much information was drawn from other sources of administrative records to the pre-census file. Fourth, information is updated more frequently, which constitutes a remarkable first step into a new (continuous) system of social statistics and at the same time significantly reduces the operational financial cost. While the 2001 census costed 200 million euros, the 2011 cost dropped to 85 million. Moreover, the 2001 census implied the mobilization of 40,000 census agents, and the 2011 required only 5,000 (Vega & Argüeso, 2016, p. 6). This is a non-negligible advantage since it is increasingly difficult to find highly qualified interviewers for such short periods of fieldwork. Fifth, IT tools purchased to help with the implementation of the interviews are suitable to use for only one census, since after ten years, technological improvements leave them obsolete, thus contributing to increasing financial burden. Lastly, Vega and Argüeso claimed that with a register-based census, better protection of privacy would be possible. In a later article, Argüeso (2021) added that administrative records of individuals residing in deprived areas were more reliable than data gathered by census agents. The same could be said about information on individuals belonging to the highest income levels. The combination of data from various administrative records provides sufficient information to assign individuals the right marital status and level of education. However, the relationship between dwellers in the same household cannot be verified with register data. Yet Argüeso argued that information from a questionnaire was hardly of better quality.

Although these advantages do not explain why controversy was practically non-existent in the Spanish transition from a traditional to a register-based census, at least they show how this transition has been legitimated. Justifications for such a transition in other cases can be expected to draw on similar arguments.

9.10.2   Challenges

In spite of the overall low level of controversy, some objections to the transition to a register-based census have indeed been made (Domingo i Valls, Bueno, & Treviño Maruri, 2021). INE officials themselves echoed some of the most notorious critiques in technical documents available on the INE website aimed at reflecting the public the process of decision-making while designing the 2021 Spanish census. To begin with, they refer to "a certain perception among users that the 'demographic reality' and the 'administrative' reality run separate paths" (INE, 2019, p. 17). Cohabitation or couple relationships that are not declared before an administrative instance, informal work not included in the labor files, or housing tenure regimes in certain cases are a few examples of phenomena of demographic interest that do not appear in the administrative records.

The flow of internal migration and the true dimension of international emigration are hard to count when relying only on administrative registers. Although the quality of municipal registers has improved, citizens of a municipality often still prefer to not unregister themselves, even if they move to another country for a period, so as not to lose the entitlement to certain services such as public health care (Lafleur et al., 2017). Estimates from INE suggest that close to a quarter of a million people left the country after the 2008 recession hit the Spanish economy, but some demographers estimate the real figure could have been three times higher (cf. Lafleur et al., 2017).

Another objection derives from the systematic imputation of certain information. Some statistical information may result from imputation, rather than from direct answers from citizens, and this may have serious implications for users, depending on their research question or practical demand for census data. A substantial number of users may not know exactly what kind of data they are dealing with if it is not explained to them adequately which variables and cases have been imputed and precisely how imputation has been applied. To cope with this, a series of, workshops and scientific meetings have been organized in order to facilitate the necessary communication between INE and the community of users of census data.

Although the INE admitted the discrepancy between reality and what is captured in administrative records, they asserted that there was also no evidence that these phenomena would be declared in traditional questionnaires (INE, 2019, pp. 21–22). According to INE, it is questionable that citizens perceive questionnaires as more confidential. A second argument tailored to cope with the objections raised is that INE realized in 2015 that the Spanish system of administrative records provided information on more than 90% of the variables required to meet most international standards. In this vein, INE (2019, pp. 12–13) argued that within the 56 UNECE countries, the current trend was to shift to register-based censuses, following the lead of the "pioneering" Scandinavian countries; referring to them on a par with "the most developed countries of the world". At first glance, this seems to be an interesting case of how a national statistical organization constructs a model case to be emulated, or a "reference society", to use Reinhard Bendix' (1979, p. 10) term.[12] Taking a second look it becomes clear, that referring to the Scandinavian model was a form of post-hoc rationalization. The INE used this argument after the decision was actually made, in order to justify it. In Spain, the INE, as an organization, became an example of a success story after contributing to the creation of the National Register. Until then, INE had been perceived as a necessary but rather costly institution that was useful only once in every ten years. The shift to a register-based census was the most obvious way to prolong the success story by adding a second mission to the organization: saving money. Also, all the investments in improving municipal registers (computerization and so on) could be better justified financially, if they served a greater purpose than merely that of municipal governments. INE and municipalities' interests complemented each other in that respect.

The aggregation of mobility data at the level of small-scale mobility areas that results from resorting to Big Data from mobile phone providers (instead of individual microdata based on census questionnaires) could be considered a drawback of the register-based census. At the same time, it is obvious that there is a trade-off between individual privacy protection and statistical precision.

The objection concerning the absence of certain information in a register-based census is admitted, but rather than a traditional census, the solution considered the best is the combination of register-based census data and a system of surveys addressing specific topics (INE, 2019). Concerning the lack of precision in capturing out-migration flows, INE officials consider collaboration with other countries' statistical offices to be the best approach. Exchanging information between countries, in a similar manner to how information is shared between Spanish municipalities, could serve to estimate more accurately the movements of the population between European countries. Hence, the objective would be to link data from INE with, for instance, data from the German Register of Foreigners (*Ausländerzentralregister*) to better count the number of Spanish citizens who moved to Germany without registering at the corresponding Spanish consulates, and vice-versa (Argüeso, 2021).

9.11   Conclusion

The process that led to the first register-based census in Spain, began earlier than the date the decision to make a register-based census was adopted. As such, there seems to be a strong case in favor of the concept of "investments in form" (Desrosières, 1991; Thévenot, 1985, 1984), which speaks also to the need for informational integration of increasingly differentiated state organizations (Schimank, 2005). Indeed, the development of a parallel system of counting inhabitants by municipal authorities based on their registers as well as the introduction of a personal identification number that worked across administrative data bases were necessary preconditions, without which a register-based census could not have been implemented. Thus, different bodies and levels of the Spanish administration had to cooperate in exchanging information and harmonizing their ways of data coding and collection in order to create the required conditions for a register-based census that complies with the high standards of statistical quality set by the EU. Thus, the methodological shift from a traditional to a register-based census in Spain was to a great extent related to the refinement of an already existing source of data, which had existed for reasons other than the planning of a register-based census. Once the mission of creating the National Register was accomplished, the exploitation of its potential advantages placed INE officials in a position where they were able to develop the concept for a register-based census, even before the technology to implement it successfully was available. The public utility of municipal registers was increased significantly by decoupling its function from policing the population (as under Franco) and extending services to irregular immigrants.

Within the framework developed by Emigh and colleagues (2016a, 2016b), the Spanish case is an example of a state-driven census, a classification that requires some further qualifications. First, the role of the actual government was secondary. The main intervention from the side of the government was the request made to INE to coordinate the efforts of municipalities to enhance the reliability of their municipal register data sets, which were to serve as a basis for the Electoral Census. However, this request was made when there was no plan to shift the methodology. In fact, INE was requested to assist municipalities because the results obtained from INE traditional censuses were considered more accurate than data gathered at the municipal registers at that time. Moreover, technologies required for the successful implementation of a register-based census were still not available. Later, when the Spanish government received the proposition from INE to approve the shift to a register-based census, the advantages associated with the lower financial costs seemed to have been sufficient to make governments comply with the initiative. Hence, a state agency seems to have been the dominant actor.

Second, the Spanish case does not seem to be consistent with a society-centered perspective, because Spanish society neither participated in the discussion nor the decision-making process (Domingo i Valls et al., 2021; Treviño Maruri & Domingo, 2020, p. 120). No relevant parliamentary debates took place to discuss the census transition, nor did the change become a matter of controversy in Spanish media. In fact, the shift from a traditional to a register-based census helped INE to become less dependent on the participation of citizens in the data-gathering process. Hence, interaction on census matters seems to be even less likely in the future. Only recently, a debate involving users of census data has been initiated (Domingo i Valls et al., 2021) The lack of debate could be partially attributed to the strong legitimacy that INE has as an independent body in charge of the production of statistical data in the eyes of the Spanish population, including media and leaders of political parties. It is noteworthy, for instance, that INE is also in charge of the production of the Spanish LFS, and that its estimations of unemployment rates have for decades been considered more accurate than the figures provided by administrative sources such as social security. In fact, as this chapter has shown, most of the initiative to foster the change from a traditional to a register-based census stems from INE officials. INE officials seem to have opted for a register-based census as a means to empower themselves. After completing the mission of creating the National Register, they were able to offer the government a means to reducing the costs of the census and becoming less dependent on the collaboration of citizens; at the same time, they were vindicating convergence with the "vanguard" Scandinavian model as a way to valorize the (improved) Spanish system of administrative data collection (INE, 2019, pp. 12–13).

Third, in the *longue durée*, a certain continuity to social categories is evident in the fact that the discrepancy between the *de jure* and the *de facto* population was one of the few issues of census taking that resurged repeatedly during recent decades. This reflects a problem of census taking that was present already during the first two modern censuses from 1857 to 1860 (Gozálvez Pérez & Martín-Serrano Rodríguez, 2016), when Spain was a country of rural-urban character as well as emigration and this distinction was of obvious practical relevance to social actors, e.g. for the alphabetization of the population. During recent decades, Spain has been converted into a country of immigration, transition and emigration at the same time, therefore, the distinction between *de jure* and *de facto* population has not lost its practical significance, especially for the calibration of municipal public services. This reading of a historical continuity specific to the Spanish case as a former country of emigration is challenged by the possible alternative view that the distinction between *de facto* and *de jure* population has probably been relevant in most countries where the number of inhabitants has any implication for the provision of municipal public services or political apportionment – independently from the level of spatial mobility relative to other countries.

The world polity perspective adopted by Ventresca (1995) and Kukutai and colleagues (2015) provides further elements for explaining the Spanish case. Their emphasis on transnational influences is difficult to place adequately within the distinction between state and society as defined by Emigh and colleagues. World polity authors point out that the development of censuses is determined to a significant extent by recommendations and regulations at the international level. Indeed, the selection of Scandinavian countries as reference societies and the further diffusion of the idea of implementing register-based census via UNECE (2007, 2018) publications as well as EU commission legislation and recommendations did play a role on shaping how the 2021 census was carried out in Spain. Although the trend at international level was certainly not a sufficient condition to provoke the shift, it is hard to imagine that the census transition in Spain might have happened without it. Hence, the Spanish case constitutes a peculiar one, which requires contributions from different theoretical frameworks for its explanation.

Different from what could theoretically be expected (Diaz-Bone & Horvath, 2021), surprisingly little criticism was directed toward the transition to a register-based census. Moreover, close to no discrepancies resulted so far from different data logics between the public and the private sector while incorporating Big Data from mobile phone companies for estimating spatial mobility. Instead of individual inhabitants on an equal base, mobile phone data represent a selective group of the population on an aggregate level. However, it might well be that controversies emerge when census data will become available for being used by different stakeholder groups.

In response to migration within the EU, administrative bodies of member states are increasingly engaged in horizontal information exchange related to the transnational realization of social citizenship rights, such as social security (Lafarge, 2018). As an outlook to the future, exchange of administrative microdata among member states might also become a relevant path for improving the quality of population and migration data (Argüeso, 2017). Furthermore, different from Spain, personal identification numbers are used to link information from public sector and private sector sources in other countries. We will have to see how privacy stakeholders in Spain respond to these imagined futures.

Notes

1 While possible connections to French pragmatism have been pointed out elsewhere (Emigh, 2022; Emigh, Riley, & Ahmed, 2020, p. 301), they still await more explicit discussions.
2 In 1989 the ministry's denomination was *Ministerio de Economía y Hacienda*; today it is *Ministerio de Asuntos Económicos y Transformación Digital*.
3 The increasing self-description of society as a world society has been captured also beyond world polity theory but without reference to census taking (Heintz, 1980; Luhmann, 1997). Therefore, it is puzzling that Emigh and colleagues stick to a dichotomous distinction between state and society that seems to be bound to a framework of methodological nationalism.
4 It is still too early to judge the impact of the Global Compact for Safe, Orderly and Regular Migration (Ferris & Donato, 2020).
5 More recently, Spain's fiscal situation was additionally affected by the Great Recession (2008–2014), which lead to some structural reforms (Royo, 2020).

6 Borders of Spain mainland have remained virtually unchanged since a war with France that ended in 1659 resulting in the loss of territory north of the Pyrenees, including the town of Perpignan. Previously, in 1640, Portugal had regained independence from Spain, this time for good as was recognized by Treaty of Lisbon 1688. Some treaties with Portugal and France (after the Napoleonic invasion) were signed during the XIX century but involved only minor changes (Capdevila i Subirana, 2017).

7 The classification system consists of Large Urban Areas, Small Urban Areas and Non-Urban Areas (MITMA, 2021, pp. 15–18). Large Urban Areas consist either of one city with at least 50,000 inhabitants or of more than one municipality with at least 1,000 inhabitants each. In 2021 there were 86 Large Urban Areas consisting of 755 municipalities (9.3%) with over 32 million inhabitants (69.42%). Of these, 19 consisted of one single urban area and 67 of several municipalities. The traditional Spanish classification has defined settlements with more than 10,000 inhabitants as urban. However, this classification was refined by the category of Small Urban Areas: On the one hand, there are 126 municipalities between 20,000 and 50,000 inhabitants that are not included in Large Urban Areas; on the other hand, there are 619 municipalities between 5,000 and 20,000 inhabitants that are not included in Large Urban Areas. Their classification as urban or non-urban requires several more sophisticated steps of analysis. The Non-Urban Areas consist of 7,046 municipalities (86.7%) containing 17.3% of the total population and extending over 79.3% of the national territory.

8 People from Latin America, North Africa and Eastern Europe (Romania, mainly) were the largest groups. Most enter regularly but then after 90 days remain irregularly until, in most cases, they find an employer that agrees to make them a legal contract, thus, making their situation regular. The process of legalization can take years.

9 After 2008, with the introduction of NIF instead of CIF, data from the cadaster could be linked to the data of individuals.

10 Sometimes, the signs-of- life suggest the person lives in a municipality different from the one that she/he is registered in, for example, individuals with two residences. Although they might spend most of the time in the second residence (especially if they are retired) they are still registered as resident in the municipality of their first residence. In such cases, INE used the data from the National Register, instead correcting it with additional administrative information.

11 There were 7.4 million unemployed according to the 2011 census vs 5.3 million unemployed according to the Labor Force Survey for the last quarter of 2011. This is, because the information given by the respondent is formally incorrect and mediated by the high prevalence of informal employment in the Spanish labor market; therefore, answers are often not in line with the official definition of "unemployed" (Vega & Argüeso, 2016, pp. 7–8).

12 Societies can also be referred to in a negative way (Waldow, 2017).

References

Alastalo, M., & Helén, I. (2022). A code for care and control: The PIN as an operator of interoperability in the Nordic welfare state. *History of the Human Sciences, 35*(1), 242–265.

Alba, C., & Navarro, C. (2011). Administrative tradition and reforms in Spain: Adaptation versus innovation. *Public Administration, 89*, 783–800. doi:10.1111/j.1467-9299.2010.01886.x

Amat i Puigsech, D., & Garcés-Mascareñas, B. (2019). *Politicisation of immigration in Spain: An exceptional case?* (Research on the Common European Asylum System No. 14). Chemnitz: CEASEVAL.

Argüeso, A. (2017). *2021 Population census and migration statistics in Spain. Why exchanging microdata?* Budapest, Hungary. Retrieved from Hungarian Central Statistical Office website: https://www.ksh.hu/dgins2017/papers/dgins2017_session1_es.pdf

Argüeso, A. (2021). El censo de población y viviendas de 2021 basado en registros administrativos: un gran paso adelante en el conocimiento estadístico de la población en España. *Revista Internacional de Sociología, 79*(1), e181a. doi:10.3989/ris.2021.79.1.19.181a

Argüeso, A., & Vega, J. L. (2014). A population census based on registers and a "10% survey" methodological challenges and conclusions. *Statistical Journal of the IAOS, 30*, 35–39. doi:10.3233/SJI-140797

Bendix, R. (1979). Why nationalism? Relative backwardness and intellectual mobilization. *Zeitschrift für Soziologie, 8*(1), 6–13.

Betts, A. (2015). The normative terrain of the global refugee regime. *Ethics & International Affairs, 29*(4), 363–375.

Bhagat, R. B. (2022). *Population and the political imagination: Census, register and citizenship in India.* London, England: Routledge.

BOE. (1989). Ley 12/1989, de 9 de mayo, de la Función Estadística Pública. *Boletín Oficial del Estado, 112*(11 mayo), 14026–14035. Retrieved from https://boe.es/boe/dias/1989/05/11/pdfs/A14026-14035.pdf

BOE. (1996). LEY 4/1996, de 10 de enero, por la que se modifica la Ley 7/1985, de 2 de abril, Reguladora de las Bases del Régimen Local, en relación con el Padrón municipal. *Boletín Oficial del Estado, 11*, 813–815. Retrieved from https://www.boe.es/eli/es/l/1996/01/10/4

BOE. (2007). Real Decreto 1065/2007, de 27 de julio, por el que se aprueba el Reglamento General de las actuaciones y los procedimientos de gestión e inspección tributaria y de desarrollo de las normas comunes de los procedimientos de aplicación de los tributos. *Boletín Oficial del Estado, 213*(05/09), 36512–36594. Retrieved from https://www.boe.es/eli/es/rd/2007/07/27/1065/con

Caldarice, O. (2018). *Reconsidering welfare policies in times of crisis.* Cham, Switzerland: Springer. doi:10.1007/978-3-319-68622-6

Capdevila i Subirana, J. (2017). *Historia del deslinde de la frontera hispano-francesa. Del tratado de los Pirineos (1659) a los tratados de Bayona (1856–1868).* Madrid, Spain: Centro Nacional de Información Geográfica. Retrieved from https://www.ign.es/resources/acercaDe/libDigPub/FronterasPirineosbaja.pdf

Caplan, J. (2001). "This or that particular person": Protocols of identification in nineteenth-century Europe. In J. Caplan & J. Torpey (Eds.), *Documenting individual identity: The development of state practices in the modern world* (pp. 49–66). Princeton, NJ: Princeton University Press.

Caplan, J., & Torpey, J. (2001). Introduction. In J. Caplan & J. Torpey (Eds.), *Documenting individual identity: The development of state practices in the modern world* (pp. 1–12). Princeton, NJ: Princeton University Press.

Chauvin, S., & Garcés-Mascareñas, B. (2020). Contradictions in the moral economy of migrant irregularity. In S. Spencer & A. Triandafyllidou (Eds.), *Migrants with irregular status in Europe: Evolving conceptual and policy challenges* (pp. 33–49). Cham, Switzerland: Springer. doi:10.1007/978-3-030-34324-8_3

Christen, P., Ranbaduge, T., & Schnell, R. (2020). *Linking sensitive data. Methods and techniques for practical privacy-preserving information sharing.* Cham, Switzerland: Springer. doi:10.1007/978-3-030-59706-1_14

Coleman, J. S. (1990). *Foundations of social theory.* Cambridge, MA: Belknap Press of Harvard University Press.

Colino, C. (2020). Decentralization in Spain. In D. Muro, I. Lago, & S. Parrado (Eds.), *The Oxford handbook of Spanish politics* (pp. 61–81). Oxford, England: Oxford University Press. doi:10.1093/oxfordhb/9780198826934.013.5

Desrosières, A. (1991). How to make things which hold together: Social science, statistics and the state. In P. Wagner, B. Wittrock, & R. Whitley (Eds.), *Discourses on society: The shaping of the social science disciplines* (pp. 195–218). Dordrecht, The Netherlands: Springer.

Desrosières, A. (2000). Measurement and its uses: Harmonization and quality in social statistics. *International Statistical Review/Revue internationale de statistique, 68*(2), 173–187. doi:10.1111/j.1751-5823.2000.tb00320.x

Diaz-Bone, R., & Horvath, K. (2021). Official statistics, big data and civil society. Introducing the approach of "economics of convention" for understanding the rise of new data worlds and their implications. *Statistical Journal of the IAOS, 37*, 219–228. doi:10.3233/SJI-200733

Domingo i Valls, A., Bueno, X., & Treviño Maruri, R. (2021). El nuevo censo de 2021 en España: un debate metodológico, epistemológico y político pendiente. *Revista internacional de sociologia, 79*(1), e181. doi:10.3989/ris.2021.79.1.19.181

Doomernik, J., & Glorius, B. (2022). The future of the common European asylum system: Dystopian or Utopian expectations? *Social Inclusion, 10*(3), 1–3. doi:10.17645/si.v10i3.5954

El-Habib Draoui, B., Jiménez-Delgado, M., & Ruiz-Callado, R. (2019). Minorities in Spanish secondary education: School segregation, between reality and official statistics. In J. Anson, W. Bartl, & A. Kulczycki (Eds.), *Studies in the sociology of population. International perspectives* (pp. 253–274). Cham, Switzerland: Springer.

Emigh, R. J. (2022). Towards interactive perspectives on information gathering: What are resolvable differences? *Journal of Cultural Economy, 15*(1), 123–126. doi:10.1080/17530350.2021.1974076

Emigh, R. J., Riley, D. J., & Ahmed, P. (2016a). *Antecedents of censuses from medieval to nation states: How societies and states count.* New York, NY: Palgrave Macmillan.

Emigh, R. J., Riley, D. J., & Ahmed, P. (2016b). *Changes in censuses from imperialist to welfare states: How societies and states count.* New York, NY: Palgrave Macmillan. doi:10.1057/9781137485069

Emigh, R. J., Riley, D. J., & Ahmed, P. (2020). The sociology of official information gathering: Enumeration, influence, reactivity, and power of states and societies. In T. Janoski, C. de Leon, J. Misra, & I. W. Martin (Eds.), *The new handbook of political sociology* (pp. 290–320). Cambridge, England: Cambridge University Press.

Esteban Romero, R. (2016). Algunos secretos del documento nacional de identidad español: una aplicación de la aritmética modular a códigos detectores de errores. *Modelling in Science Education and Learning, 9*(2), 59–66. doi:10.4995/msel.2016.6338

Faist, T. (2010). Towards transnational studies: World theories, transnationalisation and changing institutions. *Journal of Ethnic and Migration Studies, 36*(10), 1665–1687. doi:10.1080/1369183X.2010.489365

Fellegi, I. P., & Holt, D. (1976). A systematic approach to automatic edit and imputation. *Journal of the American Statistical Association, 71*(353), 17–35.

Ferris, E. G., & Donato, K. M. (2020). *Refugees, migration and global governance: Negotiating the global compacts.* London, England: Routledge.

Fishman, R. M. (2020). Spain in comparative perspective. In D. Muro, I. Lago, & S. Parrado (Eds.), *The Oxford handbook of Spanish politics* (pp. 13–31). Oxford, England: Oxford University Press. doi:10.1093/oxfordhb/9780198826934.013.2

Frohman, L. (2020). *The politics of personal information: Surveillance, privacy, and power in West Germany.* New York, NY: Berghahn Books.

Galdon Clavell, G., & Ouziel, P. (2014). Spain's documento nacional de identidad: An e-ID for the twenty-first century with a controversial past. In K. Boersma (Ed.), *Histories of state surveillance in Europe and beyond* (pp. 135–149). London, England: Routledge.

García Pérez, M. S. (2007). El padrón municipal de habitantes: origen, evolución y significado. *Hispania Nova, 7,* 79–88. Retrieved from http://hispanianova.rediris.es/7/HISPANIANOVA-2007.pdf

Goerlich Gisbert, F. J. (2007). ¿Cuántos somos? Una excursión por las estadísticas demográficas del Instituto Nacional de Estadística (INE). *Boletín de la Asociación de Geógrafos Españoles, 45,* 123–156.

Goerlich Gisbert, F. J. (2012). Estimaciones de la población actual (ePOBa) a nivel municipal. Discrepancias censo-padrón a pequeña escala. *Boletín de la Asociación de Geógrafos Españoles, 58,* 83–104.

Gozálvez Pérez, V., & Martín-Serrano Rodríguez, G. (2016). El censo de la población de España de 1860. Problemas metodológicos. Inicio de la aportación social en los censos. *Boletín de la Asociación de Geógrafos Españoles, 70,* 329–370. doi:10.21138/bage.2174

Grommé, F., Ruppert, E., & Ustek-Spilda, F. (2021). Usual residents: Defining and deriving. In E. Ruppert & S. Scheel (Eds.), *Data practices: Making up a European people* (pp. 49–88). London, England: Goldsmith Press.

Hannah, M. G. (2001). Sampling and the politics of representation in US census 2000. *Environment and Planning D: Society and Space, 19*(5), 515–534. doi:10.1068/d289

Heintz, P. (1980). The study of world society: Some reasons pro and contra. In H. H. Holm & E. Rudeng (Eds.), *Social science - for what?: Festschrift for Johan Galtung* (pp. 97–100). Oslo, Norway: Universitetsforlaget.

Herzog, B. (2021). Managing invisibility: Theoretical and practical contestations to disrespect. In G. Schweiger (Ed.), *Migration, recognition and critical theory* (pp. 211–227). Cham, Switzerland: Springer. doi:10.1007/978-3-030-72732-1_10

Howard, C. (2021). *Government statistical agencies and the politics of credibility.* Cambridge, England: Cambridge University Press.

INE. (2001). *Censos de Población y Viviendas 2001. Proyecto.* Madrid, Spain. Retrieved from Instituto Nacional de Estadistica website: https://www.ine.es/censo2001/procen01.pdf

INE. (2011). *Proyecto de los Censos Demográficos 2011.* Madrid, Spain. Retrieved from Instituto Nacional de Estadistica website: https://www.ine.es/censos2011/censos2011_proyecto.pdf

INE. (2019). *Censos de Población y Viviendas 2021. Proyecto técnico.* Madrid, Spain. Retrieved from Instituto Nacional de Estadistica website: https://www.ine.es/censos2021/censos2021_proyecto.pdf

INE. (2020). *Estudio EM-1 de movilidad a partir de la telefonía móvil. Proyecto técnico.* Madrid, Spain. Retrieved from Instituto Nacional de Estadistica website: https://www.ine.es/experimental/movilidad/exp_em1_proyecto.pdf

Kukutai, T., Thompson, V., & McMillan, R. (2015). Whither the census? Continuity and change in census methodologies worldwide, 1985–2014. *Journal of Population Research, 32*(1), 3–22. doi:10.1007/s12546-014-9139-z

Lafarge, F. (2018). EU citizens and public services: The machinery behind the principles. In E. Ongaro & S. van Thiel (Eds.), *The Palgrave handbook of public administration and management in Europe* (pp. 689–706). London, England: Palgrave Macmillan.

Lafleur, J.-M., Stanek, M., & Veira, A. (2017). South-North labour migration within the crisis-affected European Union: New patterns, new contexts and new challenges. In J.-M. Lafleur & M. Stanek (Eds.), *South-North migration of EU citizens in times of crisis* (pp. 193–214). Cham, Switzerland: Springer. doi:10.1007/978-3-319-39763-4_11

Loveman, M. (2014). *National colors. Racial classification and the state in Latin America.* Oxford, England: Oxford University Press.

Luhmann, N. (1997). Globalization or world society: How to conceive of modern society? *International Review of Sociology, 7*(1), 67–79. doi:10.1080/03906701.1997.9971223

Mann, M. (2012). *The sources of social power. Volume 2: The rise of classes and nation-states, 1760–1914* (2nd ed.). Cambridge, England: Cambridge University Press. doi:10.1017/CBO9781139381314 (Original work published 1993).

Melón, A. (1951). Los censos de la población en España. *Estudios Geográficos, 12*(43), 203–281.

Meyer, J. W. (2010). *World society: The writings of John W. Meyer* (G. Krücken & G. S. Drori, Eds.). Oxford, England: Oxford University Press.

Ministerio del Interior. (n.d.). Historia de los documentos de identidad. España 1820–2016. Retrieved from https://www.dnielectronico.es/PDFs/Historia_de_los_documentos_de_identidad.pdf

MITMA. (2021). *Áreas urbanas en España 2021*. Madrid, Spain. Retrieved from Ministerio de Transportes, Movilidad y Agenda Urbana website: https://apps.fomento.gob.es/CVP/handlers/pdfhandler.ashx?idpub=BAW087

Muñoz-Cañavate, A., & Hípola, P. (2011). Electronic administration in Spain: From its beginnings to the present. *Government Information Quarterly, 28*(1), 74–90. doi:10.1016/j.giq.2010.05.008

OJ. (2008). Regulation (EC) No 763/2008 of the European parliament and of the council of 9 July 2008 on population and housing censuses (Text with EEA relevance). *Official Journal of the European Union*, (L 218), 14–20. Retrieved from http://data.europa.eu/eli/reg/2008/763/oj

OJ. (2017). Commission Implementing Regulation (EU) 2017/543 of 22 March 2017 laying down rules for the application of Regulation (EC) No 763/2008 of the European Parliament and of the Council on population and housing censuses as regards the technical specifications of the topics and of their breakdowns (Text with EEA relevance). *Official Journal of the European Union*, (L 78), 13–58. Retrieved from http://data.europa.eu/eli/reg_impl/2017/543/oj

Pardos-Prado, S. (2020). Migration politics: The end of Spanish exceptionalism? In D. Muro, I. Lago, & S. Parrado (Eds.), *The Oxford handbook of Spanish politics* (pp. 445–464). Oxford, England: Oxford University Press. doi:10.1093/oxfordhb/9780198826934.013.27

Parrado, S. (2020). Challenges in the face of diversities: Public administration in Spain as an example. In G. Bouckaert & W. Jann (Eds.), *European perspectives for public administration* (pp. 189–206). Leuven, Belgium: Leuven University Press.

Piketty, T. (2014). *Das Kapital im 21. Jahrhundert* [Le Capital au XXIe siècle]. München, Germany: C. H. Beck.

Pino Rebolledo, F. (1991). *Tipología de los documentos municipales (siglos XII–XVII)*. Valladolid, Spain: Universidad de Valladolid.

Pollitt, C., & Bouckaert, G. (2017). *Public management reform: A comparative analysis - into the age of austerity* (4th ed.). Oxford, England: Oxford University Press.

Presidencia del Directorio Militar. (1924). Estatuto municipal, de 8 de marzo de 1924. *Gaceta de Madrid, 264*(69), 1218–1302. Retrieved from https://www.b oe.es/datos/pdfs/BOE//1924/069/A01218-01302.pdf

Prévost, J.-G. (2019). Politics and policies of statistical independence. In M. J. Prutsch (Ed.), *Science, numbers, and politics* (pp. 153–180). Cham, Switzerland: Springer.

Radermacher, W. J. (2020). *Official statistics 4.0: Verified facts for people in the 21st century*. Cham, Switzerland: Springer. doi:10.1007/978-3-030-31492-7

Riley, D., Ahmed, P., & Emigh, R. J. (2021). Getting real: Heuristics in sociological knowledge. *Theory and Society, 50*(2), 315–356. doi:10.1007/s11186-020-09418-w

Royo, S. (2020). The causes and legacy of the great recession in Spain. In D. Muro, I. Lago, & S. Parrado (Eds.), *The Oxford handbook of Spanish politics* (pp. 115–131). Oxford, England: Oxford University Press. doi:10.1093/oxfordhb/9780198826934.013.8

Ruppert, E., & Ustek-Spilda, F. (2021). Refugees and homeless people: Coordinating and narrating. In E. Ruppert & S. Scheel (Eds.), *Data practices: Making up a European people* (pp. 89–124). London, England: Goldsmith Press.

Salas-Vives, P., & Pujadas-Mora, J. M. (2021). Bottom-up nation-building: National censuses and local administration in nineteenth-century Spain. *Journal of Historical Sociology, 34*(2), 287–304. doi:10.1111/johs.12323

Schimank, U. (2005). *Beiträge zur akteurszentrierten Differenzierungstheorie: Vol. 1. Differenzierung und Integration der modernen Gesellschaft*. Wiesbaden, Germany: VS Verlag für Sozialwissenschaften.

Silva, R., Snow, R., Andreev, D., Mitra, R., & Abo-Omar, K. (2019). *Strengthening CRVS systems, overcoming barriers and empowering women and children*. Ottawa, Canada. Retrieved from International Development Research Centre website: http://hdl.handle.net/10625/60225

Simon, P. (2017). The failure of the importation of ethno-racial statistics in Europe: Debates and controversies. *Ethnic and Racial Studies, 40*(13), 2326–2332. doi:10.1080/0141987 0.2017.1344278

Sotiropoulos, D. A. (2018). Public administration in Europe North and South: Enduring differences and new cleavages? In E. Ongaro & S. van Thiel (Eds.), *The Palgrave handbook of public administration and management in Europe* (pp. 881–898). London, England: Palgrave Macmillan.

Starr, P. (1987). The sociology of official statistics. In W. Alonso & P. Starr (Eds.), *The politics of numbers* (pp. 7–57). New York, NY: Russell Sage Foundation.

Starr, P., & Corson, R. (1987). Who will have the numbers? The rise of the statistical services industry and the politics of public data. In W. Alonso & P. Starr (Eds.), *The politics of numbers* (pp. 415–447). New York, NY: Russell Sage Foundation.

Suero Salamanca, J. A. (1999). Estudio sobre el Padrón Municipal de habitantes. *Actualidad Administrativa, 15*, 3–19.

Sullivan, T. A. (2020). *Census 2020: Understanding the issues*. Cham, Switzerland: Springer. doi:10.1007/978-3-030-40578-6

Szreter, S., & Breckenridge, K. (2012). Registration and recognition: The infrastructure of personhood in world history. In K. Breckenridge & S. Szreter (Eds.), *Registration and recognition: Documenting the person in world history* (pp. 1–36). Oxford, England: Oxford University Press.

Thévenot, L. (1984). Rules and implements: Investment in forms. *Social Science Information, 23*(1), 1–45. doi:10.1177/053901884023001001

Thévenot, L. (1985). Les investissements de forme. In L. Thévenot (Ed.), *Conventions économiques* (pp. 21–72). Paris, France: Presses Universitaires de France.

Trein, P., Meyer, I., & Maggetti, M. (2019). The integration and coordination of public policies: A systematic comparative review. *Journal of Comparative Policy Analysis: Research and Practice, 21*(4), 332–349.

Treviño Maruri, R., & Domingo, A. (2020). Goodbye to the Spanish census? Elements for consideration. *Revista Española de Investigaciones Sociológicas, 2020*(171), 107–124. doi:10.5477/cis/reis.171.107

UNECE. (2007). *Register-based statistics in the Nordic countries*. Geneva, Switzerland: United Nations Economic Commission for Europe.

UNECE. (2018). *Guidelines on the use of registers and administrative data for population and housing censuses.* New York, NY: United Nations.

Vega, J., & Argüeso, A. (2016). Spain 2021. Why will this census have more quality than the previous one? European Conference on Quality in Official Statistics, Instituto Nacional de Estadistica, Madrid, Spain. Retrieved from https://www.ine.es/q2016/docs/q2016Final00025.pdf

Vega Valle, J. L., Argüeso Jiménez, A., & Pérez Julián, M. (2020). Moving towards a register based census in Spain. *Statistical Journal of the IAOS, 36,* 187–192. doi:10.3233/SJI-190516

Veira, A., Stanek, M., & Cachón, L. (2011). Los determinantes de la concentración étnica en el mercado laboral español. *Revista Internacional de Sociología, 69*(Abril), 219–242.

Ventresca, M. J. (1995). *When states count: Institutional and political dynamics in modern census establishment, 1800–1993* (PhD thesis). Stanford University, Stanford, CA.

Waldow, F. (2017). Projecting images of the 'good' and the 'bad school': Top scorers in educational large-scale assessments as reference societies. *Compare, 47*(5), 647–664.

Wallgren, B., & Wallgren, A. (2022). *Register-based statistics: Statistical methods for administrative data* (3rd ed.). Hoboken, NJ: John Wiley & Sons (Original work published 2014).

Waluyo, B. (2022). The tides of agencification: Literature development and future directions. *International Journal of Public Sector Management, 35*(1), 34–60. doi:10.1108/IJPSM-04-2020-0105

Whitley, E. A., Martin, A. K., & Hosein, G. (2014). From surveillance- by-design to privacy-by-design. Evolving identity policy in the United Kingdom. In K. Boersma (Ed.), *Histories of state surveillance in Europe and beyond* (pp. 205–219). London, England: Routledge.

10 Towards a register-based census in Germany

Objectives, requirements and challenges[1]

Thomas Körner and Eva Grimm

10.1 Introduction

The census in Germany is currently undergoing major changes. In 2022, a combined census will be conducted that is similar to the model developed for the census 2011. Meanwhile, the Federal Statistical Office, together with the State Statistical Offices of the Länder, are already preparing the establishment of a register-based census for the time after the census 2022. The factors guiding design and implementation of this new census model are manifold. This paper first examines the trajectory of the German census, describing historical and recent developments both at the national and the international level that constitute the background for the change-over to the new census model. In the second step, we present the requirements and potentials of a future register-based census model and outline its prerequisites and implementation options. In particular, questions related to quality assurance (e.g. how to deal with issues of over- and undercoverage in population registers), data provision (e.g. creation of a register on buildings and dwellings) and possibilities of linking data from different registers (e.g. establishment of record linkage procedures in the absence of a constant person identification number) are discussed as selected examples.

10.2 The trajectory of the German census

It has often been argued that the development of census methodologies can only be fully understood by taking into account the specific historical and political context (see, e.g. Desrosières, 2010; Kukutai, Thompson, & McMillan, 2015). This also holds true for the census in Germany, which has been characterised by a specific trajectory over the last decades. Thus, in order to understand the specific context of the current German census model, the major topics of this trajectory have to be considered (for the following, see Körner & Dittrich, 2017; for a short history of censuses in Germany before the 1980s see Scholz & Kreyenfeld, 2016).

At the national level, the most obvious feature is that the census that was initially planned for 1981 was postponed twice before it finally took place in 1987. The reasons for the first postponement were related to the federal structure of Germany, as no agreement could be reached regarding the distribution of the cost

DOI: 10.4324/9781003259749-14

between the Federal Government and the governments of the federal states before the end of the parliamentary term in 1980. After having been rescheduled for 1983, the census was postponed again. This time, data collection was stopped by the Federal Constitutional Court, just a few days prior to the census day. The decision was taken after strong public criticism and concerns relating to the transmission of census microdata and privacy challenging the 1983 census law. The judgement of the constitutional court had a major impact on the further development of census methodologies in Germany (and was very influential for the regulation of data protection in general). It stated that the right of informational self-determination directly follows from the fundamental right of personal freedom, guaranteed by Article 2 of the German constitution. Therefore, any data collection required from the public is only considered constitutional if it is justified by a legal basis that meets specific requirements. In particular, this legal basis needs to be specific and clear as well as commensurate compared to the public interest at stake. While data for administrative purposes may only be collected for specific, well justified and commensurate purposes, collection for official statistics, given its specific role, is allowed for a certain stock of information that can be used for multiple purposes. However, this requires that data collected for statistical purposes must be used for statistical purposes only and under no circumstances can be transferred to other public bodies (*Rückspielverbot*). At the same time, the constitutional court defined restrictions regarding the use of identification numbers (BVerfG, 1983). The census, initially planned for 1981, was finally conducted in 1987 applying the traditional model of a full enumeration. Although the strict data protection rules laid down in the judgement of the Federal Constitutional Court were taken into account, the census 1987 continued to be subject to strong public criticism.

After the controversial discussions of the census during the period 1983–1987, the Federal Government was reluctant to engage in a traditional census again. As a consequence of this historical background, a combined census model was developed, tested in 2001 and finally implemented in 2011. So, instead of carrying out a full census in the 2001 census round, a large-scale census test was conducted to assess the viability of a register-assisted approach, that combined data obtained from registers with a number of primary data collections. This combined census model integrated elements of register use, a conventional full enumeration census (on housing), and a sample survey. The results from the sample survey were used in order to correct the data for errors due to over- and undercoverage as well as to collect such variables that were not available from registers (for more details, see Bechtold, 2016; Szenzenstein, 2005). The combined approach was successfully applied in 2011 and has been implemented in a similar way in the census 2022 (Dittrich, 2019).[2] Although this method was again challenged by a complaint, the Federal Constitutional Court decided in 2018 (BVerfG, 2018) that it was in line with the German constitution. At the same time the judgement defined the basis for decisions on future methodological developments, stating that census results need to reliably portray reality (*realitätsgerecht*), while using state-of-the-art statistical methodology and minimising impacts upon the fundamental right of informational self-determination (Radermacher, 2021).

As already shown by Kukutai et al. (2015), the development of censuses is increasingly determined by recommendations and regulations at the international level. Regarding international influences on the German census, several regulations and guidelines are of particular interest. The basis for the conceptual design of population and housing censuses in Europe is provided by the Recommendations of the Conference of European Statisticians, which are updated for every census round (UNECE, 2015). The recommendations have also been used as the general framework for the European Union programme for the 2021 round of population and housing censuses from which data are to be provided to Eurostat by Member States under the provisions of EU Regulation (EC) 763/2008 (the current basic legal act regarding the census in the EU member states). The work in the context of the United Nations Economic Commission for Europe (UNECE) at the same time guided important methodological developments, amongst others by preparing guidelines on the use of registers and administrative data for population and housing censuses (UNECE, 2018) and on the assessment of the quality of administrative sources for use in censuses (UNECE, 2021). At the level of the European Union, conducting a decennial census is required in member states since the EU Regulation (EC) 763/2008 has been endorsed.

On both the national and the EU level, the requirements are currently facing profound changes. As shown by recent discussions on European (Eurostat, 2017, 2018) as well as German national level (German Data Forum, 2016; German Statistical Council, 2018), users are requesting census data more frequently, more timely, and in more differentiated regional breakdown. On the level of EU legislation, these requirements are reflected in the current preparations of a new EU framework regulation for census and population statistics (European Statistics on Population and Housing– ESOP). This framework regulation is intended to establish the compulsory census data set of the European Union after 2025 and to replace the current regulations on the census (Regulation No. 863/2008), population statistics (Article 3 of Regulation No. 862/2007) and migration statistics (Article 3 in Regulation No. 862/2007). One of the main objectives of ESOP is to provide population data broken down by grid cells (based on the 1 km square reference grid) on an annual basis from the reference year 2025 onward. This intends also to take into account the increased use of data from administrative sources in national census data collection (Eurostat, 2017). Although the current methodology of a combined census in Germany was already successfully implemented in 2011, it is not going to fulfil the future requirements on population and housing censuses that will be regulated by ESOP following the 2022 census.

In addition to these changing requirements at the international level, at the national level there are efforts towards the digitalisation of public administration and an increased use of administrative data in order to reduce the cost of data collection as well as the burden on citizens. This transition is facilitated by new national legislation strengthening the use of registers for administrative purposes. For example, the Onlinezugangsgesetz (2020) lays down the law that all administrative services must be available online from the end of 2022. The Registermodernisierungsgesetz, endorsed by the legislator in April 2021, aims to introduce a personal identification number for the purpose of the implementation of the Onlinezugangsgesetz as well as the register-based census.

In response to these developments, the Federal Statistical Office of Germany is currently preparing a transition of its census model towards a fully register-based census. The implementation of a register-based census aims at providing census data in the areas of population, housing and dwellings, households and families, labour market, and educational attainment without conducting additional surveys. Further, it could provide grid-based results and results in higher frequencies to respond to changing user needs more flexibly. At the same time, cost as well as burden on the population could be reduced through an increased use of already existing administrative data compared to the combined approach. To facilitate the test of the new methodology, the *Registerzensuserprobungsgesetz* (Register Census Testing Act) entered into force in June 2021, providing the basis for the development of a number of census topics, in particular in the field of population figures.

10.3   Requirements and potentials of a future register-based census model

A census design has to consider a multitude of requirements, ranging from quality standards, cost issues and user needs to the burden it imposes on citizens. The implementation of a register-based census in Germany is driven by several goals related to these requirements which are described in the following paragraphs (Körner, Krause & Ramsauer, 2019).

Grid-based results: The full potential of the census as a basis for local, regional, national as well as EU-level decision-making can only be tapped if data are provided in grid-based form. Up to now, census data on population and housing in Germany refer to administrative areas (e.g. NUTS or LAU areas). However, if results are provided at the level of grid cells, this will allow more flexible tabulations for non-administrative areas. As mentioned above, the annual provision of population data broken down by grid cells from the reference year 2025 onwards is one of the core requirements of the future EU framework regulation ESOP.

Higher frequency: The conduction of surveys in the combined census model is a large-scale operation. For this reason, the combined census can only be carried out at long intervals and provides results only every ten years. However, in all topics of the census, there are variables for which the current frequency of ten years is considered too long by many users. In a register-based census system that is constantly maintained and does not require the conduction of additional surveys, results can be provided more frequently and thus help to remedy the concern of the users.

Higher flexibility: With data needs changing over time, data programmes will have to be more flexible in the future. In the current German system, intercensal population updates are used in order to determine the annual population figures in between the census years (Kaus & Mundil-Schwarz, 2015). As the method of intercensal population updates always requires a census as a baseline, it can only provide results for variables and breakdowns which were included in the preceding census. New variables or breakdowns cannot be introduced before the next census provides a new baseline including these variables. The potential of intercensal population updates to analyse specific variables is thus relatively fixed between two census rounds. In contrast, in a register-based census approach, new data sources can be added and analysed more flexibly.

Timeliness: The size of the operations that are used for data collection in the combined census approach goes along with rather long production times. Due to this reason, results can only be produced with some delay. The provision of more timely results is another core requirement of the planned EU framework regulation on population statistics. The use of a register-based approach can contribute to shortening the time required for field and data processing operations and thus achieve shorter production times.

Reduction of cost and burden on citizens and administration: Interviewer-administered data collection is not popular due to its cost and the burden it imposes on respondents. Compared to traditional censuses, the combined census model in Germany has already led to significant reductions in cost and response burden. However, it still includes components related to traditional data collection which are associated with rather high costs. The supplementary household sample survey, the enumeration of buildings and dwellings and the survey at addresses with collective living quarters included in the combined census model are demanding in terms of time and effort, as more than 10 million inhabitants and 25 million owners of buildings and dwellings have to be surveyed. Several institutions have questioned whether the conduct of surveys is still justified for the purposes of the census. For example, the National Regulatory Control Council (2016, 2019) has advised increasing the use of existing administrative data sources in the census. The German Federal Constitutional Court has highlighted that collecting data from already available administrative sources is preferable to conducting surveys due to its lower intervention with the fundamental constitutional rights of the population (Leischner & Bierschenk, 2019). The basic idea of a register-based census is to replace the direct collection of data from the population with the use of administrative data held in existing administrative and statistical registers. It thus includes a reduction of burden on the population. In addition, experiences from countries with purely register-based censuses indicate that important cost reductions can be achieved (UNECE, 2014).

10.4   Basic prerequisites and challenges

Introducing a register-based census model in Germany is not straightforward for several reasons.

The introduction of a fully register-based census presupposes that a number of preconditions are met regarding data quality, register access, and the possibility of linking the records from several registers. These requirements are discussed in the following paragraphs, referring to challenges that are imposed by tight legal restrictions, an incomplete register infrastructure and coverage issues in administrative data in Germany.

Data quality: Providing results on census and population statistics belongs to the most prominent tasks of official statistics. These results serve as a basis for decision-making in various political and administrative domains, including the number of representatives in the Federal Council (*Bundesrat*) and the financial distribution between federal, state and local governments. Due to their particular importance, the results of the census are subject to very high quality requirements (Körner et al., 2019). As outlined before, the Federal Constiutional Court (BVerfGE, 2018) highlighted in its decision regarding the census 2011 drawn in September 2018 that the use of population figures in particular for the allocation of resources requires that the legislator makes sure that population figures are accurate (*relitätsgerecht*). In order to use register data for census purposes, they need to be of sufficient quality and approaches that correct the data for errors are necessary. This is of particular importance in the case of administrative registers, which are usually not primarily kept for statistical purposes. For example, as demonstrated by the results of the census 2011, the population registers which constitute the main data source for the German population figures are subject to significant over- and undercoverage (Bechtold, 2013; Bund-Länder-Arbeitsgruppe "Einwohnerzahlen", 2016). Methods that make it possible to identify over- and undercoverage in the population registers thus constitute an essential precondition of the development of a future-proof census model in Germany. In this contribution, the so-called *signs-of-life-approch* as a possible method to identify over- and undercoverage in the population registers is discussed as an example.

Data access in the context of an incomplete register infrastructure: In order to move towards a register-based census, the necessary data in the fields of demography, buildings and dwellings, household and family types, as well as labour market and education must exist and access of official statistics to data has to be ensured. Compared to countries, which have already implemented register-based approaches, the German register landscape is fragmented and lacks registers that provide comprehensive nationwide information on certain census areas. There are fewer suitable registers available in Germany than in other, e.g. in the Nordic countries which have already adopted a register-based approach many decades ago. While a significant part of the information of interest is already available in registers in Germany, some still need to be created. For instance, neither nationwide registers on buildings and dwellings nor on the educational attainment of the population currently exist. Presently, the Federal Statistical Office, in cooperation with the State Statistical Offices is preparing the establishment of a statistical education register. This education register would link information on educational achievements over time and thus provide an important data source to assess the variable educational attainment in a register-based census. Challenges related to the establishment of an education register include legal issues related to the federal structure of Germany and the fact that not all necessary data is currently available at the individual level (Gawronski, 2020). In this contribution, we discuss the possibility of creating a register on buildings and dwellings as another example to illustrate the topic of data access in more detail.

Procedures to link registers: The development of suitable approaches to link data from different registers is another basic precondition for a register-based census. On the one hand, this is necessary as not all the variables required for a population and housing census are available in one single register. On the other hand, reliable and easily implementable linkage procedures are also required for quality assurance (e.g. in order to check for administrative signs of life) as well as to analyse the consistency of information included in more than one register. For a register-based census, two objectives of record linkage are of particular importance: First, at the level of persons, it must be possible to link the data of persons across several registers, ideally supported by a constant identifier. Second, for the production of data on housing, household and family types, approaches need to be developed to link the persons to the dwellings they live in by the use of a building and dwelling identifier. Further units requiring record linkage procedures include addresses as well as institutions. In this contribution, the possibility of linking register data without a unique ID number is discussed as an example.

10.5   Selected issues

10.5.1   *Treatment of over- and undercoverage in the population registers*

Providing accurate data on the size and demographic structure of the population is one of the main tasks of a census. In Germany, the municipal population registers provide the main data source for register-based population figures as each person residing in the country is legally obligated to be registered in the population register of the municipality he or she lives in. However, accurate population figures cannot be derived by simply counting persons that are registered, because the population registers are subject to significant over- and undercoverage issues. As the population registers are updated by reported demographic events such as births or deaths or by changes of residence, errors can arise, e.g. if people do not comply with their obligation to register and deregister when changing their place of residence. In the 2011 census, about 2.1 million cases of overcoverage and 1.3 million cases of undercoverage were detected by a sample survey as well as about 600,000 cases of duplicates (Bechtold, 2013; Bund-Länder-Arbeitsgruppe "Einwohnerzahlen", 2016).

As depicted in Table 10.1, different types of over- and undercoverage in the population registers can be distinguished. First, overcoverage can relate to persons who have died or who have moved their main residence to another country without reporting it to the German authorities. Overcoverage can also be associated with duplicate records of persons who are registered more than once or in several municipalities. Without correction, such cases would be erroneously added to the population. Second, undercoverage can arise if persons who were born in or have immigrated to Germany do not report these events to the authorities. Undercoverage can also occur if persons have moved to another municipality within Germany without registering in the new municipality but are deregistered "to unknown" (e.g. due to returned election documents) by the

Table 10.1 Types of over- and undercoverage in the population registers

Type of coverage error	Description	Consequence if not corrected
Overcoverage in the population registers ("pure" cases of overcoverage)	Omitted deregistration after death	Persons would be counted although they do not live in the country
	Omitted deregistration when emigrating to another country	Persons would be counted in Germany as well as in another country
	Omitted deregistration when moving the main residence to another country	
	Registration at more than one sole or main place of residence	Persons would be counted several times within the country (duplicates)
Complementary over- and undercoverage	No change of registration when moving within the country	Persons would be counted in the wrong municipality
	No change of registration when moving the place of residence within the country (e.g. exchange of main residence with secondary residence)	
Undercoverage in the population registers ("pure" cases of undercoverage)	Omitted registration after birth	Persons would not be counted although they live in the country
	Omitted registration when immigrating from another country	
	Omitted registration of persons who are deregistered „to unknown" although still living in the country	
	Registration only with secondary residence	

municipality of their former residence. Moreover, undercoverage can also result if persons are solely registered with secondary but with no main residence in the population registers. Without correction, these persons would not be counted in the population although their actual place of residence is located in Germany. Third, there are cases of complementary over- and undercoverage that can arise if persons move their main residence to another municipality within Germany but do not report this change of residence to the authorities. Without correction, these persons would be counted to the population of a municipality in which they do not live anymore (complementary overcoverage) while they would not be counted to the population in the municipality they actually live in (complementary undercoverage).

10.5.1.1 Sample surveys

In Germany, there is no central population register. Instead, there are local population registers in the 12,000 municipalities which are maintained separately by the respective municipalities. For the purpose of the combined census, the population register data from all municipalities are merged into a single nationwide statistical population database (*Referenzdatenbestand*). Based on this database, duplicates as well as persons that are registered solely with secondary but no main residence can be determined by means of a duplicate detection methodology (Diehl, 2012). As this procedure does not necessarily require any surveys, it will be possible to adopt this duplicate detection methodology in the register-based census.

Still, the detection of the remaining cases of over- and undercoverage requires a different approach. As mentioned above, the combined census model uses a household sample survey in order to detect over- and undercoverage. Persons that are included in the population registers but not encountered by the interviewers conducting the survey are identified as cases of overcoverage. Persons that are identified as residents of a certain address but are not registered in the respective population register are identified as cases of undercoverage (Berg & Bihler, 2011). This information constitutes the basis for the estimation of over- and undercoverage for the total German population in a procedure of mathematic-statistical extrapolation. Hence, for the transition to a fully register-based census after 2022 it is necessary to develop a new methodology which is able to correct these errors without using a sample survey.

10.5.1.2 *Signs-of-life-approach*

In order to identify overcoverage in the register-based census model, the so-called signs-of-life-approach is considered, which is already applied or prepared in other countries (e.g. Austria, Estonia, Spain). The basic idea of this approach is illustrated in Figure 10.1. It is to identify potential errors by linking the data sets in the population registers with administrative signs of life in other registers (comparison registers). Comparison registers are administrative data that contain information about administrative contacts that a person might have and which indicate that the person has the main residence in Germany. If a person has had such an administrative contact in a given period, this is represented by a record in the administrative

	Population register data	Comparison register 1	Comparison register 2	Potential case of overcoverage
Person 1	✓	✓	✓	✗
Person 2	✓		✓	✗
Person 3	✓			✓
⋮	⋮	⋮	⋮	⋮

Figure 10.1 Basic idea of detecting potential cases of overcoverage in the signs-of-life-approach

register and interpreted as an administrative sign of life. Data sets included in the population registers that do relate to records in comparison registers can be counted as part of the population. Personal records that do not relate to any sign of life in other administrative registers are considered potential cases of overcoverage. For further clarification, persons who are considered potential cases of overcoverage can be contacted in order to verify their actual residence. Subsequently, a statistical correction procedure can be used to decide whether or not a person who is registered in the population registers should be counted according to his or her presence in different administrative registers (Agrüeso, 2017; Gumprecht & Wanek-Zajic, 2014; Tiit & Maasing, 2016).

In addition to the detection of overcoverage, the signs-of-life approach might also be used to detect cases of undercoverage as well as cases of complementary over- and undercoverage. By linking data from administrative registers with the population register data, potential cases of undercoverage might be identified if persons have left signs of life in administrative registers at a different place than in the population registers. A main challenge of this approach is to assign identified cases of undercoverage to their corresponding municipalities and, as geocoded population figures will be required after 2024, ideally to the correct grid cells. This requires the matching of local area codes or addresses. In order to achieve an efficient correction, local area codes or addresses of high quality and timeliness have to be available in the administrative registers which provide information on signs of life. In order to determine the potential of the signs-of-life approach to detect undercoverage as well as overcoverage, a large-scale pilot test is currently being carried out, in which the results of the methodology can be directly compared with the results obtained with the current method applied in the combined census.

A crucial element in the preparation of the signs-of-life approach is the identification of suitable registers. Comparison registers have to be selected in such a way that each person having his or her main residence in Germany is included in at least one of the selected registers. Therefore, all ages and circumstances of life need to be covered so that the information needs to be obtained from a range of different registers, e.g. pension insurance data or vehicle holder data. Moreover, the signs of life need to be specified in such a way that they can reasonably be associated with the presence of a person in Germany during the reference period. For example, persons with a main residence in other countries must not be considered and thus need to be distinguished. In preparation for the pilot test, suitable comparison registers and signs of life have been identified together with a wide range of authorities in charge of keeping these registers. The pilot study will be used to validate the choice of the registers and to assess the results obtained with the signs-of-life approach. This includes a test of the record linkage procedures used to link the population register data with the data stemming from the comparison registers. These cases can be assessed in comparison to the cases of over- and undercoverage that will be found in the household sample survey of the 2022 census.

There are two options currently discussed as to where the signs-of-life approach could be implemented. The first option is to implement the signs-of-life approach within the administrative system. This idea refers to the above-mentioned recent discussions of a modernisation of the German register system which include the implementation of a basic administrative register of personal data. This register would be used as basic data stock for the purpose of identity management in various administrative processes. If the signs-of-life approach were implemented as a regular method of quality assurance within this basic register, its results could be used for the purpose of statistics as well as the purpose of directly correcting overcoverage and undercoverage in the original data sources, including the administrative population registers. However, since it is not clear at which point in time such a system could be fully implemented, an implementation within the statistical domain becomes necessary in order to be able to comply with the coming requirement to provide annual population data at the level of geographical grid cells. Due to statistical confidentiality requirements, this option would not allow any feedback about identified cases of over- and undercoverage to the administrative population registers (*Rückspielverbot*; see above). However, it might be a transitional solution until the implementation of the administrative base register.

10.5.2   *Creation of a register on buildings and dwellings*

While in some areas, e.g. in the field of population data, existing administrative registers can be used as a basic data source for the construction of a register-based census, this is not the case in others. For the purpose of the present contribution, we discuss the issue of data on buildings and dwellings as a case in point. Besides the provision of data on the size and demographic structure of the population, providing data on buildings and dwellings as well as the housing situation of the population is the second main task of a census. Data on buildings and dwellings are the basis for generating data on the structure of the housing stock, the ownership structure of buildings and dwellings and nationwide estimates on nationwide vacancy and ownership rates in detailed regional breakdown. They are also required for the generation of data regarding household and family types, for which a linkage of dwellings and persons needs to be established.

Currently, data on buildings and dwellings are collected in a complete enumeration among all 25 million building and dwelling owners in each decennial census. Collecting these data in a register-based approach would not only allow reducing the burden considerably but may also create the potential to provide this type of information more frequently than every ten years. However, to date, there is no nationwide register on buildings and dwellings in Germany that the census may use. There are several types of administrative data sources available, yet none of these sources covers the entire population of interest and all the variables required for census purposes. For example, the land registers (*Grundbuch*) lack information on rented dwellings,[3] as well as several variables required for census purposes, while the data of the land surveying offices cover only buildings (not dwellings) and also lack important variables. Some municipalities keep municipal building registers, which are however not harmonised and quite different in design. Also, the data of the tax authorities used to determine the land property tax (*Grundsteuer*) cannot be used, since also here no rented dwellings are included and a nationwide harmonisation cannot be guaranteed since the regulation of the land property tax can be determined by the Federal States individually.

As the necessary information cannot be obtained from existing data sources, a building and dwelling register would have to be created in order to be in a position to provide the necessary information on buildings, dwellings and the housing situation of households and families using a register-based methodology. The use of the building register for census purposes requires the creation of an administrative register (and not a statistics register) because of two reasons:

• First, linking the buildings and dwellings with the information regarding the persons living in these buildings and dwellings makes it necessary to keep the respective building and dwelling ID in the population registers. In Germany, this is legally only possible in the case of an administrative register, since otherwise data would have to be transmitted from the statistical domain to the domain of administration.
• Second, a sufficient quality of the data kept by the building and dwelling register can only be guaranteed if the data are equally used for administrative purposes.

Discussions with potential users show that there is a large demand for such a register, including for administrative purposes which could also play the role of a basic real estate register in the context of the general register modernisation. The administrative purposes for which such a register could equally be used include, e.g., city and regional planning processes, the monitoring of land and dwelling market developments as well as the monitoring of the energy consumption of buildings. Such a register could finally also provide an opportunity to improve the data availability regarding commercial buildings.

When conceptualising a building and dwelling register, the following three fundamental questions need to be addressed:

1 Collection of the information on the stock of buildings and dwellings: A procedure needs to be defined on how the data on the initial stock of buildings and dwellings can be obtained. As no existing data source covers either all units or all variables, the inital stock would probably have to be determined by contacting the building and dwelling owners, similar to the processes developed for the decennial census.
2 Updating and maintenance of the register: Once the initial stock is in place, a procedure needs to be established for an ongoing update of the register information. The update process needs to be in place directly following the collection of the stock data. In doing so it is important that the different concerned authorities comply with the definition of data transmission standards in order to allow an automatic updating of the register. The different data sources to be taken into consideration for the updating and maintenance of the register include in particular the data from the building supervisory authorities, the data of the land surveying office as well as the data of the tax authorities.

3 Development of interfaces: Technical interfaces need to be provided to establish a regular data exchange as well as the integration of data provided from different sources.

In addition to these aspects, a building and dwelling identification number is required for easy identification and updating of each unit in the register. Such ID number needs to follow a common logic and should ideally at least roughly indicate the position of the different dwellings inside a building, e. g. the floor. Furthermore, building and dwelling IDs need to be linked to the persons in the population register in order to allow statistical household generation as well as deriving variables regarding the housing situation of the population (e.g. floor space per person).

10.5.3   *Linking register data without a unique ID number*

Establishing appropriate procedures to link data from different registers is a basic precondition of a register-based census. It is not only required to integrate information from different data sources at the person level since not all of the variables of interest can be found in one single register. It is also necessary as the quality assurance using the signs-of-life approach will have to rely upon comparisons between the registers and requires reliable record linkage procedures. A linkage of records from different registers can principally be achieved either by the use of a constant identifier available in all registers under consideration or by the use of identifying variables (e.g. name and date of birth) which have to be available in a sufficient number across the registers to be linked.

Constant identifiers are suited best to make sure that different registers can be linked in a unique way and that identifiers are unchangeable over time (Schnell, 2019). Assuming that a central coordination is established to assign the identifiers, it is also highly likely that constant identifiers achieve the most reliable linkage, minimising false negatives as well as false positives. To account for the requirements of IT security and data protection, constant identifiers can be encrypted (e.g. by using cryptographic hash functions) in order to limit the possibilities to generate profiles of the units available in the registers.

In Germany, the availability of constant identifiers is presently restricted to distinctive administrative spheres and their use is legally limited to purposes within these spheres. For example, the tax identification number is a constant identifier used for various taxation purposes, but it must not be used for other purposes and is only kept in registers of the tax administration (except for the population registers, in which its usage is also limited to taxation purposes). Although the use of the tax identification number as a constant identifier for the digitalisiation of administrative services and the register-based census is discussed (see Introduction), the time of its full implementation in all registers required for the register-based census is not yet fully clear. Therefore, using a constant identifier to link registers for a register-based census is not possible in a short-term perspective. Therefore, the Federal Statistical Office is currently exploring the possibilities to link registers based on personal variables. Problems going along with this alternative include the following: Links might be equivocal if the data available do not allow a sufficient differentiation of the units included in the registers under consideration (e.g. if more than one person has the same name and date of birth). The reliability of the record linkage is limited by the fact that person data available might be subject to changes over time. Even if most of them are stable, names, dates of birth, and also sex can be changed over time. Also, the harmonisation of identical variables kept in different registers will often be limited, e.g. if different standards are applied to transcribe foreign scripts or different standards of character encoding are applied. Finally, the record linkage procedures have to be applied at every single data transmission required for the register-based census, leading to an increased effort at the statistical office or the unit in charge of linking the registers.

In an expert evaluation (Schnell, 2019) it was concluded that applying record linkage procedures based on personal variables is generally feasible provided that a sufficient number of identifying variables is available. The first evaluation was further validated in a simulation study that particularly focused on the effort required for clerical checks. The results of the simulation study show that clerical checks can be necessary in a considerable number of cases, depending on the quality of the data in the registers to be linked. It therefore recommends that a combination of constant identifiers and variables identifying person variables could be envisaged ("hybrid approach"). The precondition of such a combination would however be that at least one of the registers includes the person variables as well as the constant identifiers to be used for record linkage. Moreover, the study emphasised the need to provide a high-performance IT infrastructure in order to allow for timely processing in particular of probabilistic record linkage procedures (which was found to be highly time-consuming in the case of the population of the size of Germany). Empirical tests regarding the record linkage procedures are being carried out in the large-scale pilot study mentioned above (see 10.5.1.2).

Besides the objective of providing population figures, censuses usually also aim at providing data on the housing stock, the housing situation of the population, as well as key variables regarding the household and family structures. In particular, for the latter two objectives it is essential to have the possibility to link the persons included in the population register with information on the dwellings they actually live in and the other persons they share their dwellings with. Furthermore, persons in collective living quarters need to be identified in the population count to be distinguished from persons living in private households. In order to achieve this requirement, a building and dwelling identification number must be introduced. While persons covered in several registers can be linked both by using a constant identifier and by using person characteristics (or a combination of both), in the case of buildings and dwellings only a constant identifier can be used: In the population register on the one hand and the building and dwelling register on the other, there is no overlap of information which a record linkage procedure could be based on. Therefore, we are currently discussing the idea of introducing a building and dwelling identifier in the future building and dwelling register and subsequently introducing the identifier in the population registers as well. By this means, all inhabitants can be assigned to one dwelling in the registration process.

10.6   Conclusions

Changing user requirements lead to profound changes in census and population statistics. The German Federal Statistical Office is currently preparing a change-over to a register-based census system in order to provide results more frequently, more timely, more flexibly and in greater regional breakdown.

Setting up a register-based census model is a challenging undertaking, even in a country like Germany, where a combined census model has been already in place since 2011. The extensive preparations that are necessary in this process include the development of appropriate quality assurance procedures, the access to existing and the creation of new registers and the creation of procedures to link register data in compliance with data protection requirements. As the three examples that were discussed in this chapter (treatment of over- and undercoverage in the population register, the creation of a register on buildings and dwellings as well as linking register data without a unique ID number) illustrated, the first considerations for the census post-2022 already had to start well in advance of the census 2022.

With further efforts to make progress in the digitisation of public administration, the register infrastructure in Germany is under constant development. This development may lead to further harmonisation and linking of registers, which could open new opportunities for the use of administrative data for purposes of official statistics and the register-based census system.

Notes

1 The views expressed in this paper are those of the authors and do not necessarily coincide with the views of the Federal Statistical Office. We are greatful for their helpful comments to Anja Krause, Robin Ostrowski and Saskia Fuchs, as well as the anonymous reviewers.
2 The census was initially planned for the year 2021, but had to be moved to 2022 due to the Covid-19 pandemic.
3 Rented dwellings are the majority of dwellings in Germany.

References

Agrüeso, A. (2017). *2021 Population census and migration statistics in Spain. Why exchanging microdata?* Retrieved from https://www.ksh.hu/dgins2017/papers/dgins2017_session1_es.pdf

Bechtold, S. (2013, September). *Analyse der durch den Zensus verursachten Abweichungen bei den Einwohnerzahlen.* Presentation at the conference "Statistische Woche", Berlin, Germany.

Bechtold, S. (2016). The 2011 census model in Germany. *Comparative Population Studies, 41*(1), D1–D9. doi:10.12765/CPoS-2016-07en

Berg, A., & Bihler, W. (2011). Das Stichprobendesign der Haushaltsstichprobe des Zensus 2011. *WISTA Wirtschaft und Statistik, 04/2011*, 317–328. Retrieved from https://www.destatis.de/DE/Methoden/WISTA-Wirtschaft-und-Statistik/2011/04/stichprobendesign-42011.pdf

Bund-Länder-Arbeitsgruppe "Einwohnerzahlen". (2016). Bericht der Bund-Länder-Arbeitsgruppe "Einwohnerzahlen" vom 31. März 2016. *Drucksache 389/16.* Retrieved from https://www.bundesrat.de/SharedDocs/drucksachen/2016/0301-0400/389-16.pdf

BVerfG. (1983). *Urteil des Ersten Senats vom 15. Dezember 1983 (Volkszählungsurteil) (No. 1 BvR 209/83, Rn. 1-215)*. Retrieved from https://www.bundesverfassungsgericht. de/SharedDocs/Downloads/DE/1983/12/rs19831215_1bvr020983.pdf

BVerfG. (2018). *Urteil des Zweiten Senats des Bundesverfassungsgerichts vom 19. September 2018–2 BvF 1/15 -, Rn. 1–357 (BVerfGE 150, 1–163)*. Retrieved from https:// www.bundesverfassungsgericht.de/SharedDocs/Entscheidungen/DE/2018/09/ fs20180919_2bvf000115.html

Desrosières, A. (2010). *La politique des grands nombres. Histoire de la raison statistique.* Paris, France: La Découverte.

Diehl, E. (2012). Methoden der Mehrfachfallprüfung im Zensus 2011. *WISTA Wirtschaft und Statistik, 06/2012*, 473–484.

Dittrich, S. (2019). Der registergestützte Zensus in 2021. *WISTA Wirtschaft und Statistik. Sonderheft Zensus 2021*, 5–11. Retrieved from https://www.destatis.de/DE/Methoden/ WISTA-Wirtschaft-und-Statistik/2019/07/sonderheft-wista-072019.pdf

Eurostat. (2017). Budapest memorandum "Population movements and integration issues – Migration statistics". *103rd conference of the directors general of the National Statistical Institutes*. Budapest, Hungary, 2017.

Eurostat. (2018). *Strategy for the post 2021 census*. Doc. DSS/2018/Mar/3.1, Meeting of the European Directors of Social Statistics, Luxembourg, 2–3 March 2018.

Gawronski, K. (2020). Konzeption eines Bildungsregisters in Deutschland. *WISTA Wirtschaft und Statistik, 02/2020*, 37–45. Retrieved from https://www.destatis.de/DE/Methoden/ WISTA-Wirtschaft-und-Statistik/2020/02/konzeption-bildungsregister-022020.pdf

German Data Forum. (2016). *Empfehlungen des RatSWD zum Zensus 2021 und zu späteren Volkszählungen*. Berlin, Germany: RatSWD. Retrieved from https://www.ratswd.de/dl/ RatSWD_Output2_AG-Zensus-Bericht.pdf [Zugriff am 19. November 2018].

German Statistical Council. (2018). *Fortentwicklung der amtlichen Statistik. Empfehlungen des Statistischen Beirats für die Jahre 2018 bis 2022*. Wiesbaden, Germany. Retrieved from https://www.destatis.de/DE/Ueber-uns/Leitung-Organisation/Statistischer-Beirat/ fortentwicklung-nov-2018-2022-teil3.pdf

Gumprecht, N., & Wanek-Zajic, B. (2014). *Registerbasierte Statistiken. Methodik. Statistik des Bevölkerungsstandes* (Schnellbericht 10.12, 02/2014). Vienna, Austria: Statistik Austria.

Kaus, W., & Mundil-Schwarz, R. (2015). Die Ermittlung der Einwohnerzahlen und der demografischen Strukturen nach dem Zensus 2011. *WISTA Wirtschaft und Statistik, 04/2015*, 18–38. Retrieved from https://www.destatis.de/DE/Methoden/WISTA-Wirtschaft-und-Statistik/2015/04/ermittlung-einwohnerzahlen-042015.pdf

Körner, T., & Dittrich, S. (2017). *The combined census model in Germany – origins, lessons learned and future perspectives*. Working Paper 25, 19th meeting of the Group of Experts on Population and Housing Censuses, Conference of European Statisticians, United Nations Economic Commission for Europe, Geneva, Switzerland, 4–6 October 2017.

Körner, T., Krause, A., & Ramsauer, K. (2019). Anforderungen und Perspektiven auf dem Weg zu einem künftigen Registerzensus. *WISTA Wirtschaft und Statistik, Sonderheft Zensus 2021*, 74–87. Retrieved from https://www.destatis.de/DE/Methoden/WISTA-Wirtschaft-und-Statistik/2019/07/anforderungen-perspektiven-registerzensus-072019.pdf

Kukutai, T., Thompson, V., & McMillan, R. (2015). Whither the census? Continuity and change in census methodologies worldwide, 1985–2014. *Journal of Population Research, 32*, 3–22. doi:10.1007/s12546-014-9139-z

Leischner, S., & Bierschenk, M. (2019). Zur Verfassungsmäßigkeit der Vorschriften über den Zensus 2011. Die Besonderheiten der statistischen Zweckbindung nach der Entscheidung des Bundesverfassungsgerichts. *WISTA Wirtschaft und Statistik 01/2019*, 11–18. Retrieved from https://www.destatis.de/DE/Methoden/WISTA-Wirtschaft-und-Statistik/2019/01/verfassungsmaessigkeit-zensus-012019.pdf

National Regulatory Control Council. (2016). *Stellungnahme des Nationalen Normenkontrollrates gem. § 6 Abs. 1 NKRG. Entwurf eines Gesetzes zur Vorbereitung eines registergestützten Zensus einschließlich einer Gebäude- und Wohnungszählung 2021 – Zensusvorbereitungsgesetz 2021* (NKR-Nr. 3821). Berlin, Germany.

National Regulatory Control Council. (2019). *Stellungnahme des Nationalen Normenkontrollrates gem. § 6 Absatz 1 NKRG. Entwurf eines Gesetzes zur Durchführung des Zensus im Jahr 2021* (NKR-Nr. 4684, BMI). Berlin, Germany

Onlinezugangsgesetz. (2020). Gesetz zur Verbesserung des Onlinezugangs zu Verwaltungsleistungen (Onlinezugangsgesetz – OZG). Onlinezugangsgesetz of 14 August 2017 *Bundesgesetzblatt I*, 57, 3122, 3138–3139, last modified by Article 1 of the law of 3 December 2020 *Bundesgesetzblatt I*, 59, 2668. Retrieved from https://www.gesetze-im-internet.de/ozg/OZG.pdf

Radermacher, W. J. (2021). *Statistics: A matter of trust.* Paper presented at the 40th Congress of the German Sociological Association, 14–24 September 2020, Berlin, Germany.

Registermodernisierungsgesetz. (2021, April 6). Gesetz zur Einführung und Verwendung einer Identifikationsnummer in der öffentlichen Verwaltung und zur Änderung weiterer Gesetze (Registermodernisierungsgesetz – RegMoG). *Bundesgesetzblatt I, 14*, 591–606.

Schnell, R. (2019). *Eignung von Personenmerkmalen als Datengrundlage zur Verknüpfung von Registerinformationen im Integrierten Registerzensus.* Research Methodology Group, Universität Duisburg-Essen. doi:10.17185/duepublico/49551

Scholz, R. D., & Kreyenfeld, M. (2016). The register-based census in Germany: Historical context and relevance for population research. *Comparative Population Studies, 41*(2), 175–204. doi:10.12765/CPoS-2016-08en

Statistik Austria. (2015). *Standard-Dokumentation Metainformationen Registerzählung 2011.* Bearbeitungsstand: 27.04.2015, Wien, Austria: Statistik Austria. Retrieved from https://www.statistik.at/web_de/dokumentationen/menschen_und_gesellschaft/Bevoelkerung/index.html

Szenzenstein, J. (2005). The new method of the next German population census. *Statistical Journal of the United Nations ECE, 22*, 59–71.

Tiit, E., & Maasing, E. (2016). Residency index and its application in censuses and population statistics. *Eesti Statistika Kvartalikiri, 3*, 53–60.

UNECE. (2014). *Measuring population and housing. Practices in the UNECE countries in the 2010 round of censuses.* Geneva, Switzerland: United Nations.

UNECE. (2015). *Conference of European statisticians recommendations for the 2020 censuses of population and housing.* Geneva, Switzerland: United Nations.

UNECE. (2018). *Guidelines on the use of registers and administrative data for population and housing censuses.* Geneva, Switzerland: United Nations.

UNECE. (2021). *Guidelines for assessing the quality of administrative sources for use in censuses.* Geneva, Switzerland: United Nations.

11 Techno-political transformation and adaptability in Ghanaian census history

Alena Thiel

11.1 Introduction

In spite of chronic funding constraints of the national statistical organizations (NSO), as well as long-term disregard for civil registration and administrative records by the colonial administration, the production of population statistics in contemporary Ghana has seen a steady development since Independence from Britain in 1957. In particular, the West African country's post-independence censuses have been heralded for their methodological rigour and ability to cover large stretches of remote, hard-to-reach, and scarcely populated territories. Preparations leading up to the 2020 Population Census in Ghana rearticulated the national statistical body's leadership. In 2019, the Ghana Statistical Service (GSS) was equipped by the Ghanaian Parliament with new co-ordinating powers, allowing the agency to play a significant role in the larger digitization and datafication agenda of the Ghanaian vice presidency (Thiel, 2020). In practice, GSS announced key innovations ranging from the use of digital technologies for real-time quality assurance to the further expansion of census methodology collecting new types of (geospatial) data.

This chapter explores the history of census innovations in the Ghanaian statistical system from its colonial past to current ambitions of developing new measurement practices based on interoperability-based population data infrastructures. It advances that the historical trajectories of national censuses need to pay close attention to the material and technological path dependences, as well as the model systems travelling into the national statistical system through the work of variously positioned experts and organizations, calling thus for an investigation into how these models play out in variable political and historical contexts of reception. Approaching Ghana's census history from the point of view of data infrastructures – that is, the arrangements of devices, software, experts, and organizational knowledge – and their global circulations, this chapter explores the continuous translations and adaptations that have shaped the Ghanaian census model since the middle of the 19th century. Particular attention will be paid to changing quantitative forms in relation to nation-building, from the first post-independence enumeration in light of Pan-Africanist ideology to the recent contestations around identity and technology in Africa.

DOI: 10.4324/9781003259749-15

Theoretically focussing on the genealogies of Ghana's census infrastructure, the article foregrounds what Emmanuel Didier (2021) has coined the "marbling method". Drawing on Latour, Didier proposes that in any society various quantitative forms are passed on between actors, settings, and organizations, thereby forming specific trajectories of movements. Shaped by the interplay of local forces, Didier argues, these moving quantities produce marbling veins, which can be "traced into the foundations of society, the bedrock itself being an aggregation of countless elements". Marbling, in this light, consists of quantitative forms "most frequently followed (...) in other words a series of procedures or mechanisms whose interactions are regulated with the aim of quantifying a given object over time. The method is the stabilization of marbling vein over time" and "will gradually consolidate the social aggregate that it is both holding together and being held by" (Didier, 2021, p. 21).

The chapter proceeds by laying out Ghana's census history throughout the succession of colonial and post-colonial power relations and the emerging configurations of experts, devices and ideas within them. Research for this publication was carried out in Ghana between January and March 2020, with particular focus on the Public Records and Archives Administration Department (PRAAD) as well as the George Padmore Research Library in Accra. Interviews were conducted with personnel from the GSS, but also with other relevant public authorities involved in identification and civil registration programmes. Additional primary sources (in particular, census manuals, reports and maps) were consulted at the African Studies Library of the University of Leiden.[1]

11.2   Population estimates and colonial rule

Prominent histories of official statistics and registration feature the cases of the United States of America (Anderson, 2015 [1988]; Anderson & Fienberg, 1999; Didier, 2009, 2020; Hannah, 2001), Brazil (Loveman, 2009, 2014), France (Desrosières, 2014, 2010; Noiriel, 2002), Great Britain (Higgs, 2003; Szreter, 2013) and Germany (Aly & Roth, 2004; Caplan, 2013; Hannah, 2009; von Oertzen, 2017a, 2017b). While studies on African registration systems have featured the cases of Rwanda (Tesfaye, 2014), Burundi (Uvin, 2002), Nigeria (Okolo, 1999), Kenya (Weitzberg, 2015), Mauritania and Burkina Faso (Samuel, 2014), and Ghana (Serra, 2018), no African population data system apart from South Africa's (Breckenridge, 2014) has received similarly detailed attention for its socio-historical and infrastructural composition and arrangement. Worse, scholarship has predominantly singularized the African continent as "poor" in numbers by describing its modes and practices of knowledge production as failing to meet international standards (Jerven, 2013). Szreter and Breckenridge (2012, p. 17) problematize this bias when they attribute the failure to represent the "diverse history of community registration" to the

> Profound bias in the documented comparative historical record, which provides us mainly with a history of registration which appears to be strongly

tied to those most powerful, persisting state-like forms of government which generated and archived most efficiently the records of their processes of registration.

Yet, as African census and civil registration systems originate largely from colonial histories, they require a particularly critical examination of the continuities of he-gemonic data histories in contemporary relations of power and subjectivity.

Historians of Africa have long argued that in precolonial times, African local au-thorities were preoccupied with control over labour – shifting to control over land with the introduction of colonial instruments of rule (Sackeyfio, 2012) – as man-power was indispensable for ensuring agricultural lifestyles, while also securing defence against warring chiefdoms. Unsurprisingly, then, de Graft-Johnson (1969, p. 3, see also Cardinall, 1931, p. 125) describes elaborate indigenous counting methods in place at the time of colonial powers' first attempts to count the popula-tion in the territory of what constitutes today the Republic of Ghana. Prior to the establishment of political control by the British colonizers, customary techniques involved the placing of cowries, palm kernels or grains of corn in calabashes as-signed to a town's different sub-divisions (lineages, families), while using different types of grains for each gender.

After the Bond of 1844 had expanded the control of the British Governor over the Gold Coast beyond the coastal forts onto local coastal chiefdoms, British interest in the local population began to grow. Then Lieutnant-Governor of the British Gold Coast settlements Winniett reported to the Colonial Office in the 1846 Blue Book that – though no census had been undertaken of the local population – population in districts under British political control was "no less than 275,000 scattered over a territory of about 6,000 square miles" (quoted in de Graft-Johnson, 1969, p. 1).

Valsecchi (2014, p. 223) provides more detailed insights into the micro-politics of quantifying populations in the early years of the colony. In 1849, Governor and Judicial Assessor of the British settlements James Coleman Fitzpatrick reported to the Colonial Office in London the first complete enumeration of the localities of Dixcove and Appolonia. This local census, Valsecchi points out, served a variety of purposes. By quantifying in each of the two localities' sub-divisions the number of slaves and pawns, Fitzpatrick directly attacked his adversary Henry Swanzy (favoured by the Colonial Office for the position as Judicial Assessor), accusing him of circumventing anti-slavery regulations (Valsecchi, 2014, p. 229). By quanti-tatively representing Dixcove - where Swanzy had effectively established political control - "as a concentration of social relationships rooted in slavery" (Valsecchi, 2014, p. 232), Fitzpatrick pointed to Swanzy as undermining the abolitionist idea and paying "lip-service to emancipation and freedom, while actually functioning as a stumbling block in the fight against unacceptable social habits and practices" (Valsecchi, 2014, p. 235). Beyond this personal conflict, this first documented local enumeration crucially served the purpose of justifying Britain's role in the Gold Coast on a larger political scale. By quantitatively representing Appolonia – a town where Fitzpatrick had been involved in removing the local king – as freed from slavery and dependency, Fitzpatrick directly linked native authority with enslavement, portrayed the capture of the local king as an "emancipatory moment" and ultimately "justified" further colonization of the Gold Coast (Valsecchi, 2014, p. 239f).

Yet, political control was not yet fully consolidated in large areas and quantification could not proceed with great accuracy. Methodological concerns were raised by Winniett who reported in 1849 that

> A Census of the Population of this Settlement was attempted to be taken this year but ... the Suspicious and jealous eye with which the natives view giving any information to Government officials especially respecting numbers caused this important measure to fail.
>
> (quoted in de Graft-Johnson, 1969, p. 1, see also Valsecchi, 2014, p. 233)

Accordingly, in 1851, Governor Hill estimated the population under British control at around 400,000 though, in 1860, Governor Andrew admitted that population estimates may have been exaggerated by as much as 200% (de Graft-Johnson, 1969, p. 2).

In the years following these first population estimates and partial enumerations, quantification quickly gained in importance, especially after its association with the establishment of a coherent system of taxation (Valsecchi, 2014, p. 225). State-led tax collection was established as early as in 1852 with the Poll Tax Ordinance (de Graft-Johnson, 1969, p. 2) and in turn "was used to estimate the number of people living in a given area" (Serra, 2018). At the same time, the production of the first partial population statistics specifically surveyed the military potential of local polities by counting the population of "men able to carry arms" (Valsecchi, 2014, p. 233).

11.2.1   Early colonial censuses 1891–1921

With the establishment of the Gold Coast colony in 1874, the need for more refined population statistics became increasingly pronounced. By 1883, population estimates reflected the new political organization of the colony, laying out figures district by districts (de Graft-Johnson, 1969, p. 2). By 1890, the colonial authorities no longer wanted to rely on population estimates. A first attempt at enumerating the entire population was made (de Graft-Johnson, 1969, p. 3) despite doubts about whether "a census could be taken of the Gold Coast settlements, the natives of which are suspicious of their numbers being counted, having, perhaps a lively recollection of the old days when the poll tax was in force" (1884 Blue Book, quoted in Serra, 2018). The period from 13th to 18th April 1891 marked the Gold Coast colony's first complete census, though in reality only a count was achieved, differentiating between males and females. Official enumerators operated in 16 towns, compiling statistics for approximately 70,000 persons (de Graft-Johnson, 1969, p. 4).

By 1901, census taking for the first time included the newly annexed territories of Ashanti and the Northern Territories. As Serra notes, the

Different geographical coverages of these [early colonial] counts closely mirrored the changing boundaries of British rule in the area (...) The 1891 count referred only to "The Colony" (roughly, today's [Greater] Accra, Western, and Eastern regions). In 1901 Ashanti and the Northern Regions were also included, while following Germany's loss of her African colonies in the First World War, Togoland was placed under the British administration and included in the 1921 census.

(Serra, 2018, p. 667)

Although the rebellion in Ashanti complicated the data collection (Cardinall, 1931, p. 123), for the first time the enumeration was detailed enough to reveal shifts in population statistics and to fuel speculations about demographic trends (de Graft-Johnson, 1969, p. 5).

11.2.2 The census of 1931

Although with each census officials claimed increasing levels of accuracy, the reliability of the early censuses remained a matter of intensive debate (de Graft-Johnson, 1969, p. 7). Whereas Chiefs were inclined to exaggerate the size of their community in the interest of gaining larger representation in the Provincial Councils and increasing their rates, e.g. for gun permits, individuals feared adverse consequences of disclosing property, thus encouraging false reporting (Cardinall, 1931, p. 123). At the same time, due to the lack of an accurate demographic baseline, margins of error were impossible to determine. In view of this, the census of 1931 marked the first individual enumeration carried out by trained enumerators (Cardinall, 1931, p. 125). Following the publication of the notification for the 1931 census on 14 October 1930, preparations began with campaigns carried out by the native authorities and some specialized clerks to "ascertain the number of houses there were in every village and to locate the smaller farm or hunting villages which being more often than not of a temporary character are not usually known to the commissioners of the districts" (Cardinall, 1931, p. 125). A preliminary estimate of the size of the population in each house and with it the expected requirements for census forms intended to be used for non-African statistics were to be added to this initial enumeration (Cardinall, 1931, p. 126).

The result of the 1931 census showed a population increase over 1921 at 37%. In line with the analysis that census taking played a crucial role in the reproduction of colonial rule, Cardinall's description of the 1931 census attributes this large increase to under-estimation in 1921 rather than questioning the accuracy of the 1931 results. Nonetheless, Cardinall (1931, pp. 167–169) raises the problem of unreliable civil registration statistics in the colony, noting that age groups could not be differentiated with great certainty while marriages were rarely certified. Licensing had proved to generate more reliable statistics on the occupations in the colony after it was introduced to "efficiently" fight "dishonesty" among the domestic workforce. Cynically, these same racist attitudes excluded local populations from civil registration, and though in 1926, the births and deaths registration introduced in

1888 was opened to register African births, offices remained limited to the urban areas so that, at the time of independence, a mere 38 offices existed across the country, with a reach of 16% of the population (Kpedekpo, 1971).

11.2.3   The 1948 post-war census

Questions around civil registration remained a central trope for the public discussion of the post-war census of 1948. With the rise of the population question in other parts of the world and British West Africa, especially Nigeria, migration and vital statistics became increasingly important to the Gold Coast colonial administration. Yet, as was remarked at the Second Conference of African Statisticians "No African Census [this far] can be classified as being in all cases universal, nominative and simultaneous" (C.C.T.A., 1957, quoted in Barbour & Prothero, 1961, p. 8).[2] According to observers at the time, up to 1948

> Only rough estimates of population were available which were in no way comparable in accuracy and detail with the assessment of population in Britain based on National Registration (…) [and] calculations of population growth were impossible owing to the absence of vital statistics.
>
> (Barbour & Prothero, 1961, p. 8)

In doing so, the British criticized the very absence of institutions (especially, CRVS) that the colonial administration had forcibly denied the local population.

In view of this, the census of 1948 – due to the Second World War, census taking was halted in 1941 – marked a turn to professionalization including the "latest advances and international standards in compiling vital statistics" (de Graft-Johnson, 1969, p. 8) as well as advances in "organization, methods and results" (Barbour & Prothero, 1961, p. 9). Already in 1911, 1921, and 1931, the preferred census model was that of a rolling census where enumerators were instructed to count "all persons whether strangers or natives of the town or village who had slept in the town or village on the night of" the census (de Graft-Johnson, 1969, p. 8). Yet, CVRS remained underdeveloped, as Serra notes (2018, p. 667), for the colonial authorities, quantification simply served the purpose of stabilizing the colonial expansion, it was not intended to convey any form of citizenship.

11.2.4   The first post-independence census of 1960: the rise of the population question

In 1957, Ghana achieved independence from colonialism and established its own government under the leadership of Dr Kwame Nkrumah's Convention People's Party. Serra (2018, p. 659f) underscores the importance of the 1960 census for Ghana's post-independence nation-building, pointing to the role of the census in Nkrumah's transformative political programmes aimed at providing health and education to the population as well as new imaginations of the post-colonial state, constructed through novel forms of census propaganda. Nonetheless, the census of 1960 in various respects represented a milestone for Ghana's statistical independence and professionalization (expressed not least in the increase of the census budget to nearly £300,000, or ten times the amount spent in 1948, Barbour & Prothero, 1961, p. 3).

With the aim to modernize census taking in Ghana, machine-based data analysis – using IBM punch card machines operated by – as job ads of the time indicate – predominantly young, female clerks – was introduced and documented meticulously. Besides the aura of computerization, Serra notes (2018, p. 666), "[f]or the Statistical Office of the United Nations, the essential features of a 'modern' census included government sponsorship, defined territory, universality, simultaneity, individual units of enumeration, compilation, and publication. Additionally, according to Serra (2018, p. 669), what set 1960 apart from its colonial predecessors was the Central Census Committee's "systematic attempts to gain people's trust" (Serra, 2018, p. 671). Intimately tied to post-colonial nation-building, Serra (2018, p. 672) shows in great archival detail, that the "census education campaign was designed to represent the state as a benevolent, inclusive, and knowledgeable entity" in need for "measurement and quantification as preconditions for economic and social advancement" (2018, p. 673).

Mobilizing statistical methods for building the personal cult around Kwame Nkrumah (Serra, 2018) did go hand in hand with the extensive professionalization of the field of population statistics. The Geographical Planning Officers at the University College were involved in training the field teams and census officers, dividing each local council into enumeration areas of 700–1,000 people, preparing maps (of localities and major features, especially the roads to them) as well as lists of settlements and descriptions of their boundaries and estimated population (Hilton, 1961). In Didier's (2021) words, along with the attainment of independence, a new marbling vein appeared in Ghana's statistical landscape shaped by a new self-understanding of statistical production as a professional and modern scientific exercise built on methods circulating in the global arena of statistical knowledge production.

The census was eventually held at midnight, 20 March 1960, which as Ghanaians were familiar was marked by "football matches, bonfires, tolling of church bells, mass religious services, and beating of drums" (Serra, 2018, p. 679). At 6 a.m. on 21 March, the enumeration started with a "questionnaire containing, in addition to NAME, 11 Statistical ITEMS: address at the place of enumeration, sex, age, place of birth, country of origin, tribe, school attendance, type of economic activity, industry, occupation, and employment status was completed for each person" (Gaisie, 1969). Following the processing of the questionnaires in the Accra Head Office of the Central Bureau of Statistics one advance volume, five complete volumes, as well as special reports on "Statistics of Large Towns" (Special Report A), "List of Localities by Local Authority (Special Report D) and "Tribes in Ghana" (Special Report E) were published from the census (Gaisie, 1969).[3] Stabilizing the recent political transformation, the reports reflected the political organization of Ghana's six regions at the time, that is, Northern, Brong Ahafo, Ashanti, Volta, Eastern, and Western regions, where Accra was legally part of the Eastern region, but was counted separately as Accra Capital District.

Two months after the census a Post-Enumeration Survey was carried out on about 5% of the total population. The main objectives of the Survey were: (1) to measure 'coverage and content errors' and (2) to inquire into additional topics which could not be covered in the main census. Among the items covered were Household size and structure, internal and external migration, religion, literacy, secondary occupation, economic characteristics of the unemployed, degree of employment, marital status and form of marriage, number of wives, locality of residence of husband, fertility and mortality.

(Gaisie, 1969, p. v)

The post-enumeration survey collected the additional information from a sample of 5% of the population "of such depth prohibitive of the whole population", including retrospective questions on births and deaths to allow new methods of demographic analyses in the absence of a vital registration system (previous censuses only estimated demographic levels based on the census' age data) (Gaisie, 1969, p. v).[4] The analysis focused on the differentiation between "high-fertility tribes", urban and rural contexts, age groups, and educational status, which at the time was identified as the focal area of population policy intervention (Gaisie, 1969, p. 37). Ultimately, the goal was to best represent the nation in an international comparison of demographic trends (Caldwell, foreword to Gaisie ca. 1969, p. vi). Yet, despite the global rise of the notion of population policy and the 1960 census results supporting the claim of large population growth, the Nkrumah government remained true to its natalist programme, e.g., by banning all imports of contraceptives (ILO, 1974, p. 19).

11.2.5   *The 1970 census: citizenship and population policy*

Ghana's first post-independence census became a model for the continent (Gaisie, 1969, p. v) and for all subsequent censuses in the West African country. As Serra shows (2018, p. 682),

[t]he heads of other African statistical offices were invited on a study tour to observe the implementation of the post-enumeration survey. In 1961 the Ghanaian government, in cooperation with the United Nations Economic Commission for Africa, set up and hosted a West African Training Centre in Population Census Techniques.

While the census had been technically supported by Dr Benjamin Gil, an Israeli population expert "on loan" from the United Nations, it was by large carried out by African experts. The census itself, in turn, enabled establishment of a range of institutions concerned with demographic data (Gaisie, 1969, p. v), particularly the Population Council, the Central Bureau of Statistics, and in 1966, with support of the Population Council, the Demographic Unit[5] at the Department of Sociology, University of Ghana (Legon).[6]

Despite this institutional setup, the 1970 census directly reflected a political climate of instability and economic distress of the preceding decade, while consolidating international population policy in Ghana. Inheriting dire economic conditions following the overthrow of Nkrumah in 1966 and the subsequent military rule of the National Liberation Council (NLC, 1966–1969), Ghana's Second Republic (1969–1972) under the government of Dr Kofi Busia's Progress Party entered into history as the government responsible for the expulsion of hundred-thousands of migrants in an attempt to address economic hardship. In December 1967, the NLC signed the "World Leaders Declaration on Population", confirming the commitment to population as the focus of national planning, as well as knowledge and means of family planning as a basic human right (ILO, 1974, p. 19). In addition to shifting the attention of population policy onto fertility, the NLC revitalized the country's migration policy by setting up an Immigration Quota Review Committee, restricting immigration of labour available locally, and policing the 1963 Aliens Act (NLC Decree 259, 1968). In 1968, the country's Manpower Board was installed and in March 1969 re-proposed focussing on "population planning for national progress and prosperity" (ILO, 1974, p. 20). One indirect effect of this was the strengthening of statistical, research, and analytical capacities in government, academic, and private entities while co-ordinating and integrating these capabilities with the Central Bureau of Statistics[7] (ILO, 1974, p. 20). The transfer of power to a civilian government under Busia in August 1969 reaffirmed the commitment to population policy to the effect that the 1970 census proved methodologically sensitive to the effects of these policies, particularly in the area of migration which had dropped from 12 to 7% following the Alien Compliance Order of 1969 (Peil, 1974).

Census innovations, accordingly, comprised of maternal and child health featuring prominently as main concerns of the Ministry of Health, and for the first time, the census stated the Infant Mortality Rate (122/1000 live births, an estimated improvement from 156/1000 in 1960, see ILO, 1974, p. 22). In addition,

> an innovation introduced in the 1970 census questionnaire (…) was that the adult population would be asked to give the number of days they worked for pay or profit during the four weeks preceding the Census Night. Previous occupation, industry and employment status of the unemployed would also be inquired into

adding thus a new focus on labour (1970 census report Vol. ii, 1972, p. xvi). Finally, the citizenship question, a further focal area of the Busia government, was also featured in a new quality in the 1970 census. Whereas the question in the 1960 census had been "restrictive" to the question of the country of origin (defined as the country of birth of the father or mother, depending on the respondent's matrilineal or patrilineal descent group) and therefore included some non-Ghanaians (de Graft-Johnson & Ramachandran, 1975, p. 256), in 1970 nationality was defined in the terms of the new Citizenship Act (1970 census report Vol. ii, 1972, p. xxiv).

11.2.6　1984: structural adjustment and decentralization

The government of Busia's Progress Party was short-lived. In January 1972, Ghana came under a new military regime of Colonel IK Acheampong's National Redemption Council (NRC) (Petchenkine, 1993, pp. 53, 60). The NLC alluded to a new path of national development that was diametrical to the policies of the Progress Party. Specifically, the efforts to promote a market-driven economy were reversed (Petchenkine, 1993, p. 64f), while in the international sphere, the western alignment of the previous regime was abandoned. Following the counter-coup of the SMC-II on 5 June 1978 led by General Akuffo (Mensa-Bonsu, 2007, p. 274), again, business interests were at the core of the political agenda. Although the SMC-II was aware that improving the economic situation was paramount to securing popular support, its policies did not yield the desired effects, and a return to civilian rule became inevitable. However, before elections could occur, on 15 May 1979 a renewed coup was attempted but suppressed. When the trial against coup leader Jerry John Rawlings neared its resumption, the army intervened in the imminent death sentence and staged the first successful coup on 4 June 1979 (Mensa-Bonsu, 2007, p. 274) and, following temporary return to civilian rule, a second coup on 31 December 1981.

In view of the failure of Rawlings' populist agenda to improve the living conditions in the country, the regime needed to reconcile its policy with liberal vectors of force stemming from the World Bank and International Monetary Fund (IMF) (Boafo-Arthur, 2007, p. 2; Nugent, 1995, p. 111; Petchenkine, 1993, p. 131).[8] Measures in line with structural adjustment, with their focus on rationalization and expenditure reduction, produced significant data needs "at all levels of authority (regional, district, local)" (GSS, 1987). The 1984 census reflected and foreshadowed these developments in important respects. For the first time, data production was specifically targeted at fulfilling the data needs of international organizations. On 8 February 1984 Government Statistician Oti Boateng was quoted in the Daily Graphic noting that "the census is important for information exchange, to enhance national interest and international co-operation to improve international response to the national situation". Largely oriented internationally, the census was to provide "international organizations including the World Bank (…) accurate figures to be able to offer assistance to the country" (Daily Graphic, 21 February 1984, "U.East bans census officials from trading"). In view of this, the 1983 Economic Recovery Programme "strengthened the statistical service through the provision of high-level manpower, vehicular and other facilities essential efficient statistical work" (Oti Boateng, 1995, p. 329). Part of the PNDC's commitment to international donors' conditionalities involved the advancement of the decentralization agenda to be,[9] as was argued, "better implemented when we have accurate and up-to-date information on the distribution of the population" (Daily Graphic, 12 March 1984, "1984 census is launched"). In that light, the analysis and publication of the census results by individual regions marked a "unique departure from the publication of past census results" (GSS, 1987, p. ix).

In contrast to its international outlook, practical organisation of the 1984 census was built largely on the political instruments of Rawlings' populist revolution, such as the mobilization of the People's Defence Committees (PDC). The military regime of the PNDC further involved itself in the census by issuing orders to the Central Bureau of Statistics "not to look at the census exercise as a sole responsibility of the CBS or even of the Ministry of Finance and Economic Planning" since "a people's government cannot structure its democratic machinery, arrive at policy decisions and implement them effectively without knowing how many are the people" (Daily Graphic, 17 February 1984, "View Census as a National Exercise") – therefore the PNDC initially planned to combine the census with a scheme for National Identification Numbering (see Breckenridge 2010 on the PNDC's attempts to introduce a first nationwide identification system). Census preparations further involved the rhetoric of "kalabule", which associated traders and other holders of private capital with corruption, profiteering, and essentially, the country's economic hardship. Accordingly, census officials were barred from trading, in addition to any political or religious "propaganda" during their service as a punishment for being tried by the military regime's Public Tribunals (Daily Graphic, 21 February 1984, "U. East bans census officials from trading"). At the same time, association of the census with the PNDC was anything but unproblematic as mistrust about the census data's confidentiality and worries about taxation, were for the first time accompanied by fears about forced conscription into the military regime's army (Daily Graphic, 7 February 1984, "Census not for conscription").

11.2.7   2000 and 2010: democratization and data for development

Building on the continuous professionalization of the GSS, Ghanaian government sources note, that the 2000 and 2010 census underscored important development initiatives and enabled two peaceful transfers of power by allowing the conduct of legitimate national elections in 2004, 2008, 2012, and 2016 (GNA 13 November 2019, "Government allocates GH¢45 million for 2020 population census"): Following the succession of various regimes of governance, ranging from liberal economic regimes to both civil and military authoritarianism, on 7 January 1993, Ghana saw the installation of the current Fourth Republic in which "neoliberal economic management strategies that had been practised for some years with the unflinching support of the Bretton Woods institutions and other donor or development partners" (Boafo-Arthur, 2007, p. 1) became partnered with new political liberalism.

While the 1979 Constitution already made provisions to establish a statistical service under a Statistical Service Board, in 1985 the Statistical Service Law (PNDCL 135) for the first time established statistical independence by setting up the Ghana Statistical Service as part of the public service "as an autonomous independent public service with a Board of Directors who report directly to the Office of the President" (GSS, 2020). Under the new law, the statistical service independently determined which data to collect in the most efficient way, and which data to publish "in whatever manner deemed suitable" (GSS, 2020). In this arrangement, the position of the Government Statistician, as head of the service, "is vested with the power to conduct surveys, including any census in Ghana" (Oti Boateng, 1995, p. 329).[10] The Statistical Service Act further bound all public services to collaboration with the GSS, giving the latter large co-ordinating powers. In order to facilitate this co-ordinating function, a National Committee of Users and Producers of Statistics (headed by the Government Statistician) was established to enable interaction between the various agencies. In addition, the GSS was equipped with an Analytical Studies and Development Division, in order to strengthen the methodological capacities of the service. To further improve data quality, the service further employed a higher number of graduates (Oti Boateng, 1995, p. 330).

In 2000, for the first time, data analysis of the Population and Housing Census proceeded to electronically scan census forms.

> Following the data collection, questionnaires were checked using automated control system known as CENTRACK to verify that data returned from all EAs have been received. Questionnaires were edited and scanned using Optical Character Reader (OCR) imaging technology. The data was then run through validation programmes.
>
> (*Report on the PHC*, 2000, pp. 3, 8)

Strategies to disseminate census data as quickly as possible included the distribution of CD ROMS, in addition to the publication of census reports. In terms of content, the 2000 PHC for the first time included a focus on housing conditions, surveying areas such as "household characteristics, geographical location and internal migration, demographic and social characteristics, fertility and mortality, economic characteristics, literacy and education" (*Report on the PHC*, 2000, p. 8).

Reports from the 2010 census indicate similar infrastructural investments in data processing capacity.

> Data capture for the main census began in July 2011, using Fujitsu document scanners. The data capture involved scanning of the questionnaire, interpretation of the scanned marks, transfer of the data and loading the scanned data into an oracle database. Periodic backups of the data and images were made on compact tapes. Three 8-hour shift groups, each with a scanning assistant and a supervisor initially worked around the clock, 7 days a week for the first 6 months. Later, the duration of work was changed to a 12-hour shift for the remaining of the period that the scanning of the census questionnaires took to complete.
>
> (GSS, 2016)

11.2.7   *The census of 2020/2021: census taking and the politicization of the health crisis*

The 2020 UN Round of Population and Housing Censuses, marks a moment of both continuity and innovation in the Ghanaian census history. In 2019, the Statistical

Service Act, 2019 (Act 1003) replaced the Statistical Service Law of 1985. With the new law, the GSS has been given further statistics-producing and co-ordinating functions in the national statistical system. The 2019 Statistical Service (Act 1003) of 24 September 2019 institutionally strengthens the GSS. In 2020, "for the first time, GSS would adopt the use of electronic data collection with other geospatial technologies, which has been recommended by the United Nations Statistics Division (UNSD) for the 2020 round of Population and Housing Censuses". This is to cover the population's access to basic necessities (e.g., water sources, housing conditions) and services (e.g., health care, educational infrastructure), and additionally new items (not included in previous censuses) such as "how people dispose of their household waste, places of convenience, bathing spaces and ICT usage" (Kombat, quoted in GNA, 2 November 2019, "What you need to know about the 2020 population census").

In view of these methodological innovations, a three-tier trial phase was adopted with each stage testing a different process. Francis Nyarko-Larbi, Head of Census Publicity, Education and Advocacy Committee estimated "that the 2020 Census would cost in the region of GH¢500 million" (quoted in the Ghanaian Times, 3 June 2019 "Trial census kicks off in 3 regions"), with 45 million allocated for preparatory activities in October 2020 (GNA 13 November 2019, "Government allocates GH¢45 million for 2020 population census"). The first trial census was carried out on May 26, 2019 (with 2nd June defined as trial census night) to "test the logic of the applications to be used for data collection" (Daily Graphic 25 October 2019 "Govt releases funds for 2020 Census"). The second trial census "to test all the processes" (Daily Graphic 25 October 2019 "Govt releases funds for 2020 Census") was carried out in the districts of Aowin Municipality (Western Region), Ekumfi District (Central Region), Kpone Katamanso Municipality (Greater Accra Regio) and Krachi Nchumuru District (Oti Region). The listing of structures started on 18 November 2019 and targeted the night of Sunday, 24 November for the trial census night. Government Statistician, Professor Anim, underscored the legal basis and benefits of the exercise as well as the international standards adhered to in order to foster support in the respective populations (GNA 19 November 2019 "GSS to Conduct Second Trial Census"). A third trial in December 2019 tested boundaries [as well as the process of conflict mediation, issues relating to internet connectivity, the effectiveness of supervision, the duration for enumeration, publicity, and functionality of the app (Daily Graphic 25 October 2019 "Govt releases funds for 2020 Census"; Interview GSS March 2020).

In January 2020, about 60,000 enumerators were hired, with a minimum senior secondary school education (WASSCE/SSCE diploma) required for the infrastructural setup of the census exercise. CAPIs not only facilitated allowed real-time supervision of the data collection by comparing against data from previous censuses as well as graphical representations of divergence from the mean This is to enable prompt correction of methodological error, and ultimately, the timely analysis and publication of results. As a senior official of the GSS mentioned, in the past, errors were only detected at the HQ upon scanning the questionnaires, hence too late to return to the field to correct them. In addition, GSS foresees a reduction in spending as census forms no longer need to be kept in the original but are simply stored on servers. Yet, digitization was accompanied by a host of other issues that required careful translation of the international agenda (Villacís, Thiel, Capistrano, & da Silva, 2022).

Digital census methods, first and foremost come with the issue of

> Internet fluctuation [which] is one of the principal challenges we antici-pate, so we have put in place structures to ensure that supervisors will be on stand-by to pick up data from the field where there are Internet challenges and send them to the head office.
>
> > (Daily Graphic 13 January 2020 "Statistical Service recruits
> > 60,000 officials for 2020 census")

In view of this, the 2020 PHC benefited from the current open data initiative of the Ghanaian government. In particular, the Ghana Post's GPS-based addressing system was mobilized in the census application, allowing census officers to iden-tify locations with adequate internet connectivity for the purpose of uploading and checking locations against the existing postal register (Interviews GSS and Ghana Post, March 2020).

However, due to a sequence of external factors, the 2020 Census plans were abruptly halted in March 2020. The provision of handheld devices for the census enumerators proved difficult as initial agreements reached on the level of the Af-rican Union to share the tablets between countries did not materialize. With the COVID-19 pandemic bringing production in China to a halt, Ghana's purchase of the devices was further delayed and the initial reference date for the census on 15 March was no longer attainable. With the first recording of COVID-19 cases on 12 March and the country's subsequent lockdown, the postponed date of 28 May could also no longer be upheld. This delay in the meticulously prepared census – three trial censuses had successfully confirmed the methodology at this point – caused a cascade of further complications. As the 2020 census coincided with the prepa-rations for the 2020 Parliamentary and Presidential elections, plans by the Elec-toral Commission (EC) to re-compile the country's voter register in a nationwide biometric mass registration raised concerns about the population's understanding and trust in the census as an independent exercise.[11] Widely broadcast allegations against the EC to follow partisan interests further exacerbated these tensions.

Changes in the Ghanaian census method have historically been accompanied by disagreements about the categories being used. New categories, such as boundaries of local authorities, community names, or professional distinctions were added and brought to the fore. The census of 2021 is no different. Public discussions and local boycotts of the enumeration in 2021 were centred on concerns about misrepre-sentation. For instance, in Ghana's Upper East region, concerns were raised about alleged discrimination in the listing of localities. In the Volta region, the focal area of contestation was the alleged failure to list subgroups of the Ewe ethnicity. The census also triggered public commentary about who was counted as Ghanaian.

Some alleged that certain groups listed in the survey represented foreigners. Examples include dismissing residents of the borderlands as Togolese, while Fulani and Hausa populations were repeatedly labelled as immigrants. And while the government framed the census as a means of enhancing development in the country, some residents said they wouldn't participate because development had eluded their communities. These debates politicized the census, reframing it as a moment of cementing the national identity, while also re-articulating the official narrative of data's developmental potential.

11.3 Conclusion

This chapter has presented the political history of Ghana through the lens of dual inscription of technological and political transformations in the census. It approached the question of how the establishment and the trajectories of national censuses can be described and explained from the theoretical angle advanced by Didier (2021), which suggests reading socio-political and technological transformations as stacked onto and woven through each other to the effect of the consolidation of certain methods, knowledges, and infrastructures into quantitative conventions "marbling" through society at large.

How then can we explain socio-material and methodological innovations of census taking in Didier's framework and the historiography of post-colonial censuses in particular? The case of Ghana shows us how population statistics are deeply entangled with the project of nation-building. Census taking in this light has been shaped by complex genealogies and marked by continuities of colonial hegemony, i.e. in the continued distrust of the Ghanaian population towards the exercise. Colonial forms of registration and the subjectivities imbued in them continue to shape data infrastructures and analytical practices through conventions of categorizing social aggregates, such as ethnic categories. Post-independence Ghana, in turn, saw the statistical expression of a series of illiberal regimes, while adapting to the reporting pressures of the international statistical system, in particular UN recommendations. To this day, colonial disinvestment in civil registration and vital statistics poses acute problems to data collection and data-driven governance in Ghana, while significant parts of the population face the very real consequences of exclusion from state registers and officially documented identity. Important innovations have accompanied the 2020/2021 Population and Housing Census along with the promise of better data for development. However, contestations around identity and belonging remained central concerns during the census. As social scholars of quantification have shown in abundance, statistical methods rarely close the issues they describe.

Locating the succession of population censuses in the various political conditions that have shaped Ghanaian history over the course of the last two centuries potentially opens new perspectives on political regimes as well, emphasizing these regimes' efforts in consolidating their position through data production on certain high-profile issues, while also carefully adapting to international statistical developments and funding logics. In particular, the adaptations required by the

Covid-19 pandemic illustrate graphically how census systems require adaptability to ever-changing political and societal contexts. Understanding this adaptability, it has been shown, requires an account of the arrangements of devices, software, experts, and organizational knowledge – and their global circulations, translations and adaptations into national contexts – all of which participate in the ongoing experimentations and the contingent realization of data infrastructure that make up the census in its specific historical moments. Tracing shifts in census methods and infrastructures, thereby allows reading the census as made and variable, writ through with failures and path dependences, rather than as fixed in its present state. Taking historical developments seriously finally opens the critical interrogation of the hegemonic continuities, such as conflicts about boundaries and ethnic classification, that continue to exert their social and political force on present day Ghana.

Notes

1 I want to thank the library staff of the African Studies Centre, PRAAD and the George Padmore Research Library in Accra for guidance and practical support in the data collection for this chapter. Research for this publication would not have been possible without funding from the DAAD PRIME (Project "*The Production of Measurement Policies in Africa*") and DFG Individual Grant schemes (Project "*How Democracies Know: Identification technologies and quantitative analyses of development in Ghana*", DFG TH2432/2–1).

2 While the 1931 census was the first to individually enumerate the population, complete enumeration was still largely limited to some major settlements and Ashanti (Barbour & Prothero, 1961, p. 9). Limitations in data quality were attributed to low levels of education, making "it difficult to find sufficient competent enumerators at the lowest but, nonetheless, the most important level of the census organization" (Barbour & Prothero, 1961, p. 10).

3 The 1970 census report lays out how IBM machines were used to compute census statistics. Automated analysis of geographical localities followed an elaborate eight-digit code system, in which the first digit denoting the region, the second the administrative district and the third the local authority (digits 4 to 8 denote the enumeration area and the locality within it, the largest locality in an enumeration being coded as 01 within each local authority, populations over 5,000 in the digits 50 to 59 and below that in the digits 60 to 99).

4 Gaisie (1969) lists among others, reverse survival method/estimations based on age composition of stable population model/Brass method, as well as comparative methods including total fertility by vital registration records, surveys on births in occurring per household as well as surveys on women's child bearing history, which were combined to increase statistical confidence.

5 The Demographic Unit has been heralded as the "most important development to date in Africa in terms of university demographic training" (by the late 1960s, over 100 students had graduated in demography, "a very substantial majority of all tropical Africans with such training") for its potential to replace foreign technical assistants (Gaisie, 1969, p. v).

6 Government Statistician J.E. Tandoh noted in the Second Volume published in connection with the 1970 census (1972, p. v) that while the model for publication is based on the work of Dr B. Gil for the 1960 census, "[f]or the 1970 census, however, there was no need to seek UN technical assistance and the whole work from the planning to publication was done entirely by Ghanaian officers", as "valuable assistance was received from various Government organizations and the Universities of Ghana". In particular,

the Universities of Ghana and Cape Coast "assisted in the revision of the Enumeration Area boundaries", while the Social Welfare and Community Development Department seconded staff.

7 The Office of the Government Statistician was established in 1948 and was expanded and renamed the Central Bureau of Statistics in 1961.

8 In 1983, Ghana received an initial US$367 million (followed by another US$840 million in 1987) in IMF loans to implement the Economic Reform Programme (ERP), which later took the shape of the Structural Adjustment Programmes (SAP) (Aryeetey, Harrigan, & Nissanke, 2000; Herbst, 1993; Kraus, 1991). In return, the PNDC regime agreed to reduce employment and other expenses within state organisations, among other liberalisation measures (Akonor, 2006; Overå, 2007; Petchenkine, 1993, p. 133).

9 The regime's decentralization agenda was stressed repeatedly with emphasis on how "data on various relevant issues within the communities" had been lacking and "therefore created planning and projection problems" (Oti Boateng, 1995, p. 329).

10 In line with this, the GSS divisions were reorganized and some new divisions created. "The Policy Planning and Co-ordination Division, for example, has been created to study the data needs of various users in the country and recommend appropriate statistical policies. It is also expected to co-ordinate all statistical activities in the country including those of Government Departments and to ensure that all other statistical departments in the country conform to the accepted statistical standards necessary to maintain comparability and quality of data" (Oti Boateng, 1995, p. 329).

11 This series of events reflects back on Serra's observations of the 1960 Census, when "the count took place when Ghana was about to embark on the plebiscite that would result in the adoption of a Republican Constitution (…) The census report argued that by postponing the plebiscite the government showed awareness that a simultaneous occurrence of the two events would have undermined the image of the census as a non-political and non-partisan operation" (Serra, 2018, p. 676). In 2020, other interests clearly overshadowed the census.

References

Akonor, K. (2006). *Africa and IMF conditionality: The unevenness of compliance, 1983–2000.* New York, NY: Routledge.

Aly, G., & Roth, K. H. (2004). The Nazi census. Identification and control in the Third Reich. Philadelphia, PA: Temple University Press.

Anderson, M. (2015 [1988]). *The American census: A social history.* New Haven, CT: Yale University Press.

Anderson, M., & Fienberg, E. (1999). *Who counts? The politics census-taking in contemporary America.* New York, NY: Russell Sage.

Aryeetey, E., Harrigan, J., & Nissanke, M. (Eds.). (2000). *Economic reforms in Ghana: The miracle and the mirage.* Oxford, England: James Currey.

Barbour, K. M., & Prothero, R. M. (1961). *Essays on African population.* London, England: Routledge.

Boafo-Arthur, K. (2007). A decade of liberalism in perspective. In K. Boafo-Arthur (Ed.), *Ghana. One decade of the liberal state* (pp. 1–20). Dakar, Senegal: Codesria.

Breckenridge, K. (2010). The world's first biometric money: Ghana's e-Zwich and the contemporary influence of South African biometrics. *Africa: The Journal of the International African Institute, 80*(4), 642–662.

Breckenridge, K. (2014). *Biometric state: The global politics of identification and surveillance in South Africa, 1850 to the present.* Cambridge, England: Cambridge University Press.

Caplan, J. (2013). 'Ausweis Bitte!' Identity and identification in Nazi Germany. In I. About, J. Brown, & G. Lonergan (Eds.), *Identification and registration practices in transnational perspective. People, papers and practices* (pp. 224–242). Basingstoke, England: Palgrave Macmillan.

Cardinall, A. (1931). *The gold coast, 1931.* Accra, Ghana: Government Printer.

de Graft-Johnson, J. C. (1969). The population of Ghana 1846–1967: A digest and discussion of the data in the official counts and censuses. *Transactions of the Historical Society of Ghana, 10,* 1–12.

de Graft-Johnson, K. T., & Ramachandran, K. V. (1975). *An evaluation of the 1970 population census results of Ghana.* African Population Series No 2. Addis Ababa, Ethiopia: United Nations Economic Commission for Africa.

Desrosières, A. (2010). *La politique des grands nombres. Histoire de la raison statistique.* Paris, France: La Découverte.

Desrosières, A. (2014). *Prouver et gouverner. Une analyse politique des statistiques publiques.* Paris, France: La Découverte.

Didier, E. (2009). *En quoi consiste l'Amérique ? Les statistiques, le New Deal et la Démocratie.* Paris, France: La Découverte.

Didier, E. (2020). *America by the numbers. Quantification, democracy, and the birth of national statistics.* Cambridge, MA: MIT Press.

Didier, E. (2021). *Quantitative marbling: New conceptual tools for the socio-history of quantification* (Anton Wilhelm Amo Lectures No. 7). Halle (Saale). Retrieved from Martin-Luther-Universität Halle-Wittenberg website: https://www.scm.uni-halle.de/amo_lecture/

Gaisie, S. K. (1969). *Dynamics of population growth in Ghana.* Ghana Population Studies, 1. Legon, Ghana: University of Ghana.

GSS. (1987). *1984 Population census of Ghana. Demographic and economic characteristics. Total country.* Accra, Ghana: Ghana Statistical Service.

GSS. (2016). *Report on the population and housing census 2010 – Ghana.* Accra, Ghana: Ghana Statistical Service.

Hannah, M. (2001). Sampling and the politics of representation in US Census 2000. *Environment and Planning D: Society and Space, 19,* 515–534.

Hannah, M. (2009). Calculable territory and the West German census boycott movements of the 1980s. *Political Geography, 28*(1), 66–75.

Herbst, J. (1993). *The politics of reform in Ghana, 1982–1991.* Berkeley, CA: University of California Press.

Higgs, E. (2003). *The information state in England. The central collection of information on citizens since 1500.* Basingstoke, England: Palgrave Macmillan.

Hilton, T. E. (1961). Population mapping in Ghana. In K. M. Barbour & R. M. Prothero (Eds.), *Essays on African population* (pp. 83–98). London, England: Routledge.

ILO. (1974). *Report of the national symposium on population, development and social progress.* Addis Ababa, Ethiopia: ILO.

Jerven, M. (2013). *Poor numbers. How we are misled by African development statistics and what to do about it.* Ithaca, NY; London, England: Cornell University Press.

Kpedekpo, G. (1971). Patterns of delayed registration of births and deaths in Ghana. *The Ghana Social Science Journal, 1*(1), 28–49.

Kraus, J. (1991). The political economy of stabilization and structural adjustment in Ghana. In D. Rothchild (Ed.), *Ghana: The political economy of recovery* (pp. 119–156). Boulder, CO: Lynne Rienner.

Loveman, M. (2009). The race to progress: Census-taking and nation-making in Brazil (1870–1920). *Hispanic American Historical Review, 89*(3), 207–234.

Loveman, M. (2014). *National colors: Racial classification and the state in Latin America.* Oxford, England: Oxford University Press.

Mensa-Bonsu, H. (2007). 'Political crimes' in the political history of Ghana: 1948–1993. In H. Mensa-Bonsu (Ed.), *Ghana law since independence: History, development and prospects* (pp. 239–305). Legon, Ghana: University of Ghana.

Noiriel, G. (2002). The identification of the citizen: The Birth of republican civil status in France. In J. Caplan & J. Torpey (Eds.), *Documenting individual identity. The development of state practices in the modern world* (pp. 28–48). Princeton, NJ: Princeton University Press.

Nugent, P. (1995). *Big men, small boys and politics in Ghana.* New York, NY: Pinter.

Okolo, A. (1999). The Nigerian census: Problems and prospects. *American Statistician, 53*(4), 321–325.

Oti Boateng, E. (1995). Population data collection and analysis. In *Ghana population policy: Future challenges* (pp. 321–342). Accra, Ghana: Ministry of Finance and Economic Planning.

Overå, R. (2007). When men do women's work: Structural adjustment, unemployment and changing gender relations in the informal economy of Accra, Ghana. *Journal of Modern African Studies, 45*(4), 539–563.

Peil, M. (1974). Ghana's aliens. *International Migration Review, 8*(3), 367–381.

Petchenkine, Y. (1993). *Ghana: In search of stability, 1957–1992.* Westport, CT: Praeger.

Report on the PHC. (2000). Accra, Ghana: Ghana Statistical Service.

Sackeyfio, N. (2012). The politics of land and urban space in colonial Accra. *History in Africa, 39*, 293–329.

Samuel, B. (2014). *La production macroéconomique du réel. Formalités et pouvoir au Burkina Faso, en Mauritanie et en Guadeloupe* (Doctoral dissertation). Sciences Po, Paris.

Serra, G. (2018). 'Hail the census night': Trust and political imagination in the 1960 population census of Ghana. *Comparative Studies in Society and History, 60*(3), 659–687.

Szreter, S. (2013). The Parish registers in early modern English history: Registration from above and below. In I. About, J. Brown, & G. Lonergan (Eds.), *Identification and registration practices in transnational perspective. People, papers and practices* (pp. 113–131). Basingstoke, England: Palgrave Macmillan.

Szreter, S., & Breckenridge, K. (Eds.). (2012). *Registration and recognition. Documenting the person in world history.* Oxford, England: Oxford University Press.

Tesfaye, F. (2014). *Statistique(s) et génocide au Rwanda: la genèse d'un système de catégorisation "génocidaire".* Paris, France: L'Harmattan.

Thiel, A. (2020). Biometric identification technologies and the Ghanaian 'data revolution'. *The Journal of Modern African Studies, 58*(1), 115–136.

Uvin, P. (2002). On counting, categorizing, and violence in Burundi and Rwanda. In D. I. Kertzer & D. Arel (Eds.), *Census and identity: The politics of race, ethnicity, and language in national censuses* (pp. 148–175). Cambridge, England: Cambridge University Press.

Valsecchi, P. (2014). Free people, slaves and pawns in the Western Gold Coast: The demography of dependency in a mid-nineteenth-century British archival source. *Ghana Studies, 17*, 223–246.

Villacís, B., Thiel, A., Capistrano, D., & Carvalho da Silva, C. (2022). Statistical innovation in the Global South: Mechanisms of translation in censuses of Brazil, Ecuador, Ghana and Sierra Leone. *Comparative Sociology, 21*(4), 419–446.

von Oertzen, C. (2017a). Die Historizität der Verdatung: Konzepte, Werkzeuge und Praktiken im 19. Jahrhundert. *Zeitschrift für Geschichte der Wissenschaften, Technik und Medizin, 25*(4), 407–434.

von Oertzen, C. (2017b). Machineries of data power: Manual versus mechanical census compilation in nineteenth-century Europe. *Osiris, 32*, 129–150.

Weitzberg, K. (2015). Unaccountable census: Colonial enumeration and its implications for the Somali people of Kenya. *The Journal of African History, 56*(3), 409–428.

12 Adoption of smartphones for data-collection during the fourth General Population and Housing Census of Cameroon

Motivations, opportunities and challenges

Teke Johnson Takwa

12.1 Introduction

Traditionally, paper-based data collection exercises have been the mainstay of data collection. The need to collect data fast and more accurately, the need for rapid availability of results, and the successful use of mobile devices for data collection in other African countries were some of the motivations for the adoption of smartphones for data collection for the first time during a census in Cameroon. The switch from paper-based data collection to mobile phones in the 2020 census round initially faced many obstacles. Many stakeholders felt that it was a risky venture considering the fact that handheld digital devices had never been used before. This group of people considered that the best technology is usually the one you already have, the one you know how to use, can maintain, and can afford. The high initial cost of buying mobile devices, training, low internet connectivity, and energy supply were other reasons to discourage the adoption of this novel approach. Other stakeholders, on the contrary, argued that since the world was going digital in data collection, Cameroon must not be left behind. In fact, while there were reasons to discourage the use of mobile phones in census data collection, there were also arguments for its adoption. The latter arguments came mostly from officials of the United Nations Population Fund (UNFPA) and a few national experts, who had taken part in seminars on the use of mobile devices for data collection purposes. They held the opinion that data accuracy and speedy collection of data would be enhanced through the use of this device.

Convinced by the arguments on the advantages of using smartphones, the National Census Council of Cameroon, the organ that has the final say in census matters, approved the new data collection method. The president of the council provided the following reasoning for this decision:

> Over the past year, we have been unable to decide whether we should use mobile devices to collect data during our forthcoming census or not. Some of us are of the opinion that using mobile devices to collect census data for the first time is risky because we do not have sufficient mastery. Another

DOI: 10.4324/9781003259749-16

argument is the fact that using this technology may instead increase the cost of data collection which we have been seeking deliberately to reduce. While these concerns are not completely false, reports at my level after many study visits to some African countries that have recently used this technology for census data collection such as Ivory Coast, Senegal and Cape Verde are encouraging. The use of mobile phones in many domains in Cameroon has become common. Many people are ready to embrace this change. We have a critical mass of young people commonly known as the "Android Generation" who can manipulate mobile phones with dexterity. They will need very brief training to be able to master and use any data collection apps. I have been reliably informed that the use of this technology will greatly help improve upon data quality and facilitate a real time availability of census result for development planning. After weighing the pros and cons of the use of smartphones for census data collection, the Government has instructed that mobile phones be used for this exercise. There is no turning back. We all have to move forward on the path of modernity. I urge both those who are against and those who are for to work together for its success. I am confident that we will succeed especially as we have the financial and technical support of the United Nations Population Fund, the United States Agency for International Development, the United States Bureau for the Census and the African Center for Statistics.

(President of the National Census Council of Cameroon, 2016, p. 3)

This decision ended the political and technical debate over the use or non-use of mobile phones for data collection for the then upcoming census. So far, data collection during the cartographic and pilot phases has used this technology. The Government of Cameroon has made available 33,000 smartphones for being used in the data collection of the remaining phases of the fourth General Population and Housing Census (GPHC). At the time of writing (November 2022) the remaining phases include the update of the outdated census cartography of 2018 and the actual implementation of the census scheduled for late 2023.

This chapter provides a detailed analysis of Cameroon's experiences of shifting from the traditional paper-based personal interview to the digital smartphone-based method of census data collection, with a focus on the motivations, advantages and disadvantages, and challenges of this technology change. The analysis is guided by the following two research questions: Firstly, why was there an initial resistance towards the use of mobile devices for data collection during the various phases of the fourth GPHC of Cameroon? Secondly, why and how was this resistance overcome?

The chapter is organized into nine sections. Focusing on countries from the Global South, I will first present the literature dealing with the use of mobile phones for large-scale data collection (Section 12.2). Subsequently, a theoretical framework is provided (Section 12.3), combining the world polity approach with the technology acceptance model. After presenting the methodology (Section 12.4), the institutional framework within which the census has been conducted in Cameroon so far is described (Section 12.5). Sections 12.6–12.8 provide a detailed analysis of Cameroon's experiences around three main axes: the motivations, the benefits, and advantages, as well as the challenges of changing to a digital smartphone-based method of census data collection. Finally, some conclusions and implications of Cameroon's digital data collection experience and lessons learnt are briefly discussed (Section 12.9).

12.2   Literature review

Although the use of mobile phones for data collection is relatively new, especially in countries, of the Global South, there is a rich literature dealing with this topic. This literature review focuses on the advantages and the limitations of using mobile phones compared to the traditional paper-based method.

Commonly, traditionally paper-based data collection has been the mainstay of data gathering; this has also been the case for the burden of obstructive lung disease study, an internationally recognized study that uses standardized methods to measure the burden of chronic obstructive lung diseases (Thriemer et al., 2012). However, automated data collection and processing methods are becoming more widespread in healthcare research (Garg & Mony, n.d.) and have many advantages (King et al., 2013). Although there are studies investigating the use of automated data collection via smartphones as a research tool in developing countries, they are not as comprehensive as in the developed world, and there are fewer studies exploring its use in large-scale and complex surveys, such as obstructive lung disease surveys (Schuster & Perez Brito, 2011).

Paper-based data collection is convenient for many researchers and data collectors. It has several potential advantages over the automated method: data extraction is not limited to a particular location, and it seems easier to produce, modify, manipulate, and implement. In addition, a long-lasting record of all modifications and an immediate evaluation of forms can be provided (Higgins et al., 2019; King et al., 2013). Moreover, data loss is potentially less likely with the paper-based method than with automated data collection (King et al., 2013). Studies from developing countries have found, however, that using paper-based methods results in a higher incidence of incomplete records, a greater potential for human errors, and a greater time burden to organize the data (King et al., 2013). In addition, data entry can be labor-intensive and may limit timely analysis (Weber, Yarandi, Rowe, & Weber, 2005).

Over the past decades, the number of mobile phone users in Africa has dramatically increased, with mobile phone subscribers now representing 83% of all telephone subscribers. This proportion is larger than in other world regions. South Africa leads the region in mobile phone ownership with 36.4 mobile phones per 100 people. They are no longer considered a luxury (Schuster & Perez Brito, 2011; Tomlinson et al., 2009). With the growth of information and communication technologies and the development of software such as the Android systems platform and many open-source applications, researchers in the health sector have begun using smartphones as tools for patient data collection, disease surveillance, clinical research, and national surveys (King et al., 2013). However, paper-based questionnaires continue to be the main data collection tool in many countries, particularly in sub-Saharan Africa (King et al., 2014).

Using smartphone technology-based tools for data collection has many advantages and can provide a broader range of options. It is economically and environmentally friendly, and it can provide faster reporting with greater accuracy (Pakhare, Bali, & Kalra, 2013). It is also more efficient. Data collection and entry can be combined into one step by using forms that provide built-in verification and reliability checks, and may have additional features, such as collecting information from a Global Positioning System (GPS) (Garg & Mony, n.d.). In addition, time stamps, alarms, automatic completions, and reminders can help in work-rate monitoring and data validation (King et al., 2013; Le Jeannic, Quelen, Alberti, & Durand-Zaleski, 2014).

In contrast, data security and connectivity can be a concern, and data collectors need to be familiar and comfortable with using an automated tool (King et al., 2014). Accidental loss of data, limited battery life, loss or theft of the device, security of the device, and network connectivity in rural areas are also major concerns (Pakhare et al., 2013; Tomlinson et al., 2009). Some studies have been carried out on the impact of different smartphone screen sizes on data collection efficiency. These studies found that there is little difference even though larger screen sizes improve data collection efficiency (Raptis, Tselios, Kjeldskov, & Skov, 2013).

12.3   Theoretical considerations

This chapter draws on two theoretical approaches in order to assess how people make decisions regarding new technology adoption: the world polity theory on the one hand, and the technological acceptance model on the other hand. Both theories are complementary in explaining the adoption of mobile phones for data collection during the various phases of Cameroon's fourth census.

12.3.1   World polity theory

World polity theory, also known as the "Stanford school" of the world society perspective, or the global neo-institutionalist approach (Boli, Gallo-Cruz, & Mathias, 2011; Meyer, 2010; Meyer, Boli, Thomas, & Ramirez, 1997), was developed, partly in response to Wallerstein's (1974, 1980, 1988) world systems theory focusing on global capitalist accumulation with the primacy of world economic structures, as a comprehensive, analytical frame for interpreting global relations, structures, and practices in the cultural sphere (McNeely, 2012). The concept of world society claims that social order has to be analyzed at a global level and cannot be reduced to or derived from the (national) level of nation-state societies.[1] Meyer's world society perspective is based on the hypothesis that societies organized as nation-states have a similar structure and change in a similar way over time.

World polity theory views world society as an encompassing social system with a cultural framework called world polity, which integrates and influences its actors, such as nations, international organizations, enterprises, and individuals (McNeely, 2012).

In other words, according to Boli et al. (2011, p. 148) "the world polity is con-stituted by distinct culture – a set of fundamental principles and models, mainly ontological and cognitive in character, defining the nature and purposes of social actors and action." World polity theory views the primary component of world society as "world polity," providing a set of cultural norms or directions that world society actors adopt in dealing with problems and general procedures (Boli et al., 2011). In contrast to other theories such as neo-realism or liberalism, world polity theory considers actors such as the states and institutions to be under the influence of global norms (Boli et al., 2011). Although it closely resembles constructivism, world polity theory has to be distinguished from it because "world-polity theorists have been far more resolute in taking the 'cultural plunge' than their constructiv-ism counterparts" (Boli et al., 2011, p. 152). In other words, world polity theory puts more emphasis on global homogenization than the other approaches. Through globalization, world polity is shaping local organizations and institutions while in return local cultures and organizations impact on the further evolution of world society (Boli et al., 2011).

World polity analysis launched in the 1970s by Meyer et al. (1997) and his research group at Stanford University initially focused on the study of inter-state relations. Simultaneously, extensive work on the international education environ-ment was conducted. However, in the 1980s and 1990s, due to the impact of glo-balization on world culture, the focus of the study shifted towards the analysis of transnational social movements that may contribute to a global polity, while at the same time attempting to better understand how global polity ideas are implemented by global and local actors (Boli et al., 2011). Mechanisms of world cultural inte-gration are based on so-called isomorphism, i.e. similarities in the design of basic national institutions, like educational or scientific institutions, and universally ap-plied procedures and practices, such as economic models, fiscal policy, health care models, but also national censuses and public statistics. International institutions and organizations are core actors that promote world cultural transmission and the standardization of global models and practices. Imitation (mimicry) and profes-sionalization are two core mechanisms of such institutional isomorphism. Another aspect addressed by the world polity approach concerns the phenomenon of "loose coupling," i.e. that world cultural principles and norms are institutionalized at the national and sub-national levels, but that they are only respected to a limited extent. This "flexibility" allows states and institutions to adopt new global standards with-out forcing its members (citizens, individuals) to fully adapt to these new practices (Schofer, Hironaka, Frank, & Longhofer, 2012).

In a series of empirical studies, Meyer and others observed that new states organ-ize themselves in a very similar way despite their different needs and backgrounds, supporting their argument that there is a set of norms for forming a new state relat-ing to the larger umbrella of world polity (Ahmed et al., 2018). An empirical study of International Non-Governmental Organizations (INGOs) shows that universal-ism, individualism, rational voluntarist authority, progress, and world citizenship are present in various INGOs. Sports, human rights, and environmental INGOs in particular tend to "reify" world polity (Boli et al., 2011). According to this study,

INGOs could instill the world-cultural principles of world polity to nation-states and national societies through lobbying, advocacy, criticism, and persuasion.

Critics, however, point to the fact that world polity theory assumes a rather flawless and smooth transfer of world polity norms to global actors, which may not always be plausible. The tendency to focus on the homogenizing effect has also been criticized (Meyer et al., 1997). World culture theory differs from world polity theory in this respect, as it recognizes that actors find their own identities in relation to the larger global cultural norm, rather than simply following what is suggested by world polity. Globalization cannot fully be explained by world polity theory. It is a phenomenon by which local values and global cultures converge to create something new (Boli et al., 2011).

12.3.2   Technology acceptance model

The technology acceptance model and its related theories are useful when focusing on the potential adoption of a new emerging technology, such as smartphones for census data collection. The goal of the technology acceptance model is to predict user acceptance and to highlight potential design issues before users of the technology interact with the system (Dillon & Morris, 1996; Mohd, Ahmad, Samsudin, & Sudin, 2011). The technology acceptance model was developed with support from International Business Machine Canada and is rooted in the basic psychological theory known as the theory of reasoned action (Ajzen, 1991; Fishbein, 1979). As shown in Figure 12.1, the technology acceptance model generates a framework for explaining behavioral intentions and actual behavior of users of new technology.

Perceived usefulness and perceived ease of use are the views users hold about the new product or system (Dillon & Morris, 1996). Perceived usefulness is defined

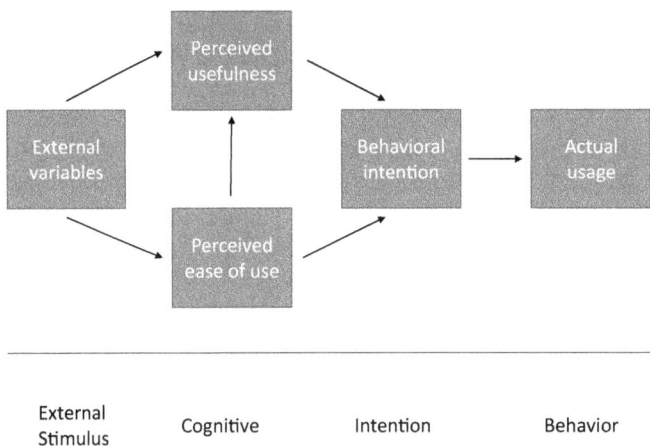

Figure 12.1 Technology acceptance model, adapted from (Davis & Venkatesh, 1996).

as the degree to which a person believes that using a particular system would enhance his or her job performance and perceived ease of use as the degree to which a person believes that using a particular system would be free of effort (Davis, 1993). The original study of the technology acceptance model generated six highly reliable items for both, perceived usefulness and perceived ease of use.

Various researchers and practitioners have validated the robustness of technology acceptance model instruments in different settings (Davis, 1993; Davis & Venkatesh, 1996; Dillon & Morris, 1996; Lee, Kozar, & Larsen, 2003). Davis and Venkatesh (1996), through four longitudinal studies, proposed a new model that was an extension of the technology acceptance model and termed it a unified theory. Various modifications have been made to the original theory (see Holden & Karsh, 2010), but its basic principles have remained unchanged.

According to the meta-analyses conducted by Yousafzai, Foxall, and Pallister (2007, 2010), the three major factors for the diffusion of the technological acceptance model are its solid theoretical foundation and the robust measurement scales, the strong empirical support and the overall explanatory power of the model, as well as its applicability across a wide range of systems and technologies. In this meta-analysis, the summary of 15 years of technological acceptance model studies revealed a high correlation between perceived use, perceived ease of use, and the intention to use various technologies.

Among the (few) weaknesses reported in the literature regarding the technology acceptance model, the most commonly mentioned limitation relates to self-reported use (Lee et al., 2003). Furthermore, since the technology acceptance model is used to predict the behavioral intention to accept technology, some researchers believe that people do not have sufficient exposure to technology prior to the assessment (Lee et al., 2003). Moreover, the original technology acceptance model did not include social influences (Ghazizadeh, Lee, & Boyle, 2012).

12.4   Sources of information and methodology

Information for this study comes from the technical report on the collection of data during the cartographic and pilot phases of the fourth GPHC of Cameroon using handheld devices (BUCREP-Cameroon, 2020). Other information comes from my personal observation as a supervisor of these operations. After introducing the use of mobile phones for data collection at the various phases of the fourth census, one of the first steps was to train future field trainers and field supervisors in the use of mobile apps. I was one of those who received training. We were then sent to the field to apply our training skills by collecting data from selected households. After a week on the field, we returned to the office to report on our experiences. Note was taken on the difficulties reported in order to improve future training. After this training, we were able to train field staff on the use of smartphones and the incorporated apps for data collection. When this field staff was deployed to collect data, national coordination staff accompanied them to observe and assess their work and the difficulties faced in using the new method of data collection in order to report to the management.

12.5   Institutional framework

Since its independence in 1960, Cameroon has carried out three population and housing censuses and is currently in the final phase of realizing the fourth.

- The first GPHC was carried out in 1976 after its institution by Decree No. 73/757 of the 6th of December 1973;
- The second GPHC, instituted by Decree No. 85/506 of the 11th of April, 1985, was carried out in 1987; and
- The third, instituted by Decree No. 2001/251 of 13th of September 2001, was conducted in 2005.
- The fourth GPHC, which is expected to be executed – after many postponements – in 2023, was instituted by Decree No. 2015/ 397 of 15th of September, 2015.

For a coherent running of the activities of the fourth GPHC, the text instituting it, just as for the past ones, specified its management organs. These bodies include the National Census Council, the Technical Committee, the Regional, Divisional, and Sub-Divisional Committees (local committees), and, as executing and coordinating body, the National Coordination Unit.

12.5.1   The National Census Council

The National Census Council is composed of all ministries concerned with population and housing issues, as well as those in charge of maintaining law and order. This body provides strategic guidance to the National Coordination Unit. It is chaired by the ministry in charge of population issues, the Ministry of Economy, Planning, and Regional Development, assisted by the ministry in charge of housing issues. The secretariat of this council is handled by the president of the Technical Committee, assisted by the National Coordinator of the fourth GPHC and the director general of the National Institute for Statistics. The UNFPA, the government's strategic partner on population issues, participates in the meetings as an observer.

12.5.2   The national Technical Committee

The Technical Committee is the operational body that ensures the smooth implementation of the fourth GPHC. Its meetings are chaired by the secretary general of the ministry in charge of population issues, assisted by the secretary general of the ministry in charge of housing issues and the director general of the National Institute for Statistics. Its secretariat is handled by the national coordinator of the fourth GPHC, assisted by the assistant national coordinator.

The Technical Committee is composed of directors and other personnel of the same rank in technical ministries in charge of population, housing, and related activities. The role of this committee is to analyze and validate technical and methodological documents prepared by the National Coordination Unit. Representatives of the UNFPA and the United Nations Children's Emergency Fund (UNICEF) also participate in these meetings as observers.

12.5.3   The local committees

Local committees for the fourth GPHC exist at the regional, divisional, and sub-divisional levels. These committees are composed of parliamentarians, senators, mayors, governors, traditional rulers, local public administrators, law enforcement officials, and religious and political party leaders. Their role is to ensure that the population is willing to participate in the census. They also work to ensure that enumerators, team leaders, controllers, and supervisors work in safe conditions at a time of growing insecurity.

12.5.4   The National Coordination Unit

The National Coordination Unit is responsible for coordinating the technical, administrative, and financial operations of the fourth GPHC. It is headed by the Director General of the Central Bureau for Censuses and Population Studies. He/she is assisted by the Assistant Director General, who may be assigned to carry out specific tasks. Several sub-units are directly attached to the National Coordination Unit. Staff working in these units are expected to advise the national coordinator on specific issues that relate to their domains of competence.

Five different departments head the activities of the fourth GPHC at the operational level. These departments include

- The Department of Cartography and Field Operations, which is in charge of the cartographic work ranging from the conception to the printing of the maps. This department also controls all personnel and material provided for cartographic activities, as well as the designation of the enumeration areas during the various phases of the census (main headcount, pilot census, post-enumeration survey).
- The Department for Methodology and Analysis which is in charge of the elaboration of the various documents on techniques and themes for census data analysis.
- The Department of Computer Activities and Archives, which is responsible for developing applications for data collection and database cleaning and management, as well as tabulating data for publication.
- The Communication and Publicity Department, which is responsible for developing and implementing advocacy and social mobilization tools for the census. Its role also includes the preparation and implementation of publication and dissemination plans.
- The Administrative and Financial Department, which is responsible for managing human resources as well as materials and equipment acquired for use during the census project.

The National Coordination Unit of the fourth GPHC has the following tasks

- Drafting and monitoring of the various census regulatory texts.
- Coordinating and monitoring the implementation of the census cartography, which is responsible for delimitating the enumeration areas.

- Preparation of various technical documents, including questionnaires and methodology documents.
- Preparation, organization, and participation in census knowledge-sharing seminars.
- Management of human, material, and financial resources allocated to the census.
- Organization of public relations and sensitization activities for the census.
- Organization and management of all phases of the census (cartography, pilot census, main headcount, post-enumeration survey, publication, and the dissemination of census results).

Due to the delicate nature of financial resource management and the huge cost of the census operation, the coordination unit has a financial controller who monitors all transactions to ensure that all expenses carried out within the framework of the fourth GPHC respect the guidelines. The National Coordination Unit also has an accountant who carries out all payments related to the census. Both the financial controller and the accountant, and are appointed by the Minister of Finance, who supervises their activities.

12.5.5   *Respect of calendar of activities of the fourth GPHC of Cameroon*

Despite the seemingly well-structured institutional arrangements to make sure that the fourth GPHC goes on smoothly, the various activities programmed witnessed significant delays. When the census was instituted on 15 September 2015, the calendar of activities stipulated that by September 2019, the whole process would be over, but up to today (November 2022), the headcount is still to be conducted. The main reason for the delay has been insufficient financing by the state and donor organizations. The census cartography that was expected to be completed in 2017 was only completed at the close of 2018. Apart from insufficient financing, insecurity in the Far North Region coming from Boko Haram attacks and violent conflicts between government forces and separatist rebels in the two English-speaking regions of Cameroon (Northwest and Southwest Regions) delayed the cartographic phase remarkably. When the census cartography was over at the end of 2018, the limited funds only permitted the realization of the pre-census survey. Funds were not immediately available to continue with the other phases and the process witnessed a halt. While funds were being mobilized for the headcount, the COVID-19 pandemic came and all attention was focused on tackling the health emergency with ramifications on all sectors of life. Now that the impacts of the pandemic have been reduced, there are renewed hopes that the headcount for this census will take place as rescheduled at the end of 2023. The long wait since the completion of the census cartography in 2018, has resulted in it becoming obsolete. Some of the funds allocated for the head count have to be used to update the obsolete cartography especially at the periphery of rapidly expanding urban areas.

While waiting for the headcount rescheduled for the close of 2023, many questions remained unanswered. Is it necessary to carry out a census at a time when some parts of the country cannot be normally covered for security reasons? Will the proposed combined method of carrying out a normal headcount in areas without security concerns and using satellite images and other methods to estimate populations in areas with security challenges be acceptable? Can the census that has been delayed by several years not wait until the security issues will have been sorted out? To what extent will the results of the census be accepted or rejected by different stakeholders? Is a questionable census better than no census at all?

12.5.6   Autonomy of the executing structure of the fourth GPHC

The fourth GPHC of Cameroon like the third, will be entirely executed by the Central Bureau for Censuses and Population Studies considered as an autonomous structure. This organization is headed by a national coordinator appointed by the president; it has an autonomous budget provided by the state. The bureau is responsible for establishing and executing the methodology used for all the phases of the census.

Despite this apparent autonomy, many questions have been raised concerning the real autonomy of this census-executing body. Critical questions raised during the last census referred to two aspects

- Why could the Central Bureau for Censuses and Population Studies based on technical and other arguments not change the method of data collection from paper-based to digital without the authorization of the National Census Council?
- Why can this body after completing the headcount, the post-enumeration survey and all the necessary verifications not publish the results of the census? Why must these results be published only by the Head of State or the Prime Minister? Can this not lead to the tailoring of the results to suit political motives?

These concerns were raised during the last censuses and could be raised again.

12.5.7   Quality assurance

Data quality is crucial as it assesses whether information can serve its purpose in a particular context. There are five traits of good quality data. These five traits are accuracy, completeness, reliability, relevance, and timeliness. The census is such a costly and essential source of data for development planning and evaluation of progress in attending to some developmental goals that special measures need to be put in place in order to ensure that its products are of goodand useable quality.

Apart from introducing digital data collection techniques for the first time, another innovation of the fourth GPHC is the introduction of a special external structure for quality assurance, the Institute for Demographic Training and Research better known by its French acronym as IFORD. IFORD was created by the United Nations Economic Commission for Africa in 1972 for training population scientists and providing technical support in the domains of data collection, analysis, and dissemination in Francophone Africa. IFORD follows up the activities of the fourth GPHC in order to ensure that data coming out from each stage of the census is of good and useable quality. Unfortunately, this organization lacks adequate financing and human resources to properly carry out this function.

12.6   Motivations for adopting mobile phones

While there were some reasons hindering the adoption of mobile phones for data collection exercises during many phases of Cameroon's fourth GPHC, there were equally strong motivations for its adoption. These motivations included its successful use in some African censuses, support from donor organizations, and finding ways to overcome the many problems associated with the paper-assisted personal interviews (PAPI) used in previous censuses.

12.6.1   *African role models for digital census taking*

While debates were heated over the use or non-use of mobile phones for data collection during the ongoing census, some countries with almost the same level of development as Cameroon had already implemented their censuses using mobile devices, including Cape Verde, Ivory Coast, and Senegal. These countries, particularly Cape Verde and Senegal, had established a Centre of Excellence for sharing knowledge and experiences on the use of mobile devices in data collection. Learning visits were organized to Senegal. At the end of these visits, it was concluded that, with proper planning and adequate technical support, census data collection using mobile devices could provide a real advantage over the traditional PAPI method.

12.6.2   *Support from donor organizations*

A key push factor for the adoption of mobile devices for data collection purposes during the fourth GPHC in Cameroon is the financial and technical support received from the UNFPA. Before the decision was made to use mobile phones for data collection during this census, UNFPA brought a team of experts and consultants from its sub-regional headquarters in Dakar, Senegal, to discuss the challenges and opportunities of collecting data using mobile devices. Due to the delicate nature of adopting a data collection method for the first time for a large-scale operation, UNFPA-Cameroon organized further consultative meetings with countries that have already implemented data collection using mobile devices. About three such meetings took place in Dakar, Senegal, with the participation of experts from UNFPA, the US Bureau of Census, and the African Center for Statistics, as well as officials from countries that have either used the device for data collection or were planning to do so.

After deciding to use mobile phones for the data collection, another key issue that came up was the development of the data collection application. Here, the UNFPA, in collaboration with the United States Agency for International Development (USAID), provided funding for experts from the US Bureau of the Census to train some officials of Cameroon's Central Bureau for Censuses and Population Studies in this domain. These applications were developed and tested, and additional training was provided. Three application developers were sent to the US

Bureau of the Census for a month's training program sponsored by UNFPA and USAID. These series of learning and training sessions convinced decision-makers that data collection using mobile devices has many advantages over traditional paper-based personal interview methods.

While the debate on whether to use smartphones or not for data collection during the fourth GPHC of Cameroon was still ongoing, the African Center for Statistics of the United Nations Economic Commission for Africa chose Cameroon's National Institute of Statistics as an experimental center for the use of mobile devices for data collection. The experience of this institute with data collection using a personal digital assistant (PDA) also supported the adoption of mobile devices for data collection.

12.6.3   *Need to solve some problems faced during previous censuses*

The last two censuses of Cameroon, the second (1987) and the third (2005), were characterized by a rather late release of results: results were released only four and five years respectively after data collection. These delays were caused by many factors, such as the need to spend time to transport tons of questionnaires from various locations to the central data processing site, long periods for coding some variables, long periods for data capture, and other financial issues. The results of the 2005 census were projected before publication since these results were not available until five years after the census. The time lapse between the 1987 census and the availability of results was equally long (four years). These delays were met with much suspicion. Some people believed that the delay occurred in order to "manipulate the results in favor of some regions and to the detriment of others". The results of the third census were released five years after the headcount, and at the time of publication, both the results from the field and projections were provided at the same time. Some people, especially politicians, who could not understand why some regions with low population counts from the field census had higher projected populations, attributed this to deliberate manipulation of results, especially if their regions did not make sufficient "gains" from the projection. Others felt that the results were outdated even before publication. The new opinion is that data collection via a mobile device will eliminate some phases such as the time lost in transporting large quantities of questionnaires, coding of some variables, and the usually lengthy process of data capture, leading to a rapid release of results, thereby increasing their acceptance.

In previous censuses in Cameroon, questionnaires were poorly handled in some localities, resulting in the loss of many of them. In some cases, some leaflets of the questionnaires were missing. Special techniques had to be used to replace the lost data. These many processes contributed to delays in the production and dissemination of the results. The use of mobile devices for data collection eliminates the need for entire questionnaires or questionnaire leaflets.

Another problem expected to be solved by the use of mobile phones in census data collection in Cameroon is inconsistent data entry. For example, the paper questionnaires showed abnormalities such as birth entries for men or for women below the age of 12, which had to be corrected in subsequent operations. These and other irregularities occurred because some enumerators did not pay sufficient attention to the various skips in the questionnaires. With the smartphone-assisted method, these skips are made automatically and enumerators cannot provide information where it is not supposed to be provided.

12.7   Advantages

Experiences with using mobile phones for many data collection operations have shown that their use provides many advantages over the traditional PAPI methods. These advantages include improvement in the quality of data collected, time and money saving in data collection, the ability to rapidly identify problems and provide real-time solutions, etc. By considering the adoption of smartphones for census data collection, Cameroon expects to benefit from these gains.

12.7.1   Data accuracy

Data collection using mobile devices reduces transcription errors that are often associated with the PAPI method. The last census of Cameroon experienced various transcription errors; most of them resulting from the difficulty of data entry agents to properly decipher the enumerator's information. With digital data capture, the possibility of enumerators leaving out key questions, which is common with the PAPI method, is almost completely eliminated. The obligation to answer some questions before proceeding with data collection ensures that all essential information is collected. This is impossible with the traditional paper-based personal interview method. During past censuses in Cameroon, key information including information on age and sex were often absent for some respondents. Another problem that was encountered during Cameroon's last census was the loss of information due to poor handling of questionnaires. Data transmission using mobile networks may ensure that no data are lost "in transit". Digital collection and transmission of data facilitates storage and access to collected data even at later points in time.

Data accuracy is enhanced during data collection due to the fact that the data collection application used in mobile devices significantly reduces the registration of inconsistent data, such as registering births for males or incorrectly reporting information for certain categories of persons who may not be concerned by certain categories of questions. Many cases of births wrongly attributed to males or girls below child-bearing ages have been found in Cameroon's past censuses. By introducing automatic filters, data collection via digital devices will significantly reduce these inconsistencies, which contributed to the generally low quality of Cameroon's census data.

12.7.2   Cost and time economy

When data is collected using conventional method, that is using printed forms, the data has to be transcribed into spreadsheets or statistical packages. Costs involved in this process are expenses in printing and transcribing. Mobile phone-based data collection saves on all these costs because data are sent directly to the database.

Data collection tools and applications have significantly reduced the cost and duration of data collection and dissemination. During the third GPCH of Cameroon, data collection using paper-based personal interviews was scheduled to last two weeks. Due to the inability of most enumerators to complete their tasks in the allotted time, this duration was extended by an additional week. This resulted in extra and non-verified costs. The pilot census for the fourth GPHC, using the digital devices, demonstrated that a slightly longer questionnaire could be completed within eleven days. This reflects a significant reduction in both, the time and cost of data collection.

The cost of large storage space for questionnaires is greatly reduced or almost eliminated with the use of mobile phones. This gain is very evident when one considers that the law requires the conservation of questionnaires for at least ten years before destruction. During the last census in Cameroon, the Central Bureau for Censuses and Population Studies, which was in charge of conducting the census, had to rent a building to store the questionnaires. This building was guarded by paid security agents for a period of ten years. This significantly increased the cost of the operation. With data collection using mobile devices, information collected can be stored more safely and for a longer time on a server, at a lower cost. This server is located within the premises of the institution, and no additional space needs to be rented to host it.

12.7.3   *Providing solutions to data collection problems in real time*

During the pilot phase of the ongoing GPHC of Cameroon, it was possible for the first time for data managers to identify problems as soon as they arose and immediately take action to fix them. With telephone-based data collection, managers can detect many errors on their monitoring boards and send signals to field supervisors to take action. The traditional paper-based data collection could not permit this, and some problems were not detected until it was too late, as with the last census.

12.7.4   *Building on familiarity*

Today, the use of mobile phones in Cameroon has become very common, especially among young people who are commonly called the "Android Generation." In urban areas of Cameroon, familiarity with smartphones has reached an important threshold. During training on the use of mobile phones for data collection during the cartographic and pilot survey phases of the ongoing fourth GPHC, there was no need to spend much time introducing the device to candidates. Extensive training was limited only to the apps that had to be used for data collection. This means that the training required to help people use mobiles for data collection is essentially about the software rather than the hardware.

12.7.5   *Improvement in the recognition of enumeration area boundaries*

A key problem of census operations in Cameroon, especially in urban areas, is how to precisely identify enumeration area boundaries. This was due to (1) the inability of some enumerators to properly read the enumeration area maps; (2) the establishment of difficult-to-identify enumeration area boundaries by cartographers; (3) the removal or modification of enumeration area boundaries after their identification; and (4) the lack of sufficient information to assist enumerators in the identification of their zones. This usually results in either omitting or double-counting parts of the enumeration areas. With the use of mobile devices for census data collection in Cameroon, a software application called Mapit, installed on mobile devices, greatly assists enumerators in identifying their work zones.[2]

12.7.6   *Potential for enriched data*

Mobile phones provide an opportunity for enriching data collection by taking pictures and registering coordinates for spatial analysis. This is not possible with the paper-based method. For the first time in the history of census-taking in Cameroon, increasing emphasis is being placed on spatial analysis. The use of smartphones for data analysis will greatly help attain this goal and improve the use of census data for decision-making.

12.8   Challenges

In shifting to data collection through the use of mobile phones in Cameroon, the advantages were overemphasized and little attention was paid to some of the inconveniences. Some of these challenges were not recognized until the use of the mobile phones for data collection was already underway. These challenges include financial constraints, limited power and internet availability, data security concerns, and technology challenges.

12.8.1   *Financial constraints*

When deciding to go digital with census data collection in Cameroon, the assessment of the financial costs was not adequately addressed. The cost of purchasing smartphones from China proved to be higher than expected. The numerous training sessions and learning visits to Senegal, Cape Verde, and the United States of America involved enormous costs. Other costs not adequately anticipated, included the costs of power banks and electricity generators for use in areas with no electricity supply. This has resulted in the cost of going digital being almost double of the traditional paper-based method. This late realization is slowing down the advancement of the process. Even though some of the expenses could be recovered through the sale of the smartphones after use, the initial costs were quite high. They amounted to nearly 4.5 billion Cameroonian Francs (about 7.96 million US dollars; BUCREP - Cameroon, 2020). This does not include other related costs such as the cost of power banks and electricity generators for use in places without electricity. For a poor country like Cameroon, this is a tremendous amount of money.

12.8.2   *Challenges linked to energy and internet availability*

In Cameroon, like in many other sub-Saharan countries, internet connection and electricity are not available everywhere. During the pilot phase of the ongoing fourth GPHC, the digital data collection process was slowed down because there were no power sources to charge phones in many small towns and rural areas. Some of the power banks used during the data collection did not work properly. Constant recharging was required for both the power banks and the phones. In some areas with electricity supply, the supply was not constant, and extended periods of electricity blackouts interrupted the work of census enumerators. In some locations, internet connection was either absent or irregular.

Even with the use of more than one SIM card in each mobile phone, there are still many areas in Cameroon where data transfer to the central server could not be carried out due to the lack of internet connections. Enumerators had to travel long distances to reach areas with internet coverage in order to be able to transfer the data collected during the pilot phase. A key problem in this regard was that some enumerators did not bother to make the extra effort to travel long distances for data transmission, so large volumes of data remained unsent to the central server. Without proper follow-up, this data can actually be lost. In all regions of Cameroon, there are areas where there is either no electricity or no internet access. To counter this problem, power generators have been purchased, which, however, contributed to increasing the cost of the operation which has been already judged to be quite high.

In order to cope with the problem of temporary or permanent absence of internet connection in some localities, the paper-based method was permitted in these localities. With this method, enumerators can administer the paper questionnaire and later transfer the responses to the mobile phone when the internet connection is re-established, or when they move to a place with an internet connection. During the pilot survey, some enumerators did not do this and had to be reminded.

12.8.3   *Concerns regarding data security and confidentiality*

During the pilot phase of the fourth GPHC of Cameroon, some respondents were afraid of giving responses registered on a mobile phone, which is an unfamiliar format for people who have been used to the PAPI method for a long time. Some respondents considered the use of the smartphone as a spying mechanism. It needed a lot of persuasion to get them to provide their responses. During the main headcount, enumerators have been advised to use paper questionnaires if respondents are suspicious of the use of mobile devices. This mixed mode can only work well if enumerators are willing or reminded to carry out the transfer of the information collected on paper to mobile phones.

12.8.4   *Getting decision-makers and implementers to buy the idea*

The switch from paper-based to computer-based face-to-face interviewing initially met with a lot of arguments among stakeholders. Some considered the PAPI method to be better because Cameroon had already used and mastered it and could afford it. The lack of experience with using mobile devices for data collection was the main reason why many thought that this method should not be used in a large-scale data collection such as the census. This was indeed a major issue of discussion, and without the timely intervention of UNFPA, it would have been very difficult to implement this method. UNFPA pointed to the example of Senegal and Cape Verde, which had just used this method with satisfactory results to collect census data, and promised to provide financial support for the learning process. UNFPA also promised to mobilize other partners, notably USAID and the US Census Bureau, as well as the World Bank, to accompany Cameroon in this new venture. This was the decisive factor that led the Government of Cameroon to adopt this novel approach.

12.8.5   *Relatively small dimensions of smartphone screens*

The screens of smartphones to be used for the data collection of the fourth GPHC of Cameroon are quite small. This reduces the visibility of information contained therein. Without appropriate zooming, the enumerator may mistakenly enter response options that are close to the correct ones. This was observed during the pilot phase of the fourth GPHC. Steps have been taken to request enumerators to fully zoom in or out, as needed, before selecting the appropriate options.

12.8.6   *Other challenges*

Other challenges in collecting data with mobile devices include common mistakes committed by enumerators in the field. These mistakes include enumerators' entering incorrect answers in a hurry and accidentally deleting part of the collected files, using the mobile phone for other purposes than data collection, accidentally logging out of the application, and mistakenly pressing the registration button when the data collection process is not yet complete. In some cases, the data collection and transmission application installed on the smartphones suffered from inappropriate manipulation of the smartphones. All of these problems were identified to some extent during the pilot phase of Cameroon's fourth GPHC. Some solutions, such as making the data collection more user-friendly and using video tutorials in order to reduce these errors, are being envisaged.

12.9   Conclusion

After many years of hesitation, Cameroon finally overcame the inertia in moving from data collection using traditional paper-based personal interviews to digital data collection of the fourth General Population and Housing Census of Cameroon. Up to now digital collection data has been used for the cartographic and pilot phases and will be used for updating the outdated census cartography that was completed in 2018 and the headcount scheduled for 2023. Cameroon's adoption of digital smartphone-based census data collection can be explained by the combination of macro and micro factors highlighted by world polity theory and the technology acceptance model. In line with world polity theory, global macro factors, notably international learning processes (mimicry), e.g. the learning opportunities offered by Ivory Coast, Cape Verde, and Senegal, the sub-Saharan African pioneers in digital data collection, but also the support of international organizations and external actors for technical assistance, financing, planning, and implementation, played a significant role.

Despite the successful adoption of digital data collection methodology, many challenges remain to be overcome for Cameroon to fully benefit from the new technology. An analysis of the use of mobile phones for census data collection during the pilot phase of this exercise has shown that the expected benefits such as improved data accuracy, reduction in real data-collection time, as well as provision of real-time solutions to problems can be offset by serious challenges if not adequate addressed. These challenges include: (i) limited availability of internet services; (ii) low electricity connectivity; (iii) low data transmission and serious difficulties in monitoring enumerators in remote rural areas; (iv) fear of some respondents of providing responses that are registered in an unfamiliar format; and (v) common field mistakes committed by enumerators when using mobile devices. Adequate solutions have been provided for maintaining battery power for long periods, theft, and malfunctioning of the apps as well as the smartphones. Respondents need to be properly sensitized in order to relieve them of the fear associated with entering information into a smartphone, which is quite new to most Cameroonians. Since digital data collection is a new approach, users need frequent follow-ups to ensure that the desired benefits of adopting this technology are realized. As it stands, despite the challenges faced, the use of digital devices for data collection in Cameroon is here to stay, and efforts are being made by all stakeholders to ensure that it works well and is offering a real improvement over the methods traditionally used. As a precaution, while the census is expected to be largely digital, a few questionnaires will be printed for use in remote areas with no internet connection. Paper captured responses are supposed to be transferred into the digital applications once the enumerator gets into a zone with internet coverage.

Notes

1 There are various concepts of world society and no unified world society theory has been developed. In addition to the "Stanford" approach of Meyer's world polity perspective, with its focus on world culture, the global diffusion of organizational models, and institutional isomorphism, the so-called "Bielefeld" approach of Luhmann based on systems theory (see Albert & Hilkermeier, 2004; Jung & Stetter, 2019; Luhmann, 1982), and the "Zurich" approach of Peter Heintz with its focus on the international development stratification system have to be mentioned (see Heintz, 1982; Suter, 2005).
2 For further information on Mapit see Mapit Spatial (https://spatial.mapitgis.com/).

References

Ahmed, R., Robinson, R., Elsony, A., Thomson, R., Squire, S. B., Malmborg, R., ... Mortimer, K. (2018). A comparison of smartphone and paper data-collection tools in the burden of obstructive lung disease (BOLD) study in Gezira state, Sudan. *PLoS One, 13*(3), e0193917. doi:10.1371/journal.pone.0193917
Ajzen, I. (1991). The theory of planned behavior. *Organizational Behavior and Human Decision Processes, 50*(2), 179–211. doi:10.1016/0749-5978(91)90020-T

Albert, M., & Hilkermeier, L. (Eds.). (2004). *Observing internatioanl relations. Niklas Luhmann and world politics*. London, England: Routledge.

Boli, J., Gallo-Cruz, S., & Mathias, M. (2011). World society, world-polity theory, and international relations. In N. Sandal (Ed.), *Oxford research encyclopedia of international studies*. Oxford, England: Oxford University Press. doi:10.1093/acrefore/9780190846626.013.495

BUCREP - Cameroon. (2020). *Report of the pilot census of the 4th GPHC*. Yaounde, Cameroon: BUCREP. Retrieved from www.bucrep.cm

Davis, F. D. (1993). User acceptance of information technology: System characteristics, user perceptions and behavioral impacts. *International Journal of Man-Machine Studies, 38*(3), 475–487. doi:10.1006/imms.1993.1022

Davis, F. D., & Venkatesh, V. (1996). A critical assessment of potential measurement biases in the technology acceptance model: Three experiments. *International Journal of Human-Computer Studies, 45*(1), 19–45. doi:10.1006/ijhc.1996.0040

Dillon, A., & Morris, M. G. (1996). *User acceptance of new information technology: Theories and models*. Medford, NJ: Information Today. Retrieved from https://repository.arizona.edu/handle/10150/105584

Fishbein, M. (1979). A theory of reasoned action: Some applications and implications. *Nebraska Symposium on Motivation, 27*, 65–116.

Garg, S. (2013). Electronic data capture for health surveys in developing countries: Use of a mobile phone based application in southern India. *Indian Journal of Medical Information, 7*, 86–95.

Ghazizadeh, M., Lee, J. D., & Boyle, L. N. (2012). Extending the technology acceptance model to assess automation. *Cognition, Technology & Work, 14*(1), 39–49. doi:10.1007/s10111-011-0194-3

Heintz, P. (1982). A sociological code for the description of world society and its change. *International Social Science Journal, 34*(1), 11–21.

Higgins, J. P. T., Thomas, J., Chandler, J., Cumpston, M., Li, T., Page, M. J., & Welch, V. A. (Eds.). (2019). *Cochrane handbook for systematic reviews of interventions*. Vancouver, Canada: Wiley. doi:10.1002/9781119536604

Holden, R. J., & Karsh, B.-T. (2010). The technology acceptance model: Its past and its future in health care. *Journal of Biomedical Informatics, 43*(1), 159–172. doi:10.1016/j.jbi.2009.07.002

Jung, D., & Stetter, S. (Eds.). (2019). *Modern subjectivities in world society. Global structures and local practices*. Cham, Switzerland: Palgrave Macmillan.

King, C., Hall, J., Banda, M., Beard, J., Bird, J., Kazembe, P., & Fottrell, E. (2014). Electronic data capture in a rural African setting: Evaluating experiences with different systems in Malawi. *Global Health Action, 7*(1), 25878. doi:10.3402/gha.v7.25878

King, J. D., Buolamwini, J., Cromwell, E. A., Panfel, A., Teferi, T., Zerihun, M., ... Emerson, P.M. (2013). A novel electronic data collection system for large-scale surveys of neglected tropical diseases. *PloS One, 8*(9), e74570. doi:10.1371/journal.pone.0074570

Le Jeannic, A., Quelen, C., Alberti, C., & Durand-Zaleski, I. (2014). Comparison of two data collection processes in clinical studies: Electronic and paper case report forms. *BMC Medical Research Methodology, 14*(1), 7. doi:10.1186/1471-2288-14-7

Lee, Y., Kozar, K. A., & Larsen, K. R. T. (2003). The technology acceptance model: Past, present, and future. *Communications of the Association for Information Systems, 12*(1), Article 50. doi:10.17705/1CAIS.01250

Luhmann, N. (1982). The world society as a social system. *International Journal of General Systems, 8*(3), 131–138.

Mapit Spatial. (2019). *MapitGIS*. Retrieved from https://spatial.mapitgis.com/

McNeely, C. L. (2012). World polity theory. In G. Ritzer (Ed.), *The Wiley Blackwell encyclopedia of globalization* (pp. 1–10). Chichester, England: Wiley-Blackwell. doi:10.1002/9780470670590.wbeog834

Meyer, J. W. (2010). World society, institutional theories, and the actor. *Annual Review of Sociology, 36*, 1–16. doi:10.1146/annurev.soc.012809.102506

Meyer, J. W., Boli, J., Thomas, G. M., & Ramirez, F. O. (1997). World society and the nation-state. *American Journal of Sociology, 103*(1), 144–181. doi:10.1086/231174

Mohd, F., Ahmad, F., Samsudin, N., & Sudin, S. (2011). Extending the technology acceptance model to account for social influence, trust and integration for pervasive computing environment: A case study in university industry. *American Journal of Economics and Business Administration, 3*(3), 552–559. doi:10.3844/ajebasp.2011.552.559

Pakhare, A., Bali, S., & Kalra, G. (2013). Use of mobile phones as research instrument for data collection. *Indian Journal of Community Health, 25*, 95–98. doi:10.1234/vol25iss2pp95

President of the National Census Council of Cameroon. (2016). Address to the National Census Council. Yaoundé, Cameroon, July 6, 2016.

Raptis, D., Tselios, N., Kjeldskov, J., & Skov, M. B. (2013). Does size matter? Investigating the impact of mobile phone screen size on users' perceived usability, effectiveness and efficiency. *Mobile HCI2013: Proceedings of the 15th International Conference on Human-Computer Interaction with Mobile Devices and Services* (pp. 127–136). Munich, Germany. doi:10.1145/2493190.2493204

Schofer, E., Hironaka, A., Frank, D. J., & Longhofer, W. (2012). Sociological institutionalism and world society. In E. Amenta, K. Nash, & A. Scott (Eds.), *The Wiley Blackwell companion to political sociology* (pp. 57–68). Chichester, England: Wiley-Blackwell.

Schuster, C., & Perez Brito, C. (2011). *Cutting costs, boosting quality and collecting data real-time: Lessons from a cell phone-based beneficiary survey to strengthen guatemala's conditional cash transfer program.* Washington, DC: World Bank. Retrieved from https://openknowledge.worldbank.org/handle/10986/10111

Suter, C. (2005). Research on world society and the Zurich school. In M. Herkenrath, C. König, H. Scholtz, & T. Volken (Eds.), *The future of world society* (pp. 377–384). Zurich, Switzerland: Intelligent Book Production.

Thriemer, K., Ley, B., Ame, S. M., Puri, M. K., Hashim, R., Chang, N. Y., ... Ali, M. (2012). Replacing paper data collection forms with electronic data entry in the field: Findings from a study of community-acquired bloodstream infections in Pemba, Zanzibar. *BMC Research Notes, 5*(1), 113. doi:10.1186/1756-0500-5-113

Tomlinson, M., Solomon, W., Singh, Y., Doherty, T., Chopra, M., Ijumba, P., ... Jackson, D. (2009). The use of mobile phones as a data collection tool: A report from a household survey in South Africa. *BMC medical informatics and decision making, 9*(1), 51. doi:10.1186/1472-6947-9-51

Wallerstein, I. (1974, 1980, 1988). *The modern world system I, II, III.* Cambridge, England: Cambridge University Press.

Weber, B. A., Yarandi, H., Rowe, M. A., & Weber, J. P. (2005). A comparison study: Paper-based versus web-based data collection and management. *Applied Nursing Research, 18*(3), 182–185. doi:10.1016/j.apnr.2004.11.003

Yousafzai, S. Y., Foxall, G. R., & Pallister, J. G. (2007). Technology acceptance: A meta-analysis of TAM Part 1. *Journal of Modelling in Management, 2*, 251–280. doi:10.1108/17465660710834453

Yousafzai, S. Y., Foxall, G. R., & Pallister, J. G. (2010). Explaining internet banking behavior: Theory of reasoned action, theory of planned behavior, or technology acceptance model? *Journal of Applied Social Psychology, 40*(5), 1172–1202. doi:10.1111/j.1559-1816.2010.00615.x

13 The global politics of census taking

Conclusions and desiderata for further research

Walter Bartl

13.1 Theorizing the politics of census taking

Since the contributions to this volume are mostly national case studies that were collected on the occasion of the global 2020 census round coordinated by the United Nations (UN), a tension between a national and a transnational perspective runs through the book, with regard to explaining the trajectory of national censuses. We will try to turn this tension into a productive force.

On the one hand, in previous work, census taking has been placed within the cultural sociology of the state, investigating how states accumulated symbolic power in the process of state formation (Loveman, 2005), how they wield symbolic power as a routine practice (Bhagat, 2022) and how census taking changes historically in interaction with society (Emigh, Riley, & Ahmed, 2016b). Together with several typologies aiming to facilitate international comparisons (Kertzer & Arel, 2006, pp. 670–673; Prévost, 2019; Rallu, Piché, & Simon, 2006), a common feature of many empirical case studies has been to operate within a framework of methodological nationalism. There are indeed good reasons why methodological nationalism might be a relevant analytical framework: Although worldwide inequality is larger between states than within states, it is national inequality that matters most to people, because the nation-state is their mundane normative and cognitive frame of reference (Schwinn, 2008). Furthermore, this frame of observation and evaluation is not arbitrary but dependent upon institutional regulations and opportunity structures that are decided upon at the national level. One could therefore expect that the relevant politics of census taking are likely to take place at the national level (Emigh et al., 2016b). On the other hand, with regard to data gathering, phenomena of transnationalization have become so manifold, that an analytical framework of methodological nationalism becomes increasingly questionable (Ruppert & Scheel, 2021a, p. 13; Villacís, Thiel, Capistrano, & Carvalho da Silva, 2022). The most far-reaching alternatives to methodological nationalism have been proposed by theories of world society (Cole, 2017; Heintz, 1982; Luhmann, 1997) with partially overlapping and partially diverging analytical premises (Meyer, 2015; Stichweh, 2015). The neoinstitutionalist world polity approach in particular argued that it is international organizations which provide the relevant cognitive scripts and standards for census taking (Ventresca, 1995). If we take

DOI: 10.4324/9781003259749-17

seriously the notion that sociology is confronted with actors that imbue their practices with first-order interpretations (Riley, Ahmed, & Emigh, 2021), we hardly can
ignore the continuous work of international organizations, such as the UN, whose
self-declared aim is establishing census standards on a global scale. In the face of
these transnational activities, it becomes all the more challenging to explain why
national and local actors comply with these standards to a varying extent.

The diagnosis that there is no coherent theory of census taking (Loveman,
2005, p. 1664), has been met with calls for more systematic research and theorizing (Emigh, Riley, & Ahmed, 2020). Typical practices of theorizing include applying existing concepts to new observations and striving for consistency among
existing concepts (Krause, 2016b). We will engage in the first form of theorizing
in a loose sense and with the aim of arriving at more precise descriptions of the
empirical material in our book in the first place. Assuming frames and scripts constituted by a World Culture (Ventresca, 1995), the notion of continuous interaction
between state and society (Emigh, Riley, & Ahmed, 2016a, 2016b), the distinction between different forms of building up symbolic state power (Loveman, 2005,
pp. 1661–1663), the idea of investments in form (Thévenot, 1984), the idea of a
statistical production chain (Desrosières, 2007), as well as comparative typologies
(Kertzer & Arel, 2006; Prévost, 2019; Rallu et al., 2006), are important analytical concepts in the sociology of census taking. Recently developed more general
concepts, such as the notion of "formalization work" still await to be applied to
census taking (Papilloud & Schultze, 2022). However, there are also significant
questions of consistency related to a possible combination of these concepts and,
at the same time, the analytical approaches chosen in the chapters of this book are
fairly diverse, which makes it difficult to apply existing concepts in a strict sense.
Recently, however, it has been suggested that theorization could also benefit from
visualization, which implies that the visual elements that are employed for theorizing don't have to represent consistent and finalized bodies of thought but could
instead invite the reader to further deliberation (Swedberg, 2016). In contrast to the
sequential logic of language, images draw their particular imaginative power from
the suggestive arrangement of visual elements on a surface (Boehm, 1994).[1] That's
why we try to map the field of census taking as an – admittedly imperfect – picture
of social space with two main dimensions (Figure 13.1).

In Figure 13.1, we bring together two exemplary theoretical approaches that
we picked up in the introduction to this book and typical actors, objects, and
institutions that have been attributed with the agency during the process of census taking in the individual chapters. As a starting point, we follow Emigh et al.
(2016a) in visualizing their theoretical approach toward state-society interactions
on different levels of aggregation (micro/meso/macro). For a visualization of the
"life cycle of statistics" see Agrawal and Kumar (2020, p. 13). Figure 13.1 depicts
the concepts of state and society as two ends of a continuum of social power
on the horizontal axis. We complement this with a vertical axis representing a
global-local continuum of spatial scales. This second dimension seems to be especially useful for being able to put the world polity approach into the picture, which
includes its typical assumption of top-down processes between the global and the

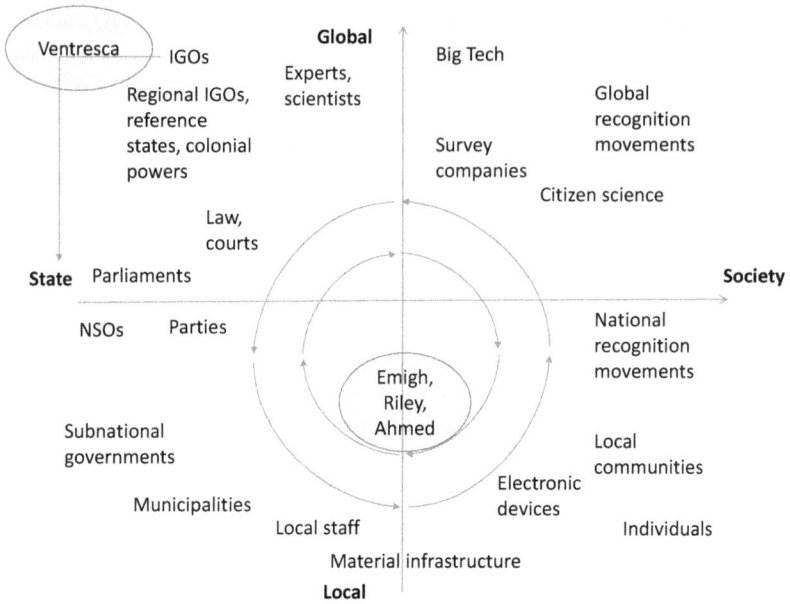

Figure 13.1 Attempt to visually theorize upon the process of census taking.

Source: Author.

national level dominating the implementation of national census taking. Following most chapters in our book, Figure 13.1 shows that beyond the national level, which would be the macro state and macro society level in the view of Emigh and colleagues (2016a), there are international actors that have to be considered as well.[2] Beyond these exemplary two analytical approaches, the chapters in our book evoked much more complex pictures with very diverse actors initiating, mediating, and translating processes in relation to census taking. For example, states are not homogeneous entities but the division of power between executive, legislative, and judicative as well as the multi-level governance processes that exist in even formally unitary states have to be taken into account. Apart from that, neither the world polity approach nor the interactive state-society approach considers material infrastructures and (technical) objects as a relevant factor in census taking, which is contrary to what research in the tradition of science and technology studies, as well as several of the chapters in our, book suggest. We placed material infrastructure at the local pole of the spatial scale, because as particular objects they occupy a unique spatial position, even if they are materially connected to other places. For example, material network infrastructures are bound up with conditions on the ground, which means that they are vulnerable locally (e.g. a data cable that is broken or a road that is impassible) although we might see the effects of damage at a much larger spatial scale (cf. Centeno, Nag, Patterson, Shaver, & Windawi, 2015). Figure 13.1 would become more dynamic if it was redrawn for particular chapters of the book including the selective configuration of actors and institutions as well as a more elaborate legend for the specific processes at work; it could also be redrawn for several points in time.

13.2   The (non-)use of ethnoracial categories in the census

We will now turn to the first more specific research questions raised in the introduction and highlight commonalities that can be observed across those chapters that addressed the use of ethnoracial categories in the census. Two chapters focused on Latin America (Chapters 1 and 3), and in both, international organizations play an important part: Mara Loveman argues that international organizations are part of a coalition of national recognition movements and scientists in the process of establishing ethnoracial census categories; after the institutionalization of these categories, international organizations also take part in pushing for the analytic production of knowledge on ethnoracial inequalities. Furthermore, mutual observation among structurally equivalent actors in different countries was important (Chapter 1). In the case of education censuses, international organizations pushed for an increase in the statistical capacity of Latin American countries (Chapter 3). Although Capistrano, Silva, and Pereira Rabelo argue that facilitating the production of market-compatible subjects was the actual motivation of international organizations pushing for statistical reforms, this impulse opened a window of opportunity for national recognition movements and the inclusion of ethnoracial categories in education censuses. In the case of India, it is mainly national state actors, politicians, and bureaucrats who continue the classificatory practice of Britain, the former colonial power (Chapter 2). Categories of caste and religion used in the census produce a more clear-cut representation of group identities than actually exist in social reality.

In terms of consequences of ethnoracial categories in census taking, Loveman argues that they also produce unintended effects because they create new political stakes, which might become highly politicized through (nationalist) political parties, court cases, and political activists (Chapter 1). In India, the redistributive effects of category-based public policy have been significant (Piketty, 2022, pp. 175–202), but at the same time they seem to come at the price of a more polarized and even divisive political culture. Although Chapter 2 is not very detailed with regard to the social actors organizing protests against the National Population Register in the run-up to the 2021 census, similar to Brazil (Chapter 1), state-society interaction among national actors seems to be crucial for bringing consequences of ethnoracial categorization to bear. Regarding possible consequences of ethnoracial categories in education censuses in Latin America (Chapter 3), there is hardly any evidence on how these categories are used for policy making. Therefore, instead of assuming that ethnoracial census categories automatically become the source of key indicators that are consequential for entire social fields (Bartl, Papilloud, & Terracher-Lipinski, 2019), this chapter could be a starting point for one of the rare studies on the non-use of actually existing indicators (cf. Lehtonen, 2013; van Dooren, Bouckaert, & Halligan, 2015).

In summary, the use of ethnoracial categories in census taking seems to have ambivalent consequences. While in the affirmative action model (Kertzer & Arel,

2006) they are crucial for redistributive public policy, they also are likely to contribute to a more politicized, sometimes even polarized climate of public debate – and sometimes it is questionable if they have any effect at all.

13.3   Institutional autonomy

How can we explain the institutional autonomy of census taking? This question implies that census taking should be free from political interference and the census takers should be able to count on sufficient financial, professional, and organizational resources to be actually able to plan and implement a census (Prévost, 2019). This engenders the National Statistical Offices (NSO) being trusted as a reliable producer of census data according to scientific standards. It does not preclude that political debate can actually enhance the instrumentality of census information (Emigh et al., 2016b).

The chapters constituting the second book section address this question, and they show that the role of international organizations and global scripts seems to be more limited compared to the introduction of ethnoracial categories. The most prominent role for transnational actors is presented in Chapter 6. Nicolás Sacco, Gabriel Mendes Borges and Byron Villacís describe a transnational project for raising the statistical literacy of census data users in Latin America. This is a rare but even more important case emphasizing that civil society networks exist not only at the national scale but increasingly also at a transnational level with a potentially global reach. Transnational problem constellations (Beck, 1996) are also a relevant factor for census taking. Data demands resulting from transnational policy problems, such as COVID-19, and the growing importance of Big Data as a competing source of information (Ruppert & Scheel, 2021b, p. 296), originating from transnational technology corporations with non-transparent conventions of quantification, are some of the most pressing challenges to NSOs according to Walter Radermacher. At the same time, the institutional autonomy of census taking is rooted to a significant extent in national institutional settings. The German case shows that robust juridical procedures allowed municipal doubts about the accuracy of the 2011 census results to be dissipated by an investigation and by decisions of the constitutional court (Chapter 4). Transnational processes are of rather secondary importance to the cases in Chapter 5, a comparison of political interventions into census taking in the USA, Brazil and Ecuador. Byron Villacís puts national actors at center stage with regard to the intervention itself. However, in the cases of Brazil and Ecuador international organizations are in an ambivalent role as propagators of neoliberal reforms leading to budget cuts on the one hand but also as sources of professional legitimacy for NSOs fighting against these budget cuts on the other hand. Overall, Villacís argues that the national institutional framework of census taking, including reliable legal procedures, as well as attention to census activities paid by civil society actors are important factors for shielding official information gathering from political interference. In the case of Nigeria, international organizations are mentioned only as financial donors benefiting census taking; the international scientific community is mentioned as repeatedly drawing attention to the politicization of census taking in the country (Chapter 7). While Temitope Owolabi attributes the manipulation of census figures by subnational governments in 2006 to the crucial role of census figures' functions for ethnic and territorial power sharing, as is not uncommon (Idike & Eme, 2015; Olorunfemi & Fashagba, 2021), it remains open why similar apportioning functions of census results do not lead to a comparable politicized outcome in other cases. Instead, the weak institutional constitution of the state itself, interethnic conflicts spilling over to state institutions, related challenges to the monopoly of violence and the practical inaccessibility of some parts of the country have to be added to the picture of a severe lack of institutional autonomy of census taking (Mohammed, Othman, & Osman, 2019). The census in the current round is scheduled for 2023; the future will show if indirect measures of population (Olorunfemi & Fashagba, 2021) or a register-based census (Mohammed et al., 2019) could be a remedy to the problems of the Nigerian census. Chapter 8, the case study on the postponement of the census in Ukraine, equally implies that the main actors undermining institutional autonomy of census taking are national politicians. While this is the case for the 2011 census, the census scheduled for the current round was in agreement with international standards and had to be postponed because of the war started by the country's imperial neighbor, Russia.

Overall, the role of global scripts and international actors regarding the institutional autonomy of census taking is rather limited. Instead resilience of census taking against political interference depends more generally upon the public integrity of state governance (cf. Mungiu-Pippidi, 2015) on the one hand and the professional statistical capacity of the NSO on the other hand. While the former director of the NSO in Greece suggested that NSOs should be extracted from policy making bodies (Georgiou, 2019), cases like Nigeria point to the fact that many of the political premises of census taking are not formally decidable but emerge from national state-society interactions. The latter is a strong point in favor of Emigh and colleagues' (2016a) argument about the crucial role of societal actors, their categorization practices, and their social power. It also points to the politics of credibility that NSOs perform when creating or defending their institutional autonomy (Howard, 2021).

13.4   Socio-technical innovation

How can socio-material and methodological innovations of census taking be explained? Two chapters in our book analyzed the (planned) transition to a register-based census in Spain and Germany (Chapters 9 and 10) and two chapters investigated the introduction of electronic data collection devices including geospatial technologies for census taking in Ghana and Cameroon (Chapters 11 and 12). In all four cases, inter- or supranational actors had an important role. Veira-Ramos and Bartl argue that the promotion of register-based censuses at the European level combined with national deliberations on cost savings were important (Chapter 9). However, less visible but even more decisive were the investments made in the quality of municipal registers that were later combined to form a national population register under the auspices of the Spanish NSO. While the building blocks of the register infrastructure are local, they acquire their emergent property only when they are connected to each other and the quality of their information is cared for continuously. The latter is assured through institutional incentives tying together the interests of municipalities and inhabitants. What is largely lacking in this case are activities of civil society; public debates about the transition to a register-based census have been practically absent. Körner and Grimm (Chapter 10) argue quite similarly for the German case with regard to the influence of international actors. In the German case, it is even more astonishing that there has been very little public debate, given that census taking was very contentious in the country's past (Frohman, 2012; Hannah, 2009). The importance of investing in the state's information infrastructure, including local material elements but also trans-local organizational routines, is even more emphasized by this case. Although efforts are being made at the moment, their actual test will follow only in 2030, when the census will be entirely based on administrative data for the first time. In particular, the required registers of education and housing are still lacking. Although Spain and Germany are part of the European Union (EU) and, hence, exposed to supranational influence, the impact of international organizations on innovations in census taking in the cases of Ghana and Cameroon is also remarkable. In both cases, innovation in the use of handheld electronic devices for data collection, including georeferential information, was facilitated by recommendations from the UN plus showcasing from neighboring countries as well as external funding and training for census officials. However, the implementation of the census was hampered by the interruption of international supply chains due to the COVID-19 pandemic, but as Teke Takwa argues (Chapter 12), also due to local gaps in technical infrastructure. Furthermore, Thiel and Takwa, both describe that the heritage of colonial enumeration practices still undermines trust in the state's census activities. Alleged political interference also led to local boycotts against census taking in Ghana (Chapter 11).

Overall, the four chapters on socio-technical innovations have in common that they are intrinsically linked to digital technologies and do not represent mere cognitive methodological changes. If we take the material dimension of census infrastructures as representing a local but nevertheless crucial element of information gathering, the relation of local material objects with elements attributed to other spatial scales should be focused more in future work.

13.5   Desiderata for future research

One characteristic of our book is that it counts on contributions from academic scholars but also practitioners of census taking. It is natural that the focus of analysis shifts accordingly between more reflective and more instrumental perspectives. Although we tried to create a productive dialogue between these different professional backgrounds, it is up to the reader to judge how well this worked or not. However, with regard to future research, it seems worthwhile to ask more explicitly what a fruitful transdisciplinary dialogue on census taking would look like.

Organizing special issues with invited contributions from both fields of activity certainly is one possible format (cf. Gong, Groshen, & Vadhan, 2022; Radermacher, 2021). But maybe future research on census taking should also include experimentation with transformative formats of knowledge production (cf. Bogusz, 2022)? Transdisciplinary formats of research would be pivotal for advancing the statistical capacity of NSOs that struggle to achieve a full census coverage of the national territory, for example (cf. Carranza-Torres, 2021). But even is full census coverage comes within reach, the limited influence of the Sustainable Development Goals and the related census-based indicators on policy making cautions us against all too ambitious expectations (Biermann, Hickmann, & Sénit, 2022). Therefore, a more systematic exploration of the translations that are required from the beginning of the statistical chain to a possible policy impact of census information seems to be worthwhile.

Indirectly related to this suggestion is the assumption put forward within the framework of the world polity approach that "theorizing" would contribute to the diffusion of practices (Strang & Meyer, 1993). "By theorization we mean the self-conscious development and specification of abstract categories and the formulation of patterned relationships such as chains of cause and effect. Without general models, cultural categories are less likely to arise and gain force" (Strang & Meyer, 1993, p. 492) With regard to census taking, we may conclude in a preliminary view that academic scholarship and debate on census taking would contribute to the diffusion of the practice of census taking. While the nature of the causal link would still have to be clarified, during the process of editing this book we gained the impression that research on census taking is very much skewed toward the Global North plus some model cases from the Global South that have been widely discussed. Is that really the case? Bibliometric evidence actually shows a global deconcentration process of scientific publications (Maisonobe, 2021). Why would research on census taking be different from the more general trend? What do the "odd and unrecognizable" cases of national census taking tell us theoretically (Krause, 2016a, p. 200)? How can we conceptualize more precisely the relationship between research and practice proposed by Strang and Meyer under the label of "theorizing"? An alternative hypothesis would be that the practice of census taking is not so much influenced by transnational "theorizing" but instead by other factors such as the institutional autonomy of NSOs (Prévost, 2019).

With regard to the use of ethnoracial categories in the census and their relation to affirmative action programs, the redistributive (and maybe unintentionally polarizing) effects of these programs should be investigated on a broader comparative base in the future (cf. Piketty, 2022). Regarding possible consequences of ethnoracial categories in education censuses in Latin America (Chapter 3), for example, it would be fruitful to analyze differential education outcomes between those countries that use these categories in a traditional framework of education planning, those countries that embed them in a more transformative postcolonial policy agenda and those countries who ignore them in policy making. Furthermore, current attempts to introduce ethnoracial categories in traditionally color-blind EU countries might offer a strategic opportunity to observe the political and professional strategies involved. Will these attempts produce more acceptance or more antagonism in formally color-blind societies that are now confronted with the self-conscious demands of third-generation descendants of former guest workers (Foroutan, 2019)? When some newcomer groups formally reclaim their recognition in the eyes of the hegemonial majority of a national society, other minority groups that perceive themselves as being denied symbolic recognition might react with particular hostility (Kalter & Foroutan, 2021).

With regard to the institutional autonomy of census taking, Big Data is considered as being in competition with the traditional services of NSOs, but might as well be coopted by them (Gbadebo, 2021; Prewitt, 2022; van der Brakel, 2022). Do we actually know in quantified terms the extent to which private data providers are involved in producing official data or have replaced traditional forms of data gathering? Furthermore, if there is cooperation between NSOs and technology companies, one can assume particular tensions due to the different conventions of quantification involved (cf. Diaz-Bone & Horvath, 2021; Vogel, 2019). These tensions have not been explored systematically in the context of census taking.

Similar tensions can also be expected when a transition to a register-based census is to occur. Formerly separate parts of public administration are suddenly forced to cooperate with each other. Although this cooperation will mostly be mediated by software solutions, the way in which these solutions can be accessed and the institutional work involved in maintaining exchange between different organizational entities has yet to be discovered. While the empirical material analyzed in Chapter 9 of this book did not uncover any significant tensions, this does not preclude that they actually existed or still exist. Furthermore, since a register-based census diminishes the required interaction with inhabitants, does it actually empower the administration vis-à-vis the population?

During the COVID-19 pandemic, the importance of timely and reliable public statistics has become evident with great clarity. At the same time, depending on the particular indicator considered, the data published often also contained a substantial amount of uncertainty. As a response to that, a coalition of academics proposed that the uncertainty involved in models of future developments should be transparently communicated (Saltelli et al., 2020). This proposal aims at increasing the public's statistical literacy and capacity to reflected decision making. With regard to demographic data, uncertainty in projections usually arises from contingent assumptions on spatial mobility and international migration or on future education trends (Barakat & Durham, 2017). However, the political salience of demographic projections does not arise from the uncertainties implied in the assumptions on migration movements but from an essentialist reading of ethnoracial indicators. Instead of treating them as an objective reality, the statistical literacy of the public and of policymakers would be substantially enhanced by transparently pointing out the contingencies that arise from the definition, categorization and characterization of ethnoracial groups (Morning & Rodríguez-Muñiz, 2016). It probably will require continuous efforts to reopen political debates in such a way, because there are also continuous efforts to essentialize them.

Notes

1 For an empirical analysis of visual constructions of "the world", see Hoggenmüller (2020).
2 Out of twelve chapters, in only four is the role of international actors more limited: Chapters 2, 5, 7, and 8. While the micro-macro distinction has its merits and limitations (Krause, 2013), we prefer to foreground the spatial scale because it is more precise than the multiple meanings that have been attached to the micro-macro distinction.

References

Agrawal, A., & Kumar, V. (2020). *Numbers in India's periphery. The political economy of government statistics*. Cambridge, England: Cambridge University Press. doi:10.1017/9781108762229

Barakat, B. F., & Durham, R. E. (2017). Future education trends. In W. Lutz, W. P. Butz, & K. C. Samir (Eds.), *World population & human capital in the twenty-first century* (pp. 397–433). Oxford, England: Oxford University Press.

Bartl, W., Papilloud, C., & Terracher-Lipinski, A. (2019). Governing by numbers: Key indicators and the politics of expectations. An introduction [Special issue]. *Historical Social Research, 44*(2), 7–43. doi:10.12759/hsr.44.2019.2.7-43

Beck, U. (1996). World risk society as cosmopolitan society? *Theory, Culture & Society, 13*(4), 1–32.

Bhagat, R. B. (2022). *Population and the political imagination: Census, register and citizenship in India*. London, England: Routledge.

Biermann, F., Hickmann, T., & Sénit, C.-A. (Eds.). (2022). *The political impact of the sustainable development goals*. Cambridge, England: Cambridge University Press. doi:10.1017/9781009082945

Boehm, G. (1994). Die Wiederkehr der Bilder. In G. Boehm (Ed.), *Was ist ein Bild?* (pp. 11–38). München, Germany: Fink.

Bogusz, T. (2022). *Experimentalism and sociology*. Cham, Switzerland: Springer. doi:10.1007/978-3-030-92478-2

Carranza-Torres, J. A. (2021). How can traditional statistical relationships be redefined through citizen to government partnerships? *Statistical Journal of the IAOS, 37*, 229–243. doi:10.3233/SJI-190578

Centeno, M. A., Nag, M., Patterson, T. S., Shaver, A., & Windawi, A. J. (2015). The emergence of global systemic risk. *Annual Review of Sociology, 41*(1), 65–85. doi:10.1146/annurev-soc-073014-112317

Cole, W. M. (2017). World polity or world society? Delineating the statist and societal dimensions of the global institutional system. *International Sociology, 32*(1), 86–104. doi:10.1177/0268580916675526

Desrosières, A. (2007). Surveys versus administrative records: Reflections on the duality of statistical sources. *Courrier des statistiques, English Series, 13*, 7–19.

Diaz-Bone, R., & Horvath, K. (2021). Official statistics, big data and civil society. Introducing the approach of "economics of convention" for understanding the rise of new data worlds and their implications. *Statistical Journal of the IAOS, 37*, 219–228. doi:10.3233/SJI-200733

Emigh, R. J., Riley, D. J., & Ahmed, P. (2016a). *Antecedents of censuses from medieval to nation states: How societies and states count*. New York, NY: Palgrave Macmillan.

Emigh, R. J., Riley, D. J., & Ahmed, P. (2016b). *Changes in censuses from imperialist to welfare states: How societies and states count*. New York, NY: Palgrave Macmillan. doi:10.1057/9781137485069

Emigh, R. J., Riley, D. J., & Ahmed, P. (2020). The sociology of official information gathering: Enumeration, influence, reactivity, and power of states and societies. In T. Janoski, C. de Leon, J. Misra, & I. W. Martin (Eds.), *The new handbook of political sociology* (pp. 290–320). Cambridge, England: Cambridge University Press.

Foroutan, N. (2019). The post-migrant paradigm. In J.-J. Bock & S. Macdonald (Eds.), *Refugees welcome?: Difference and diversity in a changing germany* (pp. 142–167). New York, NY: Berghahn Books.

Frohman, L. (2012). "Only sheep let themselves be counted" privacy, political culture, and the 1983/87 West German census boycotts. *Archiv für Sozialgeschichte, 52*, 335–378.

Gbadebo, B. M. (2021). Old, contemporary and emerging sources of demographic data in Africa. In C. O. Odimegwu & Y. Adewoyin (Eds.), *The Routledge handbook of African demography* (pp. 111–128). London, England: Routledge.

Georgiou, A. V. (2019). Extracting statistical offices from policy-making bodies to buttress official statistical production. *Journal of Official Statistics, 35*(1), 1–8. doi:10.2478/jos-2019-0001

Gong, R., Groshen, E. L., & Vadhan, S. (2022). Harnessing the known unknowns: Differential privacy and the 2020 census [Special issue]. *Harvard Data Science Review, 4*(2). doi:10.1162/99608f92.cb06b469

Hannah, M. G. (2009). Calculable territory and the West German census boycott movements of the 1980s. *Political Geography, 28*(1), 66–75. doi:10.1016/j.polgeo.2008.12.001

Heintz, P. (1982). Introduction: A sociological code for the description of world society and its change. *International Social Science Journal, 34*(1), 11–21.

Hoggenmüller, S. W. (2020). *Globalität sehen: Zur visuellen Konstruktion von »Welt«.* Frankfurt am Main, Germany: Campus Verlag . doi:10.12907/978-3-593-44235-8

Howard, C. (2021). *Government statistical agencies and the politics of credibility.* Cambridge, England: Cambridge University Press.

Idike, A., & Eme, O. I. (2015). Census politics in Nigeria: An examination of 2006 population census. *Journal of Policy and Development Studies, 9*(3), 47–72.

Kalter, F., & Foroutan, N. (2021). Race for second place? Explaining East-West differences in anti-Muslim sentiment in Germany. *Frontiers in Sociology, 6*, 735421. doi:10.3389/fsoc.2021.735421

Kertzer, D. I., & Arel, D. (2006). Population composition as an object of political struggle. In R. E. Goodin & C. Tilly (Eds.), *The Oxford handbook of contextual political analysis* (pp. 664–677). Oxford, England: Oxford University Press. doi:10.1093/oxfordhb/9780199270439.003.0036

Krause, M. (2013). Recombining micro/macro. *European Journal of Social Theory, 16*(2), 139–152.

Krause, M. (2016a). 'Western hegemony' in the social sciences: Fields and model systems. *The Sociological Review, 64*(2_suppl), 194–211. doi:10.1111/2059-7932.12008

Krause, M. (2016b). The meanings of theorizing. *The British Journal of Sociology, 67*(1), 23–29. doi:10.1111/1468-4446.12187_4

Lehtonen, M. (2013). The non-use and influence of UK energy sector indicators. *Ecological Indicators, 35*, 24–34. doi:10.1016/j.ecolind.2012.10.026

Loveman, M. (2005). The modern state and the primitive accumulation of symbolic power. *American Journal of Sociology, 110*(6), 1651–1683. doi:10.1086/428688

Luhmann, N. (1997). Globalization or world society: How to conceive of modern society? *International Review of Sociology, 7*(1), 67–79. doi:10.1080/03906701.1997.9971223

Maisonobe, M. (2021). Regional distribution of research: The spatial polarization in question. In R. Ball (Ed.), *Handbook bibliometrics* (pp. 377–396). Berlin, Germany: de Gruyter.

Meyer, J. W. (2015). Commentary: Theories of world society. In B. Holzer, F. Kastner, & T. Werron (Eds.), *From globalization to world society: Neo-institutional and systems-theoretical perspectives* (pp. 317–328). New York, NY: Routledge.

Mohammed, I. S., Othman, M. F., & Osman, N. (2019). Nigerian national population and housing census, 2018 and sustainable development: The issues at stake. *Journal of Techno-Social, 11*(1), 16–28. doi:10.30880/jts.2018.11.01.003

Morning, A., & Rodríguez-Muñiz, M. (2016). *Race in the demographic imaginary: Population projections and their conceptual foundations* (NYU Population Center Working Paper). New York, NY: New York University.

Mungiu-Pippidi, A. (2015). *The quest for good governance: How societies develop control of corruption.* Cambridge, England: Cambridge University Press.

Olorunfemi, J. F., & Fashagba, I. (2021). Population census administration in Nigeria. In R. Ajayi & J. Y. Fashagba (Eds.), *Nigerian politics* (pp. 353–368). Cham, Switzerland: Springer.

Papilloud, C., & Schultze, E.-M. (2022). *Skizze einer Theorie der Relation.* Wiesbaden, Germany: Springer VS. doi:10.1007/978-3-658-37922-3

Piketty, T. (2022). *A brief history of equality.* Cambridge, MA: Harvard University Press.

Prévost, J.-G. (2019). Politics and policies of statistical independence. In M. J. Prutsch (Ed.), *Science, numbers, and politics* (pp. 153–180). Cham, Switzerland: Springer.

Prewitt, K. (2022). 2030: A sensible census, in reach. In A. L. Carriquiry, J. M. Tanur, & W. F. Eddy (Eds.), *Statistics in the public interest* (pp. 321–336). Cham, Switzerland: Springer.

Radermacher, W. J. (2021). Governing-by-the numbers – Résumé after one and a half years. *Statistical Journal of the IAOS, 37*(2), 701–711. doi:10.3233/SJI-210819

Rallu, J.-L., Piché, V., & Simon, P. (2006). Demography and ethnicity: An ambiguous relationship. In G. Caselli, J. Vallin, & G. J. Wunsch (Eds.), *Demography: Analysis and synthesis. A treatise in population studies* (Vol. 3, pp. 531–549). Boston, MA: Elsevier.

Riley, D., Ahmed, P., & Emigh, R. J. (2021). Getting real: Heuristics in sociological knowledge. *Theory and Society, 50*(2), 315–356. doi:10.1007/s11186-020-09418-w

Ruppert, E., & Scheel, S. (2021a). Introduction: The politics of making up a European people. In E. Ruppert & S. Scheel (Eds.), *Data practices: Making up a European people* (pp. 1–28). London, England: Goldsmith Press.

Ruppert, E., & Scheel, S. (2021b). The politics of data practices. In E. Ruppert & S. Scheel (Eds.), *Data practices: Making up a European people* (pp. 269–304). London, England: Goldsmith Press.

Saltelli, A., Bammer, G., Bruno, I., Charters, E., Di Fiore, M., Didier, E., ... Vineis, P. (2020). Five ways to ensure that models serve society: A manifesto. *Nature, 582*(7813), 482–484. doi:10.1038/d41586-020-01812-9

Schwinn, T. (2008). Nationale und globale Ungleichheit. *Berliner Journal für Soziologie, 18*(1), 8–31.

Stichweh, R. (2015). Comparing systems theory and sociological neo-institutionalism: Explaining functional differentiation. In B. Holzer, F. Kastner, & T. Werron (Eds.), *From globalization to world society: Neo-institutional and systems-theoretical perspectives* (pp. 23–36). New York, NY: Routledge.

Strang, D., & Meyer, J. W. (1993). Institutional conditions for diffusion. *Theory and Society, 22*(4), 487–512.

Swedberg, R. (2016). Can you visualize theory? On the use of visual thinking in theory pictures, theorizing diagrams, and visual sketches. *Sociological Theory, 34*(3), 250–275. doi:10.1177/0735275116664380

Thévenot, L. (1984). Rules and implements: Investment in forms. *Social Science Information, 23*(1), 1–45. doi:10.1177/053901884023001001

van der Brakel, J. (2022). New data sources and inference methods for official statistics. In A. L. Carriquiry, J. M. Tanur, & W. F. Eddy (Eds.), *Statistics in the public interest* (pp. 411–431). Cham, Switzerland: Springer.

van Dooren, W., Bouckaert, G., & Halligan, J. (2015). Non-use. In *Performance management in the public sector* (2nd ed., pp. 154–173). London, England: Routledge.

Ventresca, M. J. (1995). *When states count: Institutional and political dynamics in modern census establishment, 1800–1993* (PhD thesis). Stanford University, Stanford, CA.

Villacís, B., Thiel, A., Capistrano, D., & Carvalho da Silva, C. (2022). Statistical innovation in the global South: Mechanisms of translation in censuses of Brazil, Ecuador, Ghana and Sierra Leone. *Comparative Sociology, 21*(4), 419–446. doi:10.1163/15691330-bja10060

Vogel, R. (2019). *Survey-Welten*. Wiesbaden, Germany: Springer VS. doi:10.1007/978-3-658-25437-7

Index

Note: **Bold** page numbers refer to tables; *italic* page numbers refer to figures and page numbers followed by "n" denote endnotes.

For Product Safety Concerns and Information please contact our EU
representative GPSR@taylorandfrancis.com
Taylor & Francis Verlag GmbH, Kaufingerstraße 24, 80331 München, Germany

www.ingramcontent.com/pod-product-compliance
Lightning Source LLC
Chambersburg PA
CBHW052118230326
41598CB00080B/3832

9 7 8 1 0 3 2 1 9 5 5 1 3